SPILL PREVENTION AND FAIL-SAFE ENGINEERING FOR PETROLEUM AND RELATED PRODUCTS

Spill Prevention and Fail-Safe Engineering for Petroleum and Related Products

J.L. Goodier
R.J. Siclari
P.A. Garrity

Pacific Northwest Laboratory
Battelle Memorial Institute

NOYES DATA CORPORATION
Park Ridge, New Jersey, U.S.A.
1983

Copyright © 1983 by Noyes Data Corporation
Library of Congress Catalog Card Number: 83-2204
ISBN: 0-8155-0944-8
ISSN: 0090-516X
Printed in the United States

Published in the United States of America by
Noyes Data Corporation
Mill Road, Park Ridge, New Jersey 07656

10 9 8 7 6 5 4 3 2

Library of Congress Cataloging in Publication Data

Goodier, J. Leslie.
    Spill prevention and fail-safe engineering for petro-
leum and related products.

    (Pollution technology review, ISSN 0090-516X ;
no. 100)
    Bibliography: p.
    Includes index.
    1. Oil spills. I. Siclari, R. J. II. Garrity, P. A.
III. Title. IV. Series.
TD427.P4G63 1983   665.5'4   83-2204
ISBN 0-8155-0944-8

Foreword

The annual losses of petroleum and related products due to accidental spillage during transportation, cargo transfer, bulk storage, and processing are tremendous. This spill prevention and fail-safe engineering manual was prepared to provide guideline information on spill prevention procedures for individuals who have the responsibility for maintaining a spill-free plant during the transportation, transfer, storage and processing of petroleum and related products.

An attempt has been made to cover every facet of spill prevention. Special emphasis is given to fail-safe engineering as an approach to preventing spills from the predominant cause—human failure. The book addresses state-of-the-art spill prevention practices and automation techniques that can reduce spills caused by human error. Whenever practical, implementation costs are provided to aid equipment acquisition and installation budgeting. To emphasize the need for spill prevention measures, historic spills are briefly described, after which remedial action is defined in an appropriate section of the manual. The section on plant security goes into considerable depth, since few security guidelines have been provided for industrial facilities that transfer, store, and process petroleum and related products.

The intent here is to provide fingertip reference material that can be used by interested parties in a nationwide effort to reduce losses from preventable spills.

The information in the book is from *Fuel Conservation by the Application of Spill Prevention and Fail-safe Engineering (A Guideline Manual)* (DOE Report TIC-11470), prepared by J.L. Goodier, R.J. Siclari, and P.A. Garrity of Battelle Pacific Northwest Laboratory for the U.S. Department of Energy, June 1981.

v

The table of contents is organized in such a way as to serve as a subject index and provides easy access to the information contained in the book.

Notice

Contents and Subject Index

1. Executive Summary

From federally accumulated statistics for oil and hazardous substance spills, the authors culled information on spills of hydrocarbon products. In 1978, a total of 1,456 oil spills were reported, compared to 1,451 in 1979. The 1978 spills were more severe: 7,289,163 gallons of oil were accidently discharged. In 1979, the number of gallons spilled was reduced to 3,663,473. These figures are derived from <u>reported</u> spills; it is highly possible that an equal amount was spilled and not reported. Spills that are effectively contained within a plant property and do not enter a navigational waterway need not be reported. Needless to say, there is a tremendous annual loss of oil products due to accidental spillage during transportation, cargo transfer, bulk storage, and processing.

As an aid to plant engineers and managers, federal workers, fire marshals, and fire and casualty insurance inspectors, this document is offered as a spill prevention guide.

The manual addresses state-of-the-art spill prevention practices and automation techniques that can reduce spills caused by human error. Whenever practical, the cost of implementation is provided to aid equipment acquisition and installation budgeting. To emphasize the need for spill prevention measures, historic spills are briefly described, after which remedial action is defined in an appropriate section of the manual.

The section on plant security goes into considerable depth since to date no security guidelines have been provided for industrial facilities that transfer, store, and process petroleum products. It is stressed, however, that the U.S. DOE/Sandia Laboratories have published extensive documentation on safeguards and security at nuclear facilities. Although this material can provide indepth knowledge on advanced security measures, it is so advanced that it exceeds the needs of a petroleum processing-marketing facility or a plant storing and using hydrocarbon fuels in bulk.

The intent of the document is to provide fingertip reference material that can be used by interested parties in a nationwide effort to reduce loss of oil from preventable spills.

The information presented herein can be summarized as follows.

Bulk Storage Tanks

1.   The majority of spillage from bulk storage results from overfills.
The spillage is generally contained within the dike enclosure that sur-
rounds a tank. It can, however, percolate down to and contaminate ground
water. The latter occurs following repeated or massive spills, since small
spillages are usually sorbed in the sediments, which limits the downward
travel of the spilled material. (Section 4.6)

Spills from this source can be controlled by (1) an attentive tank
gager who closely monitors the "topping" action, (2) the use of high
liquid level alarms to warn the gager when the fill operation is nearing
its maximum limit, and (3) a fill level alarm connected to a pump control
that automatically shuts down the pump when a preset liquid level is
reached. The latter system is the most desirable of the three because it
eliminates human failure.

2.   The lowermost head of a storage tank is exposed to corrosive action
from condensate water that accumulates in the lower reaches of the tank
due to water being heavier than most petroleum products. Unless carefully
inspected and gaged for thinning of metal, leakage can occur and go undetected until it seeps out from under the tank foundation. Remedial action
involves (1) frequent draining of condensate water from the tank, (2) an
established frequency of checking on plate thickness by mechanical or non-
destructive testing, and (3) cleaning the corroded metal down to a bright
steel finish by sandblasting and applying a non-corrosive epoxy coating
to the metal. (Section 4.2)

3.   Considerable quantities of petroleum products are lost by vaporization of the product during tank filling and from solar heat exchange and
vaporization during the summer months. This can be controlled by using
floating roof tanks, or by a more recent engineering development that
employs the use of a floating membrane within a fixed roof tank. Acquisition

and maintenance costs can be reduced by employing the floating roof design
for petroleum product storage. (Section 4.4)

4.   The deterioration of steam heating coils in heavy oil tanks (used to
maintain the oil in a viscous state in cold weather) from internal corrosion can result in product leakage when oil drains through a corroded
coil to discharge into a nearby watercourse. The remedies are to (1) fully
drain and dry the internal side of the coil when idle during the summer
months, (2) determine the average life span of the coil from past maintenance records and renew the coil about one year before anticipated failure,
and (3) use external heating coils and insulate the sides of the tank for
heat retention. A somewhat more expensive remedy is to circulate the oil
through an external heater. This procedure does, however, demand energy
costs for a circulating pump. (Section 4.5)

5.   In conjunction with Item 2, condensate water draw procedures can
result in product loss when the attendant worker is distracted and oil
is permitted to drain through the water draw.

    The approaches to prevent loss from this source are (1) use a
replaceable imbiber bead valve on the drain that permits water flow but
closes the flow upon oil impact, (2) pipe the water draw to an oil and
water separator that will retain and permit recovery of any oil inadvert-
ently released from the tank, and (3) station an attendant at the water
draw valve for the entire drainage period. (Section 4.8)

Secondary Containment Systems

6.   Dike failure and uncontrolled water drainage systems from within
diked areas (drain valve left open, no drain valve provided, broached
dike) can release tank spillage into the plant's surface drainage system
for release to a nearby watercourse.

    The controls available to prevent losses of this nature are to (1)
engineer impermeable dikes that will not release spillage from within the
diked area, (2) closely supervise and lock water drain valves from diked
areas, (3) eliminate drains from diked areas and rely on solar evaporation

for removal of rainfall or pump collected rainwater to a treatment pond or other disposal vehicle, and (4) incorporate a bed of imbiber valves into the final drainage system. (Section 4.10)

Buried Tanks and Lines

7.   Buried pipelines and tanks are potential spill problems due to external corrosion of the shell metal from exposure to the wet earth and electrolysis. Under some local fire codes, tanks containing highly flammable liquids must be buried. When the law or operational process demands burial, the following precaution should be exercised: Coat and wrap the external surfaces of the tank to retard physical contact of the metal with water-saturated soil and install a cathodic protection system for tank protection.

Tank Truck and Tank Car Loading Racks

8.   Tank truck and tank car loading racks are both potential spill sites due largely to the exposure from overfilling the tanks. The basic designs in this area should include (1) the provision of high liquid level alarms, (2) engineered drainage to direct spillage into a suitable spill retention sump or an oil and water separator, (3) computerized flow monitoring systems that measure the quantity of product transferred during a fill and stop flow when a predetermined quantity of material has been loaded into the tank. (Sections 8 and 9)

Loading/Unloading Piers

9.   Flexible transfer hose failure is the predominant cause of spillage during ship-to-shore transfer and vice versa. Losses can be controlled by two spill prevention techniques: (1) frequent hydrostatic testing of hose lines in accordance with U.S. Coast Guard requirements, with replacement and prompt removal of defective hoses from the dock area; (2) replacement of flexible hose lines with articulated all-metal arms for loading and unloading liquid cargoes. (Section 10)

Making and breaking of flexible hose to ship/barge manifolds is another spill cause. When the connection is broken, product remaining in the hose can accidently be released to spill material on the deck. There is one proved method to control spillage of this type: Provide a butterfly closure valve on the dockside of the line with a drain between the butterfly valve and marine manifold. Upon completion of a transfer operation, the butterfly valve and the manifold valves are closed and the drain is opened to drain entrapped product into a catch drum or back into the cargo compartment. (Section 10)

Spill Detection Systems

10. During the night a spill can go undetected until daylight. The fast detection of a spill can aid in stemming a leak and reducing the product loss. Currently, there are eight known systems that can detect a spill of floatable material on water and activate an audible or visible alarm to indicate a spillage. The report describes the characteristics and costs of the eight systems to aid in the selection of an appropriate unit. (Section 11).

Wastewater Treatment Ponds

11. Considerable loss of reclaimable waste materials and contamination of groundwater resources has occurred through improperly designed, permeable holding ponds or lagoons. The report describes a variety of sealants and liners that can be incorporated in new and existing ponds to prevent seepage and effectively contain wastes until reclaimed and reprocessed. (Section 12)

Personnel Qualifications and Training

12. An analysis of some 70,000 accidents has indicated that 88 percent resulted from human failure, 10 percent were caused by mechanical deficiencies, and only 2 percent were unpreventable. Based on this finding, an educational program to indoctrinate workers into the principles of spill prevention demands and procedures is recommended. The educational

program should be strongly supported by a program of failsafe engineering to control catastrophes occurring from the single, unintentional unsafe act of a worker. (Section 13)

Plant Security

13. Most plants can improve security procedures to prevent illegal entry with resulting theft, sabotage, and acts of vandalism. For instance, many waterfront facilities are fenced on only three sides, affording easy entry into the premises from the waterfront. Fencing and lighting systems can be substantially improved, and passive detection systems are available to alert security guards that an intruder has entered the premises. There are many degrees of security, including some high degree systems used for nuclear facilities. The report provides guidance to a degree suited to petroleum marketing terminals and waterfront facilities. It is stressed that planned sabotage could result in a casualty of massive proportion. This would be possible if high octane fuel were released by a saboteur by opening an unlocked master flow control valve on a bulk storage tank. (Section 14)

2. Introduction and Background

From a series of nationwide plant surveys dedicated to spill prevention, containment, and countermeasure evaluation, coupled with spill response action activities, the authors determined a need for a spill prevention guideline manual.

Within the past ten years, the technologies of spill prevention, spill containment, and failsafe engineering have advanced considerably. The Federal government through the U.S. Environmental Protection Agency (U.S. EPA) has published spill prevention criteria developed from historic spill investigations and field surveys of industrial and Federal installations. To date, however, there has been no attempt to define and publish information that will guide plant management and engineers in the application of spill prevention and failsafe engineering. This guideline manual is intended to extend cryptic criteria statements into a "how" and "why" to do it publication. To this end, the U.S. Department of Energy (U.S. DOE) sponsored development of this publication by the Battelle Northwest Laboratories. The U.S. DOE's major concern in oil spill prevention is to reduce the needless loss of energy products from accidental discharges. There is also, of course, a national desire to prevent polluting incidents that contaminate the inland and coastal waters of the United States.

U.S. Coast Guard records indicate that a grand total of 3,663,473 gallons of oil and other substances was spilled in the United States and immediately adjacent waters in 1979. In recent years the quantity of materials spilled has been markedly reduced. It is reasonable to assume that the reduction was brought about by recently enacted Federal regulations, coupled with an intense effort by oil handlers, processors, and users to prevent and otherwise contain oil spills. It is worthy of note that much of the oil spilled was refined product. Naturally, the more advanced the refining of crude oil becomes, the more expensive is the spill. The cost from exploration, recovery, transportation, and processing is lost with the product.

Additionally, unique spill prevention measures introduced in sur-
veyed plants are recorded for the benefit of other plant facilities.
Whenever practical, a recommended or suggested practice is supported
by a brief description of a spill incident to support and stress the
need for corrective action.

A wide variety of illustrations was collected from equipment suppliers for inclusion in the manual to clarify the written word. Photo-
graphic material and illustrations were provided with the full knowledge of intended use with the authority to reproduce. A thank you is due
to illustration suppliers, and regrets are due to some whose illustrations
were not used because of space limitations.

It is possible that this manual will eventually be revised to include
new technological advances. In this respect suggestions, comments, and
even technological input are solicited from any possible source.

It is stressed that although the names of manufacturers and products
are used freely the document is not intended as an endorsement of any
product or group of products.

Tables 2.1 and 2.2 (derived from U.S. Coast Guard data) provide
condensed information on the sources and causes of reported oil spills
during the years 1978 and 1979. Wherever practical, the spill causes
are included in the manual along with guidelines for preventing further
spills from a known cause.

TABLE 2.1.

U.S. Oil Spills, 1978

	NO. OF SPILLS	GALLONS SPILLED
Other Transportation Related Marine Facility:		
Tank Rupture or Leak	3	210,025
Transportation Pipeline Rupture or Leak	1	4
Dike Rupture or Leak	1	200
Other Structural Failure	3	126
Pipe Rupture or Leak	6	175
Hose Rupture or Leak	2	350
Pump Failure	3	52
Other Equipment Failure	2	66
Tank Overflow	4	93
Improper Equipment Handling or Operation	8	3,372
Other Personnel Error	7	210
Intentional Discharge	1	50
Natural or Chronic Phenomenon	3	13
Unknown Cause	5	400
Total	49	215,136
Onshore Bulk Cargo Transfer:		
Tank Rupture or Leak	5	4,525,440
Transportation Pipeline Rupture or Leak	14	58,020
Dike Rupture or Leak	4	460
Other Structural Failure	6	469
Pipe Rupture or Leak	21	2,066
Hose Rupture or Leak	11	4,427
Manifold Rupture or Leak	3	2,800
Loading Arm Failure, Rupture or Leak	3	28
Valve Failure	3	12,022
Pump Failure	2	23
Flange Failure	2	1,001
Gasket Failure	9	366
Other Equipment Failure	9	13,143
Tank Overflow	11	5,446
Improper Equipment Handling or Operation	14	1,112
Intentional Discharge	3	517
Other Personnel Error	11	2,159
Railroad Accident	1	3
Natural or Chronic Phenomenon	10	10,472
Unknown Cause	10	989
Total	152	4,460,063
Onshore Fueling		
Hull Rupture or Leak	1	4
Transportation Pipeline Rupture or Leak	5	2,216
Container Lost Intact	1	20
Other Structural Failure	1	100
Pipe Rupture or Leak	7	177
Hose Rupture or Leak	1	25
Valve Failure	1	20
Pump Failure	1	500
Other Equipment Failure	1	40
Tank Overflow	1	59
Total	25	3,375

Table 2.1. (Continued)

	NO. OF SPILLS	GALLONS SPILLED
Onshore Non-Bulk Cargo Transfer		
Tank Rupture or Leak	1	25
Pipe Rupture or Leak	1	50
Gasket Failure	1	1
Other Personnel Error	1	5
Total	4	81
Other Land Vehicle		
Tank Rupture or Leak	4	909
Other Structural Failure	1	2
Hose Rupture or Leak	2	60
Other Equipment Failure	1	10
Tank Overflow	2	1,650
Improper Equipment Handling or Operation	2	60
Other Personnel Error	2	29
Intentional Discharge	1	20
Railroad Accident	2	2,000
Highway Accident	3	1,052
Unknown Cause	1	5
Total	21	5,792
Rail Vehicle Liquid Bulk		
Tank Rupture or Leak	1	300
Other Structural Failure	1	200
Pipe Rupture or Leak	1	5,117
Other Equipment Failure	1	10
Railroad Accident	10	32,300
Highway Accident	1	700
Total	15	38,627
Rail Vehicle Dry Bulk		
Railroad Accident	1	20
Rail Vehicle General Cargo		
Other Structural Failure	1	500
Other Equipment Failure	1	1,000
Railroad Accident	6	29,075
Unknown Cause	2	152
Total	10	30,727
Rail Vehicle Transfer		
Valve Failure	1	300
Highway Vehicle Dry Bulk		
Tank Rupture or Leak	1	150
Other Personnel Error	1	3
Intentional Discharge	1	4,000
Railroad Accident	1	100
Highway Accident	11	8,220
Total	15	12,476

Table 2.1.   (Continued)

	NO. OF SPILLS	GALLONS SPILLED
Highway Vehicle Liquid Bulk		
Tank Rupture or Leak	14	43,222
Transportation Pipeline Rupture or Leak	1	50
Dike Rupture or Leak	1	3,000
Other Structural Failure	1	50
Pipe Rupture or Leak	2	310
Hose Rupture or Leak	11	725
Manifold Rupture or Leak	1	100
Valve Failure	4	4,636
Flange Failure	3	915
Gasket Failure	2	350
Other Equipment Failure	2	120
Tank Overflow	27	4,275
Railroad Accident	2	1,235
Highway Accident	93	184,484
Aircraft Accident	1	15
Unknown Cause	3	215
Total	180	255,605
Highway Vehicle General Cargo		
Tank Rupture or Leak	1	250
Other Structural Failure	1	3
Pipe Rupture or Leak	1	500
Hose Rupture or Leak	2	105
Valve Failure	1	100
Pump Failure	1	20
Tank Overflow	1	30
Railroad Accident	1	40
Highway Accident	9	675
Total	19	1,783
Highway Vehicle Passenger		
Tank Rupture or Leak	2	133
Other Equipment Failure	1	100
Tank Overflow	1	50
Intentional Discharge	1	2
Highway Accident	5	52
Aircraft Accident	1	20
Total	11	357
Other Land Transportation Facility		
Tank Rupture or Leak	2	80
Other Structural Failure	2	7,800
Pipe Rupture or Leak	3	1,301
Hose Rupture or Leak	3	126
Manifold Rupture or Leak	1	150
Loading Arm Failure, Rupture or Leak	1	500
Valve Failure	1	75
Other Equipment Failure	12	29,890
Tank Overflow	7	602
Improper Equipment Handling or Operation	4	120
Other Personnel Error	6	206
Intentional Discharge	3	2,700
Aircraft Accident	1	150
Natural or Chronic Phenomenon	2	16
Unknown Cause	26	3,664
Total	74	47,380

Table 2.1.   (Continued)

	NO. OF SPILLS	GALLONS SPILLED
Other Onshore Non-Transportation Related Facility		
Tank Rupture or Leak	13	13,645
Transportation Pipeline Rupture or Leak	3	8,027
Container Lost Intact	1	30
Other Structural Failure	5	2,223
Pipe Rupture or Leak	18	10,986
Hose Rupture or Leak	6	282
Manifold Rupture or Leak	2	1,420
Valve Failure	8	17,461
Pump Failure	4	22
Other Equipment Failure	12	2,248
Tank Overflow	17	4,012
Improper Equipment Handling or Operation	8	1,131
Other Personnel Error	9	7,176
Intentional Discharge	7	398
Railroad Accident	1	2
Natural or Chronic Pehnomenon	4	122
Unknown Cause	14	3,366
Total	132	72,551
Onshore Refinery		
Transportation Pipeline Rupture or Leak	1	200
Dike Rupture or Leak	1	2
Other Structural Failure	1	1,000
Pipe Rupture or Leak	2	11
Hose Rupture or Leak	1	2
Valve Failure	1	2
Pump Failure	1	12
Gasket Failure	2	105
Other Equipment Failure	9	454
Tank Overflow	3	2,053
Improper Equipment Handling or Operation	4	287
Intentional Discharge	1	3
Natural or Chronic Phenomenon	2	31
Unknown Cause	5	105
Total	34	4,267
Onshore Bulk Storage Facility		
Tank Rupture or Leak	7	20,750
Transportation Pipeline Rupture or Leak	3	135
Other Structural Failure	1	100
Pipe Rupture or Leak	7	836
Hose Rupture or Leak	1	4,200
Manifold Rupture or Leak	4	831
Loading Arm Failure, Rupture or Leak	3	10,922
Valve Failure	11	30,213
Pump Failure	3	251,265
Gasket Failure	1	1,000
Other Equipment Failure	15	16,046
Tank Overflow	36	167,641
Improper Equipment Handling or Operation	5	30,343
Other Personnel Error	5	13,540
Intentional Discharge	4	1,780
Railroad Accident	1	500
Highway Accident	1	200
Natural or Chronic Phenomenon	5	2,634
Unknown Cause	4	14,100
Total	117	567,036

Table 2.1.   (Continued)

	NO. OF SPILLS	GALLONS SPILLED
Onshore Industrial Plant or Processing Facility		
Tank Rupture or Leak	6	17,400
Dike Rupture or Leak	5	1,420
Container Lost Intact	1	22
Other Structural Failure	6	100,381
Pipe Rupture or Leak	17	26,505
Hose Rupture or Leak	2	20
Valve Failure	11	6,801
Pump Failure	5	137
Flange Failure	1	50
Gasket Failure	2	600
Other Equipment Failure	88	20,772
Tank Overflow	21	9,299
Improper Equipment Handling or Operation	15	9,547
Other Personnel Error	12	4,498
Intentional Discharge	5	120,485
Railroad Accident	1	1,400
Natural or Chronic Phenomenon	5	415
Unknown Cause	18	592
Total	221	320,344
Onshore Oil or Gas Production Facility		
Tank Rupture or Leak	3	3,310
Dike Rupture or Leak	3	1,368
Other Structural Failure	1	1
Pipe Rupture or Leak	7	1,270
Hose Rupture or Leak	2	200
Valve Failure	3	1,880
Other Equipment Failure	6	32,742
Tank Overflow	3	1,240
Intentional Discharge	3	2,318
Unknown Cause	1	200
Total	32	44,529
Power Plant		
Tank Rupture or Leak	1	20
Transportation Pipeline Rupture or Leak	3	605
Dike Rupture or Leak	1	15
Other Structural Failure	1	40
Pipe Rupture or Leak	6	830
Hose Rupture or Leak	2	25
Valve Failure	1	5
Pump Failure	1	2
Gasket Failure	3	260
Other Equipment Failure	11	9,321
Tank Overflow	3	703
Improper Equipment Handling or Operation	4	14
Other Personnel Error	5	129
Intentional Discharge	1	30
Natural or Chronic Phenomenon	1	3
Unknown Cause	4	295
Total	48	12,297
Pipeline Within Non-Transportation Related System		
Tank Rupture or Leak	1	200
Other Structural Failure	1	10
Pipe Rupture or Leak	5	8,780
Valve Failure	2	16
Other Equipment Failure	1	50
Improper Equipment Handling or Operation	2	1,505
Other Personnel Error	3	130
Natural or Chronic Phenomenon	1	10
Unknown Cause	1	550
Total	17	11,251

Table 2.1.   (Continued)

	NO. OF SPILLS	GALLONS SPILLED
Other Transportation Related Marine Facility		
.Hull Rupture or Leak	1	42
Tank Rupture or Leak	2	65
Transportation Pipeline Rupture or Leak	1	3
Other Structural Failure	3	57
Pipe Rupture or Leak	6	69
Hose Rupture or Leak	4	1,557
Valve Failure	1	400
Pump Failure	2	700
Gasket Failure	1	1
Other Equipment Failure	4	234
Tank Overflow	5	131
Improper Equipment Handling or Operation	6	45
Other Personnel Error	10	142
Intentional Discharge	3	251
Aircraft Accident	1	10
Natural or Chronic Phenomenon	1	5
Unknown Cause	4	207
Total	55	3,919
Onshore Bulk Cargo Transfer		
Hull Rupture or Leak	2	13
Tank Rupture or Leak	1	150
Transportation Pipeline Rupture or Leak	9	18,467
Container Lost Intact	3	9,000
Other Structural Failure	8	111,067
Pipe Rupture or Leak	16	4,116
Hose Rupture or Leak	16	657
Manifold Rupture or Leak	1	200
Loading Arm Failure, Rupture or Leak	6	791
Valve Failure	12	3,301
Pump Failure	1	60
Flange Failure	4	212
Gasket Failure	9	12,688
Other Equipment Failure	5	36
Tank Overflow	9	6,062
Improper Equipment Handling or Operation	12	6,040
Other Personnel Error	12	154
Intentional Discharge	2	505
Highway Accident	1	100
Natural or Chronic Phenomenon	5	216
Unknown Cause	3	22
Total	137	173,857
Onshore Fueling		
Tank Rupture or Leak	3	515
Transportation Pipeline Rupture or Leak	4	10,227
Other Structural Failure	2	29
Pipe Rupture or Leak	8	529
Hose Rupture or Leak	7	434
Valve Failure	1	50
Pump Failure	1	1
Flange Failure	2	22
Gasket Failure	4	82
Other Equipment Failure	3	311
Tank Overflow	3	55
Improper Equipment Handling or Operation	4	104
Other Personnel Error	5	1,204
Intentional Discharge	3	1,024
Natural or Chronic Phenomenon	1	125
Unknown Cause	1	1
Total	52	14,713

Table 2.1.   (Continued)

	NO. OF SPILLS	GALLONS SPILLED
Onshore Bulk Storage Facility		
Tanks Rupture or Leak	3	7,843
Transportation Pipeline Rupture or Leak	3	30,025
Container Lost Intact	2	601
Other Structural Failure	1	5
Pipe Rupture or Leak	3	162
Hose Rupture or Leak	4	601
Valve Failure	4	35,848
Pump Failure	1	700
Flange Failure	1	100
Gasket Failure	1	2,878
Other Equipment Failure	11	60,690
Tank Overflow	25	10,357
Improper Equipment Handling or Operation	6	5,170
Other Personnel Error	5	807
Intentional Discharge	17	147,814
Natural or Chronic Phenomenon	7	2,978
Unknown Cause	2	1,510
Total	96	308,089
Onshore Industrial Plant or Processing Facility		
Tank Rupture or Leak	5	2,150
Other Structural Failure	3	257
Pipe Rupture or Leak	8	2,316
Hose Rupture or Leak	5	31
Valve Failure	2	2,075
Pump Failure	4	485
Flange Failure	1	1
Gasket Failure	4	372
Other Equipment Failure	26	14,832
Tank Overflow	20	16,987
Improper Equipment Handling or Operation	9	957
Other Personnel Error	8	193
Intentional Discharge	9	19,957
Natural or Chronic Phenomenon	6	2,799
Unknown Cause	25	9,263
Total	135	72,675
Onshore Oil or Gas Production Facility		
Transportation Pipeline Rupture or Leak	2	11
Other Structural Failure	5	482
Pipe Rupture or Leak	23	21,171
Hose Rupture or Leak	10	413
Loading Arm Failure, Rupture or Leak	1	42
Valve Failure	2	44
Pump Failure	2	521
Other Equipment Failure	4	157
Tank Overflow	4	115
Other Personnel Error	2	35
Highway Accident	3	4,920
Aircraft Accident	1	4,000
Natural or Chronic Phenomenon	2	19
Unknown Cause	4	2,780
Total	65	34,710

Table 2.1. (Continued)

	NO. OF SPILLS	GALLONS SPILLED
Highway Vehicle Liquid Bulk		
Tank Rupture or Leak	9	33,250
Transportation Pipeline Rupture or Leak	1	2
Other Structural Failure	3	240
Pipe Rupture or Leak	5	2,590
Hose Rupture or Leak	12	632
Manifold Rupture or Leak	1	1
Valve Failure	5	2,590
Pump Failure	2	2,193
Gasket Failure	4	1,210
Other Equipment Failure	3	7,175
Tank Overflow	19	11,075
Improper Equipment Handling or Operation	7	302
Other Personnel Error	6	876
Intentional Discharge	4	156
Highway Accident	71	189,125
Aircraft Accident	5	12,120
Natural or Chronic Phenomenon	1	10
Unknown Cause	5	255
Total	163	263,802
Highway Vehicle Dry Bulk		
Tank Rupture or Leak	2	52
Railroad Accident	1	10
Highway Accident	5	190
Total	8	252
Highway Vehicle General Cargo		
Tank Rupture or Leak	2	90
Transportation Pipeline or Leak	1	10
Container Lost Intact	1	120
Hose Rupture or Leak	1	20
Improper Equipment Handling or Operation	1	30
Other Personnel Error	2	15
Highway Accident	13	849
Aircraft Accident	1	100
Unknown Cause	1	1
Total	23	1,135
Highway Vehicle Passenger		
Tank Rupture or Leak	3	517
Other Equipment Failure	2	30
Improper Equipment Handling or Operation	1	2
Other Personnel Error	1	1
Intentional Discharge	2	2
Highway Accident	5	140
Unknown Cause	1	1
Total	15	693
Unknown Type of Land Vehicle		
Tank Overflow	4	450
Railroad Accident	1	2,500
Highway Accident	5	8,649
Total	10	11,599

TABLE 2.2.

U.S. Oil Spills - 1979

	NO. OF SPILLS	GALLONS SPILLED
Other Land Transportation Facility		
Hull Rupture or Leak	1	35
Tank Rupture or Leak	2	4,100
Other Structural Failure	3	16
Pipe Rupture or Leak	7	24,623
Hose Rupture or Leak	2	1,025
Manifold Rupture or Leak	1	800
Valve Failure	1	5
Other Equipment Failure	5	80
Tank Overflow	10	2,816
Improper Equipment Handling or Operation	3	23
Other Personnel Error	8	2,118
Intentional Discharge	7	11,596
Highway Accident	1	40
Aircraft Accident	1	450
Natural or Chronic Phenomenon	6	438
Unknown Cause	6	1,031
Total	64	49,196
Railway Cargo Transfer		
Other Structural Failure	1	1
Pipe Rupture or Leak	1	3
Valve Failure	1	250
Tank Overflow	1	2
Natural or Chronic Phenomenon	1	50
Total	5	306
Railway Fueling Facility		
Other Equipment Failure	1	10
Tank Overflow	1	2
Improper Equipment Handling or Operation	1	2,000
Intentional Discharge	1	32,460
Unknown Cause	2	40
Total	6	34,512
Highway Cargo Transfer		
Tank Rupture or Leak	2	115
Hose Rupture or Leak	1	1
Pump Failure	1	50
Tank Overflow	4	1,155
Improper Equipment Handling or Operation	1	50
Other Personnel Error	1	1,000
Intentional Discharge	1	1,800
Total	11	4,171
Highway Fueling		
Tank Rupture or Leak	6	4,395
Pipe Rupture or Leak	5	2,064
Hose Rupture or Leak	1	10
Pump Failure	1	50
Tank Overflow	5	505
Improper Equipment Handling or Operation	2	371
Other Personnel Error	3	260
Highway Accident	2	350
Unknown Cause	1	300
Total	26	8,305

<u>Table 2.2.</u>   (Continued)

	NO. OF SPILLS	GALLONS SPILLED
Other Pipeline		
Transportation Pipeline Rupture or Leak	1	300
Other Structural Failure	1	1,000
Pipe Rupture or Leak	12	5,495
Hose Rupture or Leak	2	51
Tank Overflow	2	600
Intentional Discharge	1	20
Natural or Chronic Phenomenon	1	20
Unknown Cause	1	84
Total	21	7,570
Onshore Pipeline		
Transportation Pipeline Rupture or Leak	13	393,526
Other Structural Failure	1	20
Pipe Rupture or Leak	227	1,672,368
Loading Arm Failure, Rupture or Leak	3	43
Valve Failure	5	1,639
Flange Failure	3	90
Gasket Failure	5	9,443
OtherEquipment Failure	4	2,335
Tank Overflow	1	150
Improper Equipment Handling or Operation	6	10,023
Other Personnel Error	1	42
Intentional Discharge	3	1,596
Natural or Chronic Phenomenon	5	5,660
Unknown Cause	3	13,160
Total	280	2,110,095
Other Onshore Non-Transportation Related Facility		
Tank Rupture or Leak	11	30,185
Dike Rupture or Leak	1	210
Container Lost Intact	1	1,800
Other Structural Failure	8	10,114
Pipe Rupture or Leak	5	776
Hose Rupture or Leak	1	1
Valve Failure	4	26
Pump Failure	1	200
Gasket Failure	2	4,600
Other Equipment Failure	25	8,167
Tank Overflow	13	2,080
Improper Equipment Handling or Operation	13	4,341
Other Personnel Error	17	1,773
Intentional Discharge	10	3,327
Highway Accident	1	3,450
Natural or Chronic Phenomenon	8	1,690
Unknown Cause	7	675
Total	128	100,388
Onshore Refinery		
Tank Rupture or Leak	1	7,000
Transportation Pipeline Rupture or Leak	2	1,360
Other Structural Failure	3	12,090
Pipe Rupture or Leak	4	640
Valve Failure	3	8,610
Gasket Failure	1	80
Other Equipment Failure	2	210
Tank Overflow	2	10,001
Other Personnel Error	2	94
Natural or Chronic Phenomenon	2	500
Unknown Factor	1	1
Total	22	40,536

Table 2.2.   (Continued)

	NO. OF SPILLS	GALLONS SPILLED
Railway Cargo Transfer		
Pipe Rupture or Leak	1	30
Tank Overflow	1	2,200
Unknown Cause	2	1,015
Total	4	3,235
Railway Fueling Facility		
Tank Rupture or Leak	1	20,000
Pump Failure	1	20
Other Equipment Failure	3	1,299
Tank Overflow	2	1,925
Other Personnel Error	1	300
Intentional Discharge	1	200
Railroad Accident	1	50
Natural or Chronic Phenomenon	2	35
Unknown Cause	1	300
Total	13	24,129
Highway Cargo Transfer		
Tank Rupture or Leak	2	800
Valve Failure	1	200
Tank Overflow	2	100
Improper Equipment Handling or Operation	2	301
Other Personnel Error	1	2
Highway Accident	1	1,000
Total	9	2,403
Highway Fueling		
Hull Rupture or Leak	1	2,000
Tank Rupture or Leak	4	6,380
Pipe Rupture or Leak	1	10
Valve Failure	1	300
Pump Failure	1	350
Tank Overflow	6	691
Improper Equipment Handling or Operation	2	51
Other Personnel Error	1	25
Highway Accident	1	250
Total	18	10,057
Other Pipeline		
Tank Rupture or Leak	1	3
Pipe Rupture or Leak	8	10,715
Hose Rupture or Leak	3	30,564
Unknown Cause	1	3
Total	13	41,285
Onshore Pipeline		
Hull Rupture or Leak	1	8,000
Transportation Pipeline Rupture or Leak	8	29,865
Pipe Rupture or Leak	200	1,037,299
Hose Rupture or Leak	1	10
Valve Failure	4	9,520
Gasket Failure	1	200
Other Equipment Failure	3	1,008
Improper Equipment Handling or Operaton	1	20
Other Personnel Error	1	1
Intentional Discharge	2	18,200
Unknown Cause	1	15
Total	223	1,104,138

Table 2.2. (Continued)

	NO. OF SPILLS	GALLONS SPILLED
Onshore Non-Bulk Cargo Transfer		
Tank Rupture or Leak	1	400
Hose Rupture or Leak	1	15
Gasket Failure	1	210,000
Improper Equipment Handling or Operation	1	2
Other Personnel Error	1	300
Total	5	210,717
Other Land Vehicle		
Tank Rupture or Leak	3	2,000
Pipe Rupture or Leak	1	70
Hose Rupture or Leak	4	49
Tank Overflow	2	930
Improper Equipment Handling or Operation	3	4,780
Intentional Discharge	2	65
Railroad Accident	1	10,000
Highway Accident	5	1,657
Aircraft Accident	1	10
Natural or Chronic Pehnomenon	1	10
Unknown Cause	4	436
Total	27	20,007
Rail Vehicle Liquid Bulk		
Tank Rupture or Leak	1	1,000
Other Equipment Failure	2	30,150
Tank Overflow	1	2,000
Railroad Accident	10	61,170
Natural or Chronic Phenomenon	1	2,000
Total	15	96,329
Rail Vehicle General Cargo		
Flange Failure	1	1,000
Railroad Accident	5	3,480
Highway Accident	1	500
Unknown Cause	1	4,000
Total	8	8,980
Rail Vehicle Transfer		
Tank Rupture or Leak	1	800
Hose Rupture or Leak	1	1,800
Railroad Accident	2	4,225
Total	4	6,825

Table 2.2. (Continued)

	NO. OF SPILLS	GALLONS SPILLED
Power Plant		
Tank Rupture or Leak	1	10,000
Transportation Pipeline Rupture or Leak	1	5
Dike Rupture or Leak	1	40
Other Structural Failure	3	703
Pipe Rupture or Leak	8	2,142
Loading Arm Failure, Rupture or Leak	1	10
Valve Failure	9	1,875
Pump Failure	2	7
Flange Failure	1	50
Gasket Failure	1	30
Other Equipment Failure	7	270
Improper Equipment Handling or Operation	4	18,022
Other Personnel Error	6	113
Natural or Chronic Phenomenon	2	2
Unknown Cause	4	585
Total	51	33,854
Pipeline Within Non-Transportation Related Facility		
Other Structural Failure	1	51
Pipe Rupture or Leak	11	43,447
Hose Rupture or Leak	1	5
Gasket Failure	1	700
Other Equipment Failure	1	1
Other Personnel Error	1	2,000
Intentional Discharge	1	3
Unknown Cause	2	30
Total	19	46,238

3. Facility Layout and Planning

3.1 <u>GENERAL</u>

Site selection has become increasingly difficult since the advent of the era of environmental concern. Numerous objections are raised when potential sites are studied in highly populated and industrial areas. Whereas workers once selected to live as close to the plant as possible, this is no longer the case. When sites are selected in isolated areas, the conservation of virgin territory becomes a predominant factor. Attempts to gain authority to construct a marine terminal on the Southeast coast (right or wrong) has resulted in an expenditure of $12 million just to overcome environmental objections and legal injunctions.

There is a dire need for a new industry to consider such factors as access to the labor market and raw materials, access to the product markets and transportation facilities, and the availability of adequate energy sources. Also of considerable importance is the accessibility to waste disposal facilities. The waste transportation and disposal sources should have federally issued permits to handle the wastes generated in the production process.

The site planners must satisfy the federal and local governments and the general public that socioeconomic goals will be met and that the natural ecosystems and microenvironments will be fully protected from plant operations, from raw material through to finished product.

One noticeable European trend is to develop industrialized port facilities that eliminate the transportation of raw materials into the hinterland and finished products back to the shipment center. There are indications that the European trend is extending to the U.S. coastal areas.

Figure 3.1 provides an insight into three decisional approaches to site selection.

22

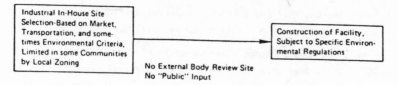

Traditional Siting Scenario (excluding nuclear facilities)

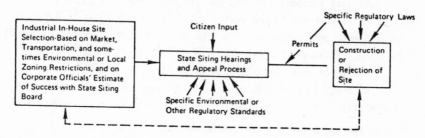

Siting Under Typical State Site Evaluation Law of the 1970s

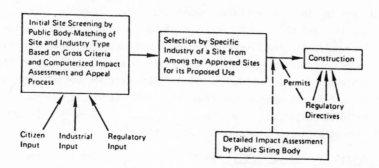

Alternative "Public" Site Selection Scenario for Major Industrial Facilities

SOURCE: D. Stever, U.S. Department of Justice, Washington, D.C.

FIGURE 3.1.

Three Decisional Approaches
to Site Selection

3.2 BULK STORAGE TANK SITING

The environmental and economic exposures developed from the design of bulk storage tank facilities warrant special consideration.

The trend toward "jumbo" storage tanks of 500,000 barrel capacity (79,000 m^3) and even more has created loss potentials equal to those of many new process units. A 500,000-barrel tank filled with crude at early 1980 prices of $25/barrel would have a content value of $12.5 million. The loss of a bulk container can increase this figure to $15 million.

A past loss incident involved two 120,000-barrel (19,000 m^3) floating roof tanks within a common diking. The common diking was a major factor in their destruction and the loss of over 50,000 barrels (8,000 m^3) of gasoline. A third 120,000-barrel (19,000 m^3) tank some 85 ft (26 m) away in a separate dike suffered only light fire damage to its upper shell. Incidents such as this give basis for considering separate diking of each storage tank.

Wide spacing between tanks and between tank fields and processing locations is most advisable. Even if closely adjacent tanks escape being ignited, they can be damaged by radiant heat. The Industrial Risk Insurers recommend that tank spacings be at least equal to the diameter of the largest tank and that large storage tanks containing flammable products be located at least 500 to 1,000 ft (150 to 300 m.) from plant processing units. The National Fire Prevention Association specifies tank separation distances by tank diameter and the combustibility and flammability of the tank's content. The recommended separation distances are not as great as the risk insurers, as can be seen from Table 3.1. As far as practical, tanks should be installed on level ground; they should never be located uphill from other tanks or processing facilities.

3.2.1 Protection Against the Elements

Site selection warrants extensive investigation into potential flooding and the susceptibility to damage from hurricanes, earthquakes, or tidal

TABLE 3.1.

Minimum Tank Spacing (Shell-to-Shell)

	Floating Roof Tanks	Fixed Roof Tanks	
		Class I or II Liquids	Class IIIA Liquids
All tanks not over 150 feet diameter	⅙ sum of adjacent tank diameters but not less than 3 feet	⅙ sum of adjacent tank diameters but not less than 3 feet	⅙ sum of adjacent tank diameters but not less than 3 feet
Tanks larger than 150 feet diameter			
If remote impounding is in accordance with 2-2.3.2	⅙ sum of adjacent tank diameters	¼ sum of adjacent tank diameters	⅙ sum of adjacent tank diameters
If impounding is around tanks in accordance with 2-2.3.3	¼ sum of adjacent tank diameters	⅓ sum of adjacent tank diameters	¼ sum of adjacent tank diameters

NFPA Table

waves. On the Texas Gulf Coast, a combination of high winds and high tide caused flooding in a plant normally protected from high water by a flood control dike. All electric motors at grade level were damaged and had to be rewound, many with replacement of bearings and almost 40 percent with replacement of motors. Remedial action after this expensive learning process resulted in motors, pumps, and so on, being raised above the flood's high water mark.

Also on the Texas Coast under hurricane conditions, the contents of waste treatment ponds were washed away and partially full product tanks floated off their foundation. Bulk storage tanks in the most recent Johnstown, Pennsylvania flood also floated clear of their foundation.

In the hurricane belts, such as the Gulf of Mexico, where the land areas are prone to flooding, special precautions are warranted. The weight of the tank itself and its partial content should not be considered a suitable means of anchorage. With adequate warning, it may also be advisable to fill tanks with water, contamination being preferable to tank damage and possible loss of the content.

Earthquakes present a number of problems, sometimes unsurmountable, depending on the severity of the tremor or tremors. Structures in earthquake prone locations are generally designed to resist the dynamic forces of earthquakes by use of equivalent static loads, a lesson learned from experience. In most areas, local building codes demand such designs. Basically, foundations become a major design factor. No structural design system is adequate unless it transmits all forces acting upon it into adequate support in the ground. In the San Francisco Bay area one plant is known to have supported tanks containing critical materials on a spring foundation.

3.3 OLD SITES

Older plants suffer greatly from lack of space. The frontage is usually bound by an access roadway, with adjacent plants on either side. They generally back up to a waterway. This situation restricts the ability to provide or expand spill containment systems. Therefore, special environmental management procedures are demanded.

At many product transfer locations, especially tank car loading racks, overfills have saturated the earth surrounding the railroad tracks. It was also a standard procedure to drain the residue from tank cars directly onto the tracks prior to a product change. Past sins are now causes of serious plant problems. Ground water contamination has occurred at many locations. At other facilities, the spilled material is continually leaching out to the shoreline to develop a sheen that can be most prominant during period of low tide. It is common to find wells drilled down to ground water along a shoreline or even in the center of a plant. These are not water wells but wells used to recover oil spilt over many years of operation.

In an attempt to terminate the problem, one Baltimore, Maryland, plant dug a deep trench and filled it with impermeable clay to control material flow of the spilled product and contain it on the plant property.

Dual drainage systems have also been constructed--one to direct surface water from "clean areas" out to a direct discharge and the second to service "dirty areas," such as bulk storage and processing locations. Surface water from these "dirty areas" is directed through concrete-lined trenches (generally covered to prevent the entry of debris) to waste-water treatment facilities or oil and water separation systems. Plants lacking space to treat wastewater to the stage now demanded by Federal law have, for a fee, connected the discharge into municipally operated wastewater treatment plants.

In many "dirty water" drainage systems, oil spill retainment sumps have become state of the art. Figure 3.2 shows a design suggested by the Chemical Manufacturers' Association (CMA).

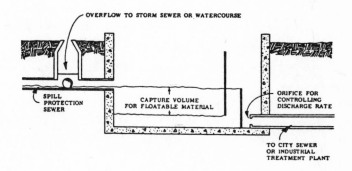

FIGURE 3.2.

Spill Material (Floatable) Containment Sump
(From CMA Guidelines)

3.4 NEW SITES

Once the protocol for obtaining a permit to construct a new hydro-carbon processing or distribution facility has been satisfied, it will be found that new facilities have operational advantages over "old sites." The new technologies can be incorporated into new plant design. Space problems can be eliminated. Modern drainage and waste treatment facilities can be provided with monitoring systems at strategic locations to detect any flow that deviates from the normal. Newly erected tanks can be internally coated while new to protect against internal corrosion. Vapor collection systems can be installed to reduce the fire and explosion hazards and to reduce the odors normally associated with handling hydrocarbon products. In short, some or all of the items contained in this guideline manual can be introduced in the facility.

References:

1. "Long Range Environmental Outlook." Proceedings of a National Academy of Sciences' Workshop.

2. "Tank Farms--Loss Prevention and Protection." Bulletin 6805M. Industrial Risk Insurers.

3. "Flammable and Combustible Liquids Code 1977." National Fire Protection Association.

4. Bulk Storage Tanks

4.1 GENERAL

Spills of hydrocarbon products from bulk storage tanks can generally be attributed to the following causes:

- Tank overfill during the filling process whereby oil is released to the environment through the tank's roof-mounted venting system.

  At some time or another, most tanks have been subjected to an overfill and the evidence can be viewed by product streakage down the side of the tank. Following a recent overfill, pools of product could be seen within the diked or otherwise spill-restrained area.

- Loss of product from vapor release during exceptionally warm weather and during a tank filling operation.

  Section 4.4 which follows details the extent of product loss from vapor emission.

- Accidental release of stored oil through bulk storage tank steam heating coils.

  The coils are used generally in northern climates to fluidize heavy oils. The coil corrodes from the inside and when idle the oil, under gravity head pressure, escapes through the coil's return or open discharge line.

- A worn seal on a floating roof tank.

  The warn seal permits rainwater to enter a tank. Frequently, water drains can become frozen shut with ice, resulting in heavy accumulations of water on the tank top. One USN fueling station experienced a problem with drifting snow that collected on one side of the tank top, causing the entire floating roof to tip and displace the jet fuel content.

29

Floating roof tanks that have worn seals can also
accumulate rainwater. A similar situation exists
for fixed roof tanks having badly corroded heads.
Flights over bulk storage tank fields will quickly
indicate that the latter water contamination source
is not an isolated condition.

- Loss of product during the condensate water drainage procedure.

Night and day temperature changes result in a buildup
of condensate water within the inner surfaces of the
storage tank. This can be further complicated with
water that has entered the product during transfer
from a ship, barge, or other transportation vehicle.

The water, being heavier than the hydrocarbon product,
collects in the lower reaches of the vessel, specifically the tank bottom. Most organizations draw the
accumulated water from the tank on a regular cycle to
reduce internal corrosion and to avoid product contamination. Water draining can be a lengthy procedure,
dictated largely by the size of the storage vessel and
the humidity associated with the geographical area.
Unless careful monitoring is exercised, massive
quantities of product can be drained from the tank
when the water and oil interface is reached. One
marketing terminal in Massachusetts spilled 250,000
gallons of #2 fuel oil when a tank attendant forgot
to close the water draw valve during a draining procedure. This oil escaped into a navigable waterway
through a partially opened rainwater drain valve within
the tank's diked enclosure. This drain valve was of
flap type design under which a stone had become lodged.

- Tank failure.

Complete tank failure is not a common occurrence, although
a major spill did occure in Europe when a new tank failed
during its first filling. The tidal wave of product spilled
from the tank washed away an earthen dike, releasing the
spilled material into a heavily traveled canal system serv-
ing a number of countries. Cleanup contractors from all
over Europe were needed to mitigate the effects of spill.

- Seam leakage.

Minor leaks from tank seams and corroded openings
are common. Visual examination of any large tank
field will provide rapid proof of this. There are
still many riveted seam tanks in service at both
industrial and federal storage facilities. Riveted
seams and riveted heads are quite prone to leakage,
especially the lowermost shell to bottom joint, which
is often underwater during periods of heavy rain and
is thereby vulnerable to corrosion.

- Tank bottom deterioration.

Tank bottoms are subject to extensive corrosive action;
many storage tanks will require tank bottom replacement
during their operating life. The detrimental effects
of bottom head and even shell plate corrosion depend
upon its rate of penetration. Corrosion affecting large
areas of steel plate is not so likely to penetrate the
metal as rapidly as localized corrosion on small areas.

Localized corrosion may be in the form of pitting.
The pits result from the formation of "rust blisters"
that eat into the metal. The pits can be small as a
pinhead or as large as a half dollar, and they may be
widely scattered or so close that a honeycomb surface
results. Obviously, scattered pitting does not weaken
the metal to the extent of closely spaced pitting.

Riveted seams are also exposed to metal deterioration
from "grooving." This is a form of deterioration that
eats into the steel plate parallel to an overlapped
seam. It results from a combination of stress concentration and localized corrosion. Dished heads at their
attachment to a tank shell are also subject to this form
of metal deterioration. Since the defect occurs at locations subjected to stress concentration, it can result
in failure unless detected and suitably repaired
by welding.

There is no doubt that lower head failure due to
metal deterioration can result in extensive spills
that go undetected for a lengthy period of time.
The leakage can continue until it seeps out from
under the tank foundation or until a low liquid
level alarm is activated.

● Tank support failure.

Another source of total loss of a tank's content
can be attributed to tank support failure. Horizontal storage tanks supported by masonry supporting walls have been exposed to total failure
of one support. Total loss of product has occurred
at three known locations where the unsupported end
of the tank ruptured upon impact with the earth or,
in one case, a poured concrete foundation.

4.2 CORROSION CONTROL

Corrosion control is one of the main approaches to spill prevention
from metal tanks. In recent years there have been great advances in the
development of synthetic paints and coatings to combat corrosive deterioration on steel surfaces. The protective coatings are of excellent quality
and durability if applied over properly prepared surfaces. It has been
quite common practice throughout the petroleum industry to leave steel surfaces unpainted for at least a year. This action permits some rusting of
the steel to occur in the form of minor pitting. Additionally, this has

been an accepted method of removing mill scale from steel plate and pipe without any extensive labor. The initial rusting roughens the steel surface, thereby increasing the cohension factor of the protective coating. Many plants continue to operate in the described manner.

The National Association of Corrosion Engineers (NACE) has, however, conducted extensive investigations and suggests that steel surface preparation can best be accomplished by abrasive blast cleaning or through the medium of water blasting prior to coating or recoating.

4.2.1 Abrasive Blast Cleaning

The NACE Technical Committee assigned to abrasive blasting investigations reported that "the blast cleaning method of surface preparation is a relatively simple procedure having only two end components: The abrasive and its velocity." There is, however, an increasingly wide range of abrasives and the selection of the right abrasive is most important to gain the optimum result.

The size of the abrasive should range between 20 to 30 mesh, as most abrasives recommended for surface preparation are used in these sizes. After cleaning tank surfaces, it is extremely rare that the abrasive can be recycled and reused economically. Under these circumstances the following abrasives are recommended for general blast cleaning where the abrasives cannot be economically reclaimed:

- Silica Abrasives
- Slag Sand
- Slag Shot
- Flint Abrasives

The cleaned surface should, according to the National Association of Corrosion Engineers, be cleaned to one of the following standards:

12.3.1 NACE No. 1 White Metal Blast Cleaned Surface
Finish: Defined as a surface with a gray-white (uniform metallic) color, slightly roughened to form a suitable anchor pattern for coatings. This surface is free of all oil, grease, dirt, mill scale, rust, corrosion products, oxides, paint, and other foreign matter (comparable to SSPC-SP 5-63, White Metal Blast Cleaning)

12.3.2 NACE No. 2 Near-White Blast Cleaned Surface

Finish: Defined as a surface from which all oil, grease, dirt, mill scale, rust, corrosion products, oxides, paint, or other foreign matter have been removed except for light shadows, streaks, or slight discolorations (of oxide bonded with metal). At least 95% of any given surface area has the appearance of NACE No. 1, and the remainder of that area is limited to slight discolorations (comparable to SSPC-SP 10-63T, Near-White Blast Cleaning).

12.3.3 NACE No. 3 Commercial Blast Cleaned Surface

Finish: Defined as a surface from which all oil, grease, dirt, rust scale, and foreign matter have been completely removed and all rust, mill scale, and old paint have been removed except for slight shadows, streaks, or discolorations caused by rust stain or mill scale oxide binder. At least two-thirds of the surface area shall be free of all visible residues, and the remainder shall be limited to light discoloration, slight staining, or light residues mentioned above. If the surface is pitted, slight residues of rust or paint are found in the bottom of pits (comparable to SSPC-SP 6-63, Commerical Blast Cleaning).

12.3.4 NACE No. 4 Brush-Off Blast Cleaned Surface

Finish: Defined as a surface from which oil, grease, dirt, loose rust scale, loose mill scale, and loose paint are removed, but tightly adhering mill scale, rust, paint, and coatings are permitted to remain if they have been exposed to the abrasive blast pattern, so that numerous flecks of the underlying metal are uniformly distributed over the entire surface.

The quality and durability of the protective coating to be applied over the cleaned metal surface is highly dependent on the cleanliness of the metal. In consideration of this fact, NACE No. 1 and No. 2 surface finishes are recommended.

The cost of abrasive blast cleaning varies geographically. In the New England area, blast cleaning costs about $1.50/sq. ft., according to one contractor. In the Washington, D.C. area, a tank cleaning organization reports that a first-class blast cleaning followed by a single application of "Glass Armor"* costs $20,000 for a 50,000-gallon capacity fuel tank.

4.2.2 Water Blast Cleaning

The NACE reports that water blasting is well suited for cleaning steel and other hard surfaces when sand blasting is not feasible. In essence, sand blast cleaning is preferable if it can possibly be used. Because water blasting has no abrasive effect on steel or other hard surfaces, it does not provide an anchor pattern for coating adhesion: its use is recommended primarily for maintenance painting programs.

It should be noted, however, that sand can generally be introduced into the water stream if a more thorough cleaning is desired. This in effect introduces a wet sand blasting techniques, which aids in dust control but develops muddy conditions at grade level.

Actually, conventional wet sand blasting requires heavy duty compressors to develop the cleaning force and large mixing tanks to store the sand and water mix. This has in recent years been circumvented by the development of water spray fingers that surround the sand discharge nozzle to "wet" the sand in jet suspension. The water supply, a hose line, is connected to an annular collar from which it enters the water supply "fingers" to mix with the dry sand some 3 in. forward of the sand outlet. This design effectively controls dust generation, and heavy-duty compessors/ pumps are not required.

---

*A product of Bridgeport Chemical Corporation, Pompano Beach, Florida.

Water blasting usually removes anything that is not tightly adhered to the surface, and it is an effective technique for cleaning irregularly shaped surfaces such as valves, flanges, back-to-back angles, grating, and floor plates.

Tables 4.1 and 4.2, taken verbatim from NACE publication "Surface Preparation of Steel and Other Hard Materials by Water Blasting Prior to Coating or Recoating," provide data on cleaning time for coated and uncoated steel.

The equipment used for cleaning consists of a trailer-mounted high pressure pump driven by gasoline, diesel, electric, or air motors or engines. The units can be operated at pressures that range from 2,000 to 10,000 psi. Experience has shown that for most practical purposes a pressure range from 2,500 to 5,000 psi is adequate. Pressures above 5,000 psi constitute a hazard because they are difficult to handle and place undue stress and strain on the operator.

Detergents and other cleaning chemicals and hot or cold water can be used to suit a particular application. However, the cleaning agents must be removed from the surface prior to the application of a protective coating.

For cleaning most ferrous metal, rust inhibitors are injected at the nozzle or at the water supply to prevent oxidation of bare metal.

Water blasting techniques are similar to those of sand blasting. The nozzle is normally held 6 to 10 in. from the surface being cleaned. In some instances a distance of 2 to 3 ft. may achieve the desired cleaning action. The best nozzle angle for cleaning purposes can be readily detected by trial and error, although for brittle substances such as dead paint or rust scale the nozzle normally is held perpendicular to the surface.

4.2.3 Internal Surface Coating - General

It has become an accepted practice to coat the complete interior of a bulk storage or production vessel. However, in the petroleum production industry coatings are sometimes restricted to the following areas.

TABLE 4.1.   Range of Time Required to Water Blast
Uncoated Steel* (Square Feet Per Hour)

Structural Classification	Conditions						
	1	2	3	4	5	6	7
A	300-500	175-350	150-300	100-200	75-150	200-400	75-150
B	450-600	325-450	275-400	150-300	100-225	350-500	100-225
C	500-800	375-625	300-525	200-450	125-375	400-700	125-375
D	600-800	450-725	400-600	250-550	150-450	500-800	150-450
E	150-400	90-275	80-250	50-150	25-100	200-400	75-150

**Structural Classification of Fabricated Items
Including Structural Steel and Piping**

A — Small size members having less than 1 square foot per lineal foot.
B — Medium size members having 1-3 square feet per lineal foot.
C — Members having greater than 3 square feet per lineal foot.
D — Flat surfaces such as vessels, tanks, checker plate, floors, and undersides of floor plates, etc.
E — Gratings.

Surface Condition

1. Mud and loose foreign matter (Note: waterblasting will not remove adherent mill scale).

2. 10% of surface covered with rust, loose scale, and loose foreign matter.
3. 30% of surface covered with rust, loose scale, and loose foreign matter.
4. 60% of surface covered with tightly adhering hard rust and pitting.
5. Greater than 60% of surface covered with tightly adhering rust and pitting.
6. Very light coating of oil and grease. No corrosion present.
Note: If corrosion exists on 6 and 7 (under the oil and grease), classify by the appropriate condition 2 through 5 above.

*All data are based on actual experience. The production was less in some isolated cases and greater in others. General conditions, environment, accessibility to work area, personnel protection, and equipment protection requirements will exert some influence on the time required.

TABLE 4.2.   Range of Time Required to Water Blast Coated Steel*
(Square Feet Per Hour)

Structural Classification	Conditions							
	1	2	3	4	5	6	7	8
A	200-400	175-350	150-300	100-200	75-150	200-400	75-150	5-25
B	350-500	325-450	275-400	150-300	100-225	350-500	100-225	5-30
C	400-700	375-625	300-525	200-450	125-375	400-700	125-375	10-30
D	500-800	450-725	400-600	350-550	150-450	500-800	150-450	12-40
E	100-300	90-275	80-250	50-150	25-100	200-400	75-150	5-25

**Structural Classification of Fabricated
Items Including Structural Steel and Piping**

A — Small size members having less than 1 square foot per lineal foot.
B — Medium size members having 1-3 square feet per lineal foot.
C — Members having greater than 3 square feet per lineal foot.
D — Flat surfaces such as vessels, tanks, checker plate, floors, and undersides of floor plates, etc.
E — Gratings.

Surface Condition

1. Finish coat of paint weathered thin; paint has chalked; a very minor amount of contaminants and other foreign matter present; and no corrosion of substrate.
2. Finish coat of paint weathered thin and some primer showing. 10% of surface covered with rust, loose scale, and loose paint film.

3. Finish coat of paint thoroughly weathered, and considerable primer shows. Approximately 30% of surface covered with paint, rust, and corrosion scale. Some surface pitting and paint blistering.
4. Finish coat of paint thoroughly weathered with almost all primer showing. Approximately 60% of surface covered with tightly adhering rust. Some surface pitting and paint blistering.
5. Finish coat of paint and most primer completely worn off. Most of surface covered with hard rust and pitting.
6. Very light coating of oil and grease.
7. Heavy coating of oil and grease.
8. Asphalt mastic and coal tar coatings to be removed between 1/32 and 1/4 inch thickness.

*All data are based on actual experience. The production was less in some isolated cases and greater in others. General conditions, environment, accessibility to work area, personnel protection, and equipment protection requirements will exert some influence on the time required.

- The bottom of the tank and 18 in. up the wall of
  the shell (including pumps, striker plates, and
  exterior drain lines). This is the collection
  point of water that has separated from the hydrocarbon product, either accumulations of condensate
  water or water that has mixed with the product during
  transport or at other periods of production or storage. The water is normally drawn off through specially designed "water draw/drain valves," which are
  described later in this report.

- The underside of the tank roof and 18 in. down the
  wall of the shell. This practice is exercised if
  hydrogen sulfide bearing crude or condensate is to
  be stored or oxygen is not excluded.

It is known that some 90 percent of the U.S. tanker fleet now have
the interior of the cargo compartments "epoxy" coated; there appears no
inventory available to determine the percentage of on-land tanks that have
been treated.

4.2.4 Coating Procedures and Products

Bulk storage tanks are most commonly constructed of iron and steel for
reasons of strength, hardness, durability, and cost control. The major
limitations to iron and steel tanks are weight and vulnerability to corrosion and rusting.

To prepare a metal surface for the application of a corrosive resistant
coating, the metal must first be cleaned of any contaminant that will interfere with the full development of adhesion of the protective paint and coating systems. Most coatings adhere properly when surface "anchor patterns"
are developed by sand blasting.

For optimum adhesion or mechanical bonding, an anchor pattern of between
1 and 2 mils in depth is required. The depth of the anchor pattern is controlled mainly by the selection of the abrasive and the pressure applied to
the cleaning process.

The Glidden Coating and Resins Division of SCM Corporation describes other cleaning systems that are alternatives to sand blasting:

- <u>Water Washing</u> - This method is used for removing water-soluble chemicals or foreign material. Care must be taken to prevent extended contact of the water with the iron or steel surface, since this may result in rust formation.

- <u>Steam Cleaning</u> - Steam cleaning is usually accomplished with a "Steam Jenny." The "Jenny" may use steam alone or in combination with cleaning compounds or detergents. The cleaning compound residue should be rinsed from the surface with water following the steam cleaning. Steam cleaning is effective for removing oils, greases, and various water-soluble chemicals.

- <u>Weathering</u> - Natural weathering is often used as one of the most economical methods of removing mill scale, dislodging it by the development of rust. Negatively, poor appearance prevails during the weathering period and heavy rust must be removed before applying finishes.

- <u>Solvent Cleaning</u> - Solvents such as mineral spirits, xylol, tuluol, and so on, may be used to remove greases or similar solvent-soluble foreign materials from the surface of the iron or steel. If rags are used for application of these solvents, they should be changed periodically so that the accumulated oil or grease is not redeposited on the metal surface.

  Solvents may also be used in vapor degreasing units. This method avoids the redeposit of oils and greases on new areas of metal surfaces being cleaned. The solvents used in such cases are usually those with non-flammable characteristics, such as perchlorethylene.

- Flame Cleaning - Flame cleaning is often used to dislodge foreign particles or mill scale on the surface of hot rolled steel. Due to differences in expansion and contraction of the mill scale as compared to the steel substrate, the mill scale is broken loose when a very hot flame is played over the surface. This process may be acceptable at the tank builder's plant but not in a refinery or petroleum storage facility.

- Acid Cleaning or "Pickling" - Acid cleaning is usually done in shops, not in the field. When properly controlled, such cleaning will remove mill scale and foreign materials while producing a very fine anchor pattern. Thorough rinsing of the surface after pickling is necessary to remove all traces of the acid, the presence of which may adversely affect the adhesion and performance of protective coatings.

- Hand-Tool Cleaning - This is a mechanical method of surface preparation involving wire brushing, scraping, chipping, and sanding. It is not the most desirable method of surface preparation, but it can be used for mild exposure conditions. Optimum performances of protective coatings systems should not be expected when hand-tool cleaning is employed.

- Power-Tool Cleaning - This mechanical method of surface preparation is widely used in industry and involves the use of power sanders or wire brushes, power chipping hammers, abrasive grinding wheels, needle guns, and so on. Although usually more effective than hand-tool cleaning, it is not considered adequate for use under severe exposure conditions or for immersion applications.

Naturally, for newly constructed or existing tanks of large dimension and capacity, abrasive blast cleaning develops a cleaner surface, is not as labor intensive as other processes, and generally is the cheaper cleaning process.

U.S. paint manufacturers produce a wide range of coatings that includes the following generic types:

Alkyd
Bituminous
Epoxy
Latex
Oil Base
Polyurethane
Silicone

The most commonly used coating, however, for the interior painting of bulk liquid storage tanks, is coal tar epoxy. This is a rugged and strongly adherent industrial coating, highly resistant to chemicals, water, abrasion, and corrosion. The coating generally consists of a combination of a coal tar epoxy resin base and a polyamide curing agent. It can be obtained in measured containers (4 gallons of base and 1 gallon of curing agent per 5-gallon unit). The unit is mixed prior to application, and when applied it possesses high film building properties. However, an 8.0-mil dry coating provides desirable coverage. A typical cost per gallon is about $10 (1980). Unlike most protective coatings, coal tar epoxy requires no special primers for most applications. Its limitations are that it should not be used for prolong contact with strongly oxidizing chemicals, dilute alkalies, ketones, esters or alcohols, or for lining tanks used to store "white" petroleum products.

The coating can be applied by brush, roller, or spray. Two coats are recommended for best results in chemical and immersion service and up to 20 hours of drying time is suggested between coats. Coverage is about 155 ft^2/gal. at 10.5 mils wet, 8.0 mils dry. Xylol is recommended for cleanup and equipment cleaning.

Another product used for lining bulk storage tanks is "Glass Armor," a product of Bridgeport Chemical Corporation, Pompano Beach, Florida. This organization contends that the product actually builds a new tank inside an existing tank and that one-quarter of a million tanks have been "Glass Armored" over the past 22 years. The material is a specially designed formulation of thermosetting epoxy resins. The epoxy is formed by mixing a resin material with an "activator."

One bucket of the mixture covers approximately 40 ft^2 of tank surface. The manufacturer says this allows for a coating about 3/16 in. thick. An indirect-fired heater is used for curing, which cures the coating in a relatively short period of time.

Once the tank is cured, a "sparking machine" is used to check for any leaks or missed spots in the coating. Since the epoxy does not conduct electricity, the spark tester is run over the walls of the tank and where a spark appears a "holiday" is found. Repairs are made with a quick-drying epoxy. This protective process is provided by a nationwide group of franchised applicators.

Another state-of-the-art product is known as "Glassflake," produced by Glassflake International, Inc., Jacksonville, Florida 32223. This material has been used extensively on marine applications, on ship's decks, tank interiors, above water hull surfaces (freeboard), riveted strakes, sea chests, stern frame, and rudders. According to the manufacturer, it is resistant to a variety of acids, organics, and salts, as listed in Table 4.3.

Glassflake is claimed by the manufacturer to have the following advantages and features.

- Corrosion. Coating provides a solid barrier 20 mils or more thick, with minimal permeability against vapor particles that cause corrosion.

TABLE 4.3.

Approved Chemical Environments
For Glassflake Coatings*

ENVIRONMENT	% BY WEIGHT	MAXIMUM TEMPERATURE LIMITATION		
		70°F.	150°F.	200°F.
Hydrochloric Acid	to 10%	R	R	R
	10 to 38%	R	R	NR
Phosphoric Acid	to 85%	R	R	R
Sulphuric Acid	to 50%	R	R	R
Sulphuric	50 to 70%	R	R	(to 160°F)
Sea Water		R	R	R
Distilled Water		R	R	NR
Mineral Oil	—	R	R	R
Vegetable Oils	—	R	R	R
Animal Fats	—	R	R	R
Calcium Hypochlorite	to 20%	R	R	NR
Sodium Hypochlorite	to 15%	R	R	NR
Chlorine Water, Sat'd.	—	R	R	R
ORGANICS				
Asphalt (uncut)		R	R	—
Crude Oil—sour		R	R	—
Crude Oil—sweet		R	R	—
Furnace Oil		R	R	—
Gasoline		R	R	—
Kerosene		R	R	—
SALTS				
Aluminum Chloride	100%	R	R	—
Aluminum Nitrate	to 10%	R	R	—
Aluminum Sulphate	100%	R	R	—
Ammonium Chloride	100%	R	R	—
Ammonium Nitrate	to 83%	R	R	(to 170°F)
Ammonium Sulphate	to 25%	R	R	—
Barium Carbonate	100%	R	R	—
Barium Chloride	100%	R	R	—
Calcium Sulphate	100%	R	R	—
Copper Chloride	100%	R	R	—
Ferric Chloride	100%	R	R	R
Ferric Nitrate	100%	R	R	—
Ferrous Chloride	100%	R	R	—
Ferrous Sulphate	100%	R	R	—
Lead Acetate	100%	R	R	—
Magnesium Carbonate	100%	R	R	—
Magnesium Chloride	100%	R	R	—
Magnesium Sulphate	100%	R	R	—
Potassium Alum	100%	R	R	—
Potassium Chloride	100%	R	R	—
Potassium Dichromate	100% (No HF)	R	R	—
Potassium Nitrate	100%	R	R	—
Potassium Sulphate	100%	R	R	—
Silver Nitrate	100%	R	R	—
Sodium Bicarbonate	100%	R	NR	NR
Sodium Bisulphate	100%	R	NR	NR
Sodium Carbonate	to 5%	R	R	(to 180°)
	to 25%	R	(to 120°F)	NR
Sodium Chloride	100%	R	R	R
Sodium Ferrocyanide	100%	R	R	—
Tridodium Phosphate	to 5%	R	R	—
Zinc Chloride	100%	R	R	—
Zinc Sulphate	100%	R	R	—

NOTE: R—Recommended for Service
NR—Not recommended for Service
— -Inadequate information available for a recommendation;
or, inapplicable.

While recommended for use as indicated, conditions vary to such
an extent from job to job that the foregoing is not to be taken
as a representation or warranty that the coating will in all
cases perform as indicated.

*Table developed by manufacturer

- <u>Erosion</u>. Glassflake has excellent resistance to erosion. It fills pits and cavities in old steel to extend the service life before plate replacement.

- <u>Electrolysis</u>. In cavitation areas, a significant reduction in the number of sacrificial metal anodes can be made.

- <u>Abrasion</u>. The effect of mechanical damage to the hard, solid coating is minimized and confined, without undercutting or loss of adhesion to the surrounding area.

- <u>Easily Repaired</u>. The coating can be easily repaired in case of heavy damage by cleaning the damaged area and feathering edges of surrounding coating, then recoating to maintain a long-lasting coating protection.

- <u>Three Colors</u>. Glassflake coating is available in three colors--white, marine grey (especially recommended for decks), and black.

- <u>Non-Skid Finish</u>.[*] A non-skid finish is available for deck coatings, which sand is added to the resin at time of application.

- <u>Can Be Machined</u>. The Glassflake coating can be machined with conventional metal working tools, which permits the repair of pumps and valves for extended service life. It can be applied in any thickness desired without cracking or crazing.

---

[*]This type of finish may also be well suited to tank tops.

The material consists of glass flake particles approximately 1/8 in. in diameter, 2 to 3 microns thick, in a 100 percent solids, polyester resin vehicle. There are no solvents to evaporate. Polymerization of the risen vehicle at the time of application makes the coating hard, solid, and of minimum permeability. There are normally about 130 overlapping layers of the flakes to make a continuous surface that is a minimum of 30 mils thickness.

The coating can be applied by plant maintenance personnel under the direct and constant supervision of a Glassflake field engineer. All materials and application equipment are furnished by the Glassflake engineer.

The metal surface is first sand blasted to a white finish, after which a Glassflake primer is applied to prevent surface oxidation before the coating is applied. The Glassflake is then sprayed (or on small sur-faces, troweled) on the metal surface while another employee immediately rolls the coating to orient the flakes before the resin starts to dry. The coating currently (1980) costs \$3.96/lb for the Glassflake and coating material (sprayed coverage 1/2 lb/ft^2), \$3.90/lb for the catalyst (used 2-3 percent by weight to the glass and the resin), and \$1.50/lb for the styrene modifer (mix 3 gallons modifier/per gallon of curative catalyst), which is also used as a cleaning agent. Figure 4.1, 4.2 and 4.3 depict application techniques for the Glassflake coating.

Another generally accepted coating is "Glid Guard," produced by Glidden Coatings and Resins Division of SCM Corporation. This organi-zation recommends a 3-mil (dry) primer coat of Glid Guard Y5251 with Y5252 activator or curing agent applied over a sand blasted finish. This would be followed by two applications of Glid Guard Epoxy Chemical Resistant Finish Y5240 mixed with Y5242 curing agent in equal proportion. Once mixed, the coating material is left to gel for about 30 minutes; it can then be applied for a period of about 8 hours (depending on heat of day). All mediums of application can be used--spray, brush, and so on. Once completed, the dry coating would be approximately 9 mils thick.

FIGURE 4.1. Spray Application of Glassflake Type
GF-202 to Tank Wall Interior (Photograph Courtesy
of Glassflake International)

FIGURE 4.2. Application of Glassflake Type 202. Finish brushing
to ensure penetration into holes, etc. and to orient glassflakes
in resin, thus providing laminar matrix and maximum resistance
to vapor penetration and abrasion. (Photograph Courtesy of Glassflake International)

FIGURE 4.3. Application of Glassflake GF-202 to Tank Floor Using Pole Gun   (Glassflake International Photo)

One gallon of the mix will coat 350 ft^2 and the cost of the primer and finish coat is comparable, being about $17.78 for the epoxy finish and $16.27 for the curing agent. In Arizona, California, New Mexico, Idaho, Montana, Nevada, Oregon, Utah, Washington, Alabama, and Hawaii the cost is increased by about 20¢ per gallon.

Woolsey Marine Industries, Inc. produces a coating under the trade name "Res-N-Glas." This is also a polyester resin vehicle, containing 120 layers of 3-micron thick glass flakes that self-laminate into a 30-mil, 98 percent solids coating. It is precatalyzed and sprayed over the bare metal surface with conventional high-capacity spray equipment. No special technique is required--just a simple cross-path spray action. The material is claimed to be corrosion, abrasion, and impact resistant and is compatible with cathodic protection systems. Coverage is approximately 52 ft^2/gallon at 30-mils thickness. At 77°F the regular cure is hard in 4 hours and full cured in one week.

Koopers Company Inc. has developed a wide range of protective coatings. Table 4.4 provides basic data on their range of coatings.

As an additional aid toward the selection of a protective coating, Tables 4.5 and 4.6 are reproduced from the paper "How To Evaluate Urethane Coatings," by J. A. Cross as published in the October 22, 1979 issue of Chemical Engineering."

4.2.5 Flexi-Liners

An additional technique for corrosion control and protection of the inner surface of a storage tank has been developed by the Flexi-Liner Corporation, Pasadena, California 91102. The organization manufacturers balloon-like containers, tailer-made to suit a given tank size. The flexi-liner is produced with built-in connections and manholes and can be installed in welded, riveted, or bolted metal tanks. The supplier claims that the liner is easy to install and that tank preparation such as sand blasting is not required. Unskilled labor can readily install the liner, which is suspended from the top of the tank. Bonding is not required and the patented process of liner installation is suitable for any vertical, horizontal, or rectangular tank, with or without tops.

TABLE 4.4.
Koppers Company Inc. Protective Coatings

	(1) 6060-5 Isophthalic Polyester Laminate	(1) 980 Series Gel Coat Off-White	(2) Bitumastic No. 300-M	(2) Koppers Inorganic Zinc No. 701	(3) Bitumastic Non-Drying Tank Bottom Coating	(4) Koppers 200 HB Epoxy	(5) Koppers 1122B Linear Polyurethane	(5) Koppers 1515 Silicone Alkyd	(6) P-527 Strontium Chromate Epoxy Primer	(7) Koppers 654 Epoxy Primer	(8) Koppers 622 Rust-Penetrating Primer
Number of coats	Laminate	Multi-pass	2	1	1	1	2	2	1	1	1
Theoretical Coverage (ft²/gallon) 1.0 mils dry film thickness	1,604	1500	1,184	640	1,344	603	675	704	460	718	792
Film build ratio: Minimum dry film required	90 (entire laminate)	15	8-10	2.0-3.0	125 mils (¼")	6.0	1.0	1.5	1.6	1.5	1.5
Film build ratio: Wet film required	Not applicable	18	10-14	5-7	149 mils	13.8	2.4	3.5	2.8	3.4	3.0
Sq foot coverage to achieve minimum film	18	60-80	90-115	257	8	90	540	375	230	385	425
Temperature limitations: Dry	225°F	200°F	250°F	800°F	110°F	200°F	150°F	250°F	250°F	250°F	250°F
Temperature limitations: Wet	160°F	N/A	120°F	100°F	100°F	100°F	N/A	N/A	100°F	100°F	N/A
Drying-Curing Time at 70°F and 50% relative humidity: To Touch	2 hrs curing time usually required to reach Barcol hardness of 30 with 1% MEKP catalyst	20 min	3-4 hrs	15 minutes	2 hrs - semi-drying material	4 hrs	1 hr	2 hrs	1 hr	1 hr	2 hrs
Drying-Curing Time: To Recoat		1 hr (dry hard)	24 hrs	Overnight to 24 hrs		Overnight	½ hr to 24 hrs	18 hrs	2 hrs to overnight	Overnight	48 hrs
Primer	P-527	N/A	Koppers 654 Epoxy Primer	N/A	N/A	Koppers Inorganic Zinc No. 701	Koppers 654 Epoxy Primer	Koppers 622 Rust Penetrating Primer	N/A	N/A	N/A
Thinner	Styrene	Styrene	Koppers Thinner 2000	Koppers Thinner 442	Koppers Thinner 1000	Koppers Thinner 10,000	Koppers 1184 Thinner	Koppers Thinner 4000	Koppers Thinner 10,000	Koppers Thinner 10,000	Koppers Thinner 4000
Cleaner	Acetone	Acetone	Koppers Cleaner 2300	Koppers Thinner 442	Koppers Thinner 1000	Koppers Cleaner 2300	Koppers 1184 Thinner	Koppers Thinner 4000	Koppers Thinner 10,000	Koppers Cleaner 2300	Koppers Thinner 4000

(1) Pit fillers, fiberglass strips for tank bottoms

(2) Steel coating general

(3) Applied on tank bottom between foundation

(4) Used primarily for JP Fuel tanks

(5) Tank exterior coating

(6) Primer for 6060-5 Isophthalic Polyester Laminate

(7) Primer for 300 M Coal Tar Epoxy

(8) Primer for 1515 Silicone Alkyd

TABLE 4.5.

Properties of eleven different industrial coatings

	*Gloss retention	Weathering	Abrasion	Heat	Water	Salts	Solvents	Alkalis	Acids
Vinyl	6	10	8	6	10	10	5	9	10
Vinyl acrylic	8	10	8	6	9	9	4	8	9
Epoxy polyamide	3	8	9	8	9	10	8	10	8
Water-base epoxy	3	8	8	6	8	8	8	7	7
Aliphatic urethane	10	10	10	8	0	10	8	10	10
Chlorinated rubber	6	7	7	5	10	10	3	9	9
Coal-tar epoxy	5	7	9	7	10	10	4	8	10
Acrylic	9	9	8	6	9	9	4	7	7
Industrial acrylic emulsion	7	8	6	5	8	7	2	6	6
Silicone alkyd	9	9	6	7	8	8	4	4	6
Alkyd	6	8	6	7	8	8	4	4	5

Ratings: 10 Excellent
9-8 Good
7-6 Fair
5-1 Poor

*Gloss retention based on Southern Florida exposures.

TABLE 4.6.

Chemical resistance of urethane and epoxy coatings

Acids	Aliphatic urethane	Polyester epoxy	Urethane epoxy	Urethane-modified lacquer
10% HCl	Pass	Pass	Soft	Soft
10% H_3PO_4	Pass	Pass	Soft	Soft
10% CH_3COOH	Pass	Pass	Soft	Soft
10% HNO_3	Slight yellow	Soft	Soft	Failed
10% H_2SO_4	Pass	Pass	Pass	Failed
Caustics				
10% KOH	Pass	Pass	Pass	Failed
10% NaOH	Pass	Pass	Pass	Failed
Household NH_3^-	Slight yellow	Soft	Slight yellow	Failed
Solvents				
Toluol	Pass	Pass	Pass	Failed
MEK	Pass	Pass	Slight soft	Failed
Butanol	Pass	Pass	Slight soft	Failed

Reproduced from "How To Evaluate Urethane Coatings," J. A. Cross, Chemical Engineering, October 22, 1979.

Liners fabricated from any of fifteen different materials are available to contain the following products:

Acetic acid	Chromic Acid	Nitric Acid
Alcohol	Fatty Acids	Paper Mill Liquors
Alum	Formal-dehyde	Phenol
Amines	Glycerine	Phosphoric Acid
Ammonium Nitrate	Glycols	Stoddard Solvent
ASTE Oils	Hexane	Sulphuric Acid
Bleach	Hydrochloric Acid	Turpentine
Caustic	Muriatic Acid	Vegetable Oil

Almost all plating solutions and combinations of chemicals can be accommodated. Liners have been produced for tanks having two million-gallon capacity. Figure 4.4 indicates the procedures for installing a Flexi-Liner in an existing storage tank.

An additional advantage of this type of inner liner results from the fact that there is no contact between deteriorated metal and product. This eliminates contamination of the tank's content.

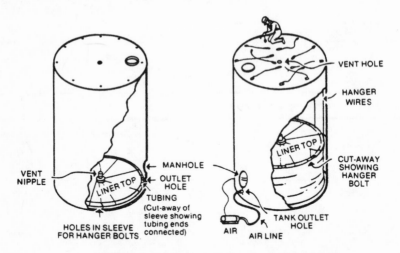

FIGURE 4.4. Procedure for Installing Flexi-Liner In Existing Storage Tank (Illustration Courtesy Flexi-Liner Corporation)

4.3 TANK SHELL THICKNESS TESTING

Many industrial facilities have tested shell thickness of bulk storage tanks for many years. There has, however, been a decided increase in
this preventive maintenance procedure over the past ten years. Three
known testing procedures are currently in use.

- Visual Examination - during regular internal tank
  cleaning (generally every five years). By this process vulnerable tank surfaces are visually examined
  for metal deterioration, and measurements are taken
  with a depth gage, equipped with a micrometer screw
  head. The depth of deterioration is then related to
  the original plate thickness as shown on the mill
  specification sheets. Records are maintained and
  the rate of plate deterioration is determined.

- Drill and Weld - With this technique the shell or
  head plate is drilled and measured with a micrometer or vernier caliper, after which the drilled
  hole is filled with weld metal. Records are maintained in line with visual examination.

- Nondestructive Testing - At least three organizations
  manufacture ultrasonic thickness measurement devices.
  Ultrasonic sound is used to penetrate the shell and
  measure the metal thickness. The result is graphically depicted on a range scale or a lighted crys-
  tal display (LCD). Measurement ranges from 0.050 in.
  to 25 in. are obtainable.

The following manufacturers can provide a range of thickness
testing devices:

Automation Industries, Inc./Sperry Products Division, Danbury, CT 06810

Model UTM-24* ($1,900) is a lightweight pocket-size instrument designed to measure thickness of steel from 0.06 in. to 4 in. This is a four-place, battery-operated unit with automated decimal location.

Model UTM-20* duplicates Model UTM-24 in cost and operation but has three-place metric readout. Figure 4.5 and 4.6 depict the UTM-24 unit.

Model NOVA-21 ($1,950) gages metal, glass, and most plastics in a range of 0.005 in. to 25 in. It has inch and metric readout and has either battery or AC energy source. The cables and transducer for this unit increase the price by $250.

UJ Reflectoscope with digital thickness gage ($5,717 plus $250 for cables and transducer) is a dual purpose unit that detects flaws in metal and can gage thicknesses from 0.010 in. to 19.99 (0.25 mm to 199.9 mm). Readout presentation is both video and digital.

Panametrics, Inc., Waltham, MA 02154

The Model 5226 Ultrasonic Thickness Gage is available from this organization. It is a lightweight portable unit that can be used on pipes, pressure vessels, and storage tanks to determine thinning of the metal due to corrosion and erosion. The specifications of the unit are as follows.

Thickness Measurement Range	0.046 to 10.000 inches or 1.00 to 199.99 mm	**Power Requirements**	100, 115, or 230 VAC, 50-60 Hz, 5W max. (Charger/AC adaptor) 6VDC, 2.6 Amp-Hr
Calibrated Accuracy	±0.004 inches or ±0.1 mm over measurement range	**Battery Type**	6V rechargeable solid gel battery
Resolution	0.001 inches or 0.01 mm or 0.1 mm	**Battery Life**	8 hours continuous operation (automatic shutoff extends battery life)
Velocity Calibration Range	0.0500 to 0.7999 in/μsec or cm/μsec		

---

*The manufacturer is in the process of replacing these units with Models UTM-100 and UTM-200, which are smaller, more accurate, and easier to use (information supplied by letter July 7, 1980).

FIGURE 4.5. Model UTM-24 Thickness Tester

FIGURE 4.6. Model UTM-24 Thickness Tester
(Photographs Courtesy of Automation Industries, Inc./Sperry Products Division)

Zero Adjustment Range	± 1.3 μsec or ± 0.15 inches (± 3.9 mm in steel)	Battery Charging Time	16 hour maximum charging time (fully discharged battery)
Display Type	4½ digit high contrast liquid crystal display with switchable backlight and fixed decimal point	Automatic Shutoff	Gage automatically shuts off power approximately 3 minutes after last reading
Display Mode	Switchable HOLD/BLANK HOLD—Display holds the last reading. Decimal point blinks to indicate hold mode BLANK—Display blanks when no reading is being made	+ Sync Output	Positive t_r < 30 nsec; Z_{out} ≈ 50Ω
		Rcvr Monitor Output	± 0-2.4V from 50Ω source, no load
Transducer Type	Dual (pitch-catch) pulse echo	Marker Output	± 4V TTL logic pulse with main bang blank pulse superimposed
Test Mode	First echo measurement	Size	2¾" × 6½" × 9½" (70mm × 165mm × 242mm)
Gage Operating Temperature Range	0° to 50°C	Weight	4¼ lbs. including battery (1.9 kg)
Transducer Operating Temperature Range	−40°C to 500°C surface temperature (intermittent use)		

This unit complete with transducer, carrying case, and cable currently (May, 1980) sells for $2,195.

NDT International, Inc., West Chester, PA 19380

NDT International conducts tank thickness testing on a contractual basis. A pricing sheet follows that provides typical costs for this type of service. The concern has developed a device that magnetically attaches to the tank and actually crawls over the tank surface to provide thickness data. The "magnetic crawler" is powered by two 12-volt motors that allow the operator to "drive" the unit on the sides of the tank.

A supply of water couplant is fed to the transducer, which is gimbel mounted under the mobile device. The design enables the operator to make continuous scans and to gage the tank shell thickness from either the top of the tank or grade level. Figure 4.7 is a view of the instrument in operation while the strip chart (Figure 4.8) illustrates a typical read-out showing thinned or corroded sections of a tank's inner shell metal.

The unit, which can also be purchased at a cost of $5,100 (June, 1980), has the following basic data.

NDT INTERNATIONAL, INC.

711 CREEK ROAD
WEST CHESTER, PA. 19380   (215) 793-1705

RATE SCHEDULE

1 Man - NDT Level II	$28.00/hr.
2 Men - NDT Level II & Helper	$40.00/hr.
Immersion Inspection	$35.00/hr.
Level III Work	$35.00/hr.
"C" Scan Inspection	$35.00/hr.
Milage	.20/mile
Expenses	Actual
Overtime	1.5 x Rate
Sundays & Holidays	2.0 x Rate
Night Shift Work	+ 15%
Minimum Charge	5 hours

Terms: Net 15 days

EQUIPMENT RENTAL

Ultrascope	$25.00/Day
Digital Hand Scanner or NDT-6D/Recorder	$45.00/Day
Magnetic Crawler	$150.00/Day
Magnetic Tapes	$15.00/ea.

In order to avoid confusion NDT International would like to specify
its policy on billing for field work. It is as follows:

> Time: All time expended is considered working time be it travel,
> report preparation, equipment clean-up,etc. and will be
> billed at normal rates for that time period.

> Expenses: Field engineers are entitled to submit expenses as follows:

> Meals: 8-10 hours   - 1 meal
>   10-12 hours   - 2 meals
>   Over 12 hours - 3 meals

All additional expenses pertaining to the job such as motel, tolls,
parking, car rental, air fare, etc. Receipts will be submitted for
all expenses over $5.00

Ultrasonic Transducers and Eddy Current Probes will be billed at
cost where unusual wear exists and causes rapid deterioration.

Effective 10/79

FIGURE 4.7. NDT International, Inc. Thickness Gage

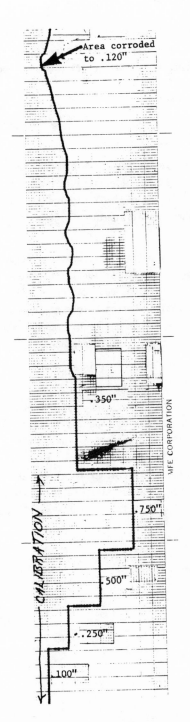

FIGURE 4.8. NDT International, Inc.
Typical Strip Chart Readout

NOTE: "Bathtub Ring" at air liquid
interface with .120" of metal
corroded away.

Required Utilities:	12 volt D.C. water at 50 psi
Required Electronics:	Ultrasonic instrument with time analog gate such as the Nortec 131 D Ultrascope
Weight:	19 lbs.
Speed:	0 to 10 ft/min.
Holding Force:	250 lbs.

4.4 BULK STORAGE TANK VAPOR EMISSIONS

When released to the atmosphere, hydrocarbon vapors develop an obvious fire hazard. In addition, the vapors generate an objectionable odor, which can be very noticeable in the area of refineries and bulk storage facilities. This odor can detract from the good neighbor relationship so eagerly sought by the oil industry.

Vapor release can also result in a heavy dollar loss, as it is actually a loss of product. The dollar loss increases in proportion to the extent of refining. Loss of fully refined product involves duplication of all costs from exploration work up to the operation where the loss of the usable liquid occurs. In short, the later the loss occurs as the oil moves from well head to the final consumer, the more costly it is to replace.

Of greater importance is the actual loss of product in a period of short, in fact, dwindling, supply.

The American Petroleum Institute (API) has calculated that one 80,000-barrel cone-roofed tank in a typical operation can be expected to lose almost 2,700 barrels of gasoline each year from evaporation. Under the federal definition, this would be considered a major spill in either coastal or inland waters if it occurred at one time rather than over the course of a year. At a current market price of about $1.30/gallon[*], the dollar loss per year would be in the vicinity of $147,420 from a single tank.

---

[*]U.S. News and World Report, Jan. 7, 1980.

Based on the number of tanks in service nationwide, although the actual
figure is not known, the total dollar loss must be astronomical.

It must also be accepted that the release of vapors is in fact a spill--
in this case a known or intentional rather than an accidental discharge.
Obviously, it behooves industry to effectively control the release of vapors
from bulk storage tanks. The intent of this section of the report is to
provide guidance toward this end.

4.4.1 Fixed Roof Tanks

The fixed or cone roof tank has become the standard container for the
storage of volatile liquids.

API reports that the amount of evaporation from a tank depends strongly
upon the true vapor pressure of the liquid at the average storage temperature. The average storage temperature for a particular site is about 5^0F
above the mean atmospheric temperature. Figures 4.9 and 4.10 depict average
storage temperatures and average daily temperature changes on a mainland
national basis. Figure 4.11 provides the true vapor pressure, using the
Reid vapor pressure, the slope of distillation curve, and the average storage temperature. Figure 4.12 provides the yearly breathing loss from the
true vapor pressure, the average tank outage (distance from liquid level to
tank top), and the tank diameter.

The inset in Figure 12 graphically shows the reduction in vapor loss
as a function of the external color of the storage tank. The reduction
resulting from painting a tank white develops the first engineering recommendation in this section of the report. Specifically, dark colors, or
company colors, should be avoided. This alone can reduce vaporized emissions
by almost 50 percent.

There is a trend to convert fixed roof tanks into internal floating
roof tanks. This can be done by installing a pipe column or guide bar
from the floor to the fixed roof. A membrane sized to suit the tank diameter is then installed within the tank, using, for example, pie sections
of stryofoam "welded" together with fiber glass and polyester resin binder.

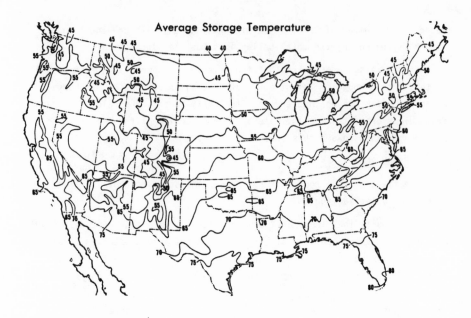

FIGURE 4.9.

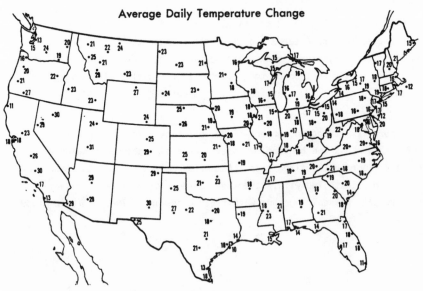

FIGURE 4.10.

(Illustrations Courtesy of
API)

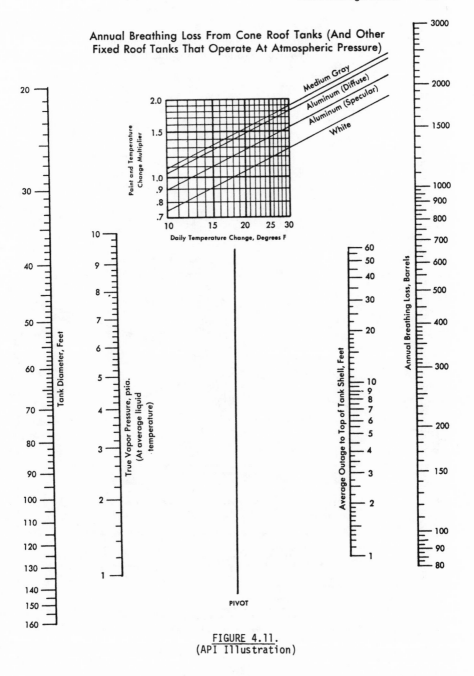

Annual Breathing Loss From Cone Roof Tanks (And Other
Fixed Roof Tanks That Operate At Atmospheric Pressure)

FIGURE 4.11.
(API Illustration)

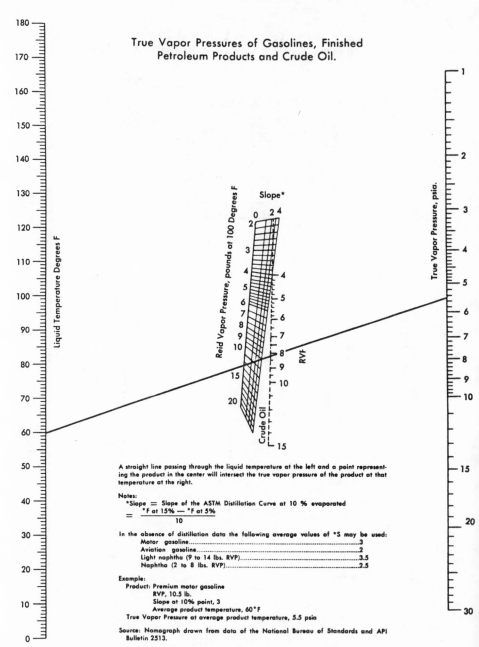

A straight line passing through the liquid temperature at the left and a point representing the product in the center will intersect the true vapor pressure of the product at that
temperature at the right.

Notes:
*Slope = Slope of the ASTM Distillation Curve at 10 % evaporated
= °F at 15% — °F at 5%
10

In the absence of distillation data the following average values of *S may be used:
Motor gasoline...3
Aviation gasoline..2
Light naphtha (9 to 14 lbs. RVP)...3.5
Naphtha (2 to 8 lbs. RVP)..2.5

Example:
Product: Premium motor gasoline
RVP, 10.5 lb.
Slope at 10% point, 3
Average product temperature, 60°F
True Vapor Pressure at average product temperature, 5.5 psia

Source: Nomograph drawn from data of the National Bureau of Standards and API
Bulletin 2513.

FIGURE 4.12.
(Figure Courtesy of Chicago Bridge and Iron Co.)

The membrane floats on the surface of the stored liquid, moving along
the central guide/pipe column with the rise and fall of the liquid level.
In this manner, the available vapor space is reduced and emissions are
reduced in proportion. Other floating membranes have been fabricated of
sheet aluminum and even suitably braced synthetic sheeting.

CBI, as one of the leading organizations in evaporation loss prevention, has developed the "Weathermaster" tank (Figure 4.13), which is a
fixed roof tank that is adapted for floating roof service. The Weathermaster tank combines the advantages of both the fixed and floating roof
tanks. The design keeps the elements--snow, ice, rain--out while keeping
the product in. These tanks are available in diameters up to 320 ft
(97.6 m).

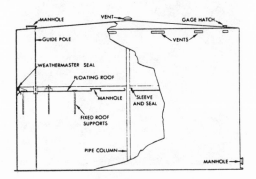

FIGURE 4.13.   Weathermaster Floating Roof Tank
               (CBI Illustration)

A fixed roof completely covers the tank, while a steel inner floating
cover, complete with a resilient foam seal, rides up and down with the
liquid level (Figure 4.13). This reduces evaporation loss and product emissions

by eliminating vapor space. The fixed roof and the inner floating seal
(Figure 4.14) effectively combat corrosion and product contamination.
In addition to a central pipe column or guide, this design incorporates
guide poles to fully control the movement of the internal floating roof.

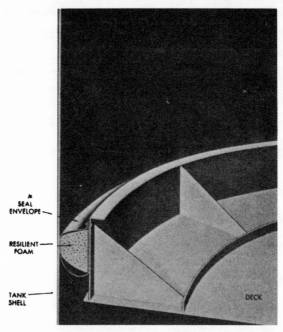

FIGURE 4.14. Horton Resilient Foam Fabric Seal for
Floating Roofs in Weathermaster Tanks   (CBI Illustration)

The design also eliminates the need for the roof drainage systems that are
required for conventional external floating roof tanks, which are described
later.

    The advantages of these designs are manifold: Maintenance costs
normally associated with floating roof tanks are reduced. There is no
snow to remove from the floating roofs, no icing on seals, and no drains
that require constant attention. Tests (CBI) have also shown that the
space between the floating roof and the fixed roof does not contain flam-
        mixtures, except for a short time immediately after the product is
        empty, or nearly empty, tank.

Internal floating roof tanks are suitable for the storage of the same products usually stored in other floating roof tanks. Precautions should be taken, however, for successful operation during butane blending or mixing procedures and storage of boiling hydrocarbons or gaseous products. Any of these conditions can cause severe agitation of the product beneath the floating roof, which could result in the product splashing onto the roof around the seals or roof support sleeves.

In summary, vapor emission losses can be drastically reduced by converting fixed or cone roof tanks to integral floating roof systems. The actual reduction in emissions cannot be readily predicted, since this is greatly dependent on knowing the extent of tank use, product throughput, number of fillings per year, and general environmental conditions (heat, cold, wind, tank color, and so on), but the reduction will be significant.

4.4.2 External Floating Roof Tanks

The standard means for controlling the release of vapor emissions is the floating roof tank. The external floating roof is in complete contact with the product and rides up and down on the surface of the liquid, thereby eliminating the vapor generation space. Roof buoyancy is maintained by annular buoyant ring known as a "pontoon." Some floating roofs are of double deck design whereby air space provides a degree of insulation from the heat of the sun and inhibits "boiling" of the stored product.

All floating roofs have some type of seal at the rim space that rides on the inner surface of the tank. There are a variety of designs that control the release of vapors. Figures 4.15 and 4.16 indicate designs used by CBI for their Horton Floating Roof Tank Seals. Figure 4.16 illustrates an advanced design whereby a secondary seal is provided for vapor containment.

All tanks of floating roof design require drains to remove rainwater from the tank top. Some floating roof drains direct the rainwater into the tank from the center of the roof. This naturally results in the accumulation of water at the bottom of the tank and a certain degree of product contamination.

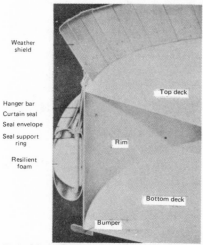

The Horton Resilient Foam Fabric Seal (SR-7) can safely accommodate rim space variations up to plus-or-minus four inches (100 mm) and can be removed from above the floating roof.

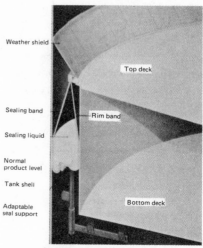

The Horton Liquid-Filled Fabric Seal (SR-5). Its exclusive adaptable seal support ring accommodates a decrease or increase in rim space of four inches (100 mm).

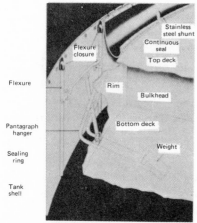

SR-1 Horton Flexure for floating roofs in tanks with butt-welded shells.

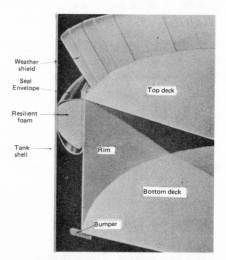

Horton Resilient Foam Fabric Seals for Horton Floating Roofs. SR-9 Seal for 8 inch (200 mm) rim space. SR-10 Seal for 12 inch (305 mm) rim space.

FIGURE 4.15. Floating Roof Seal Details
(Illustrations Courtesy of CBI)

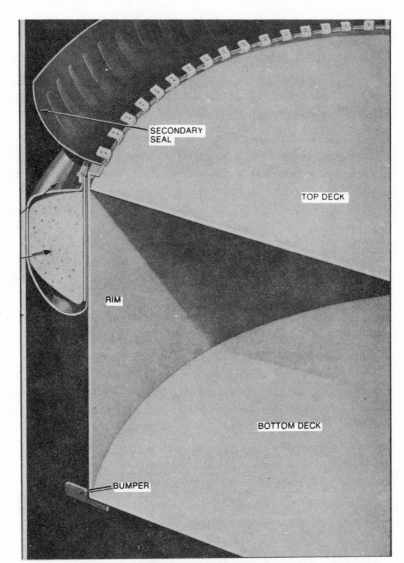

PRIMARY
SEAL
ENVELOPE

RESILIENT
FOAM

TANK
SHELL

SECONDARY
SEAL

TOP DECK

RIM

BOTTOM DECK

BUMPER

HORTON RESILIENT FOAM FABRIC PRIMARY SEAL WITH SECONDARY SEAL FOR HORTON FLOATING ROOFS
SR-9 For 8 inch or 200 mm Rim Space
SR-10 for 12 inch or 305 mm Rim Space

HORTON ® RESILIENT FOAM FABRIC PRIMARY SEAL WITH SECONDARY SEAL FOR HORTON FLOATING ROOFS
SR-9 For 8 inch or 200 mm Rim Space
SR-10 for 12 inch or 305 mm Rim Space

FIGURE 4.16.   (Illustration Courtesy of CBI)

Other drains employ flexible or articulated drain lines that
direct the water through the tank shell into the diked containment area
or directly to a disposal area.

Drains with swing or articulated joints project down into the
bottom of the tank. The articulated joints compensate for the rise and
fall of the tank's contents and are connected to a drain valve.

Flexible hose drains provide a means of handling accumulated rain
water, directing it either into the tank or to a water drain connection.
This design incorporates a specially fabricated hydrocarbon-resistant
synthetic rubber hose, reinforced with fabric and protected against
kinking and collapse by helically wound reinforcing wire. To prevent
airlock, the hose is weighted with a lead core rod that is reinforced
with a steel cable. These drains can become blocked and drainage can
become restricted in freezing weather. In fact, should the hose become
ice bound in the bottom of the tank, it could be physically damaged when
the roof floats in an upward direction.

Figure 4.17 graphically depicts the various drainage systems that
are available for CBI Horton Floating Roof Tanks.

Considering that conventional floating roofs' water drainage problems
can result in a contaminated product and the possibility of ice blockage
and freezing that could interfere with drain operation, the internal floating roof system offers the best advantage of both the fixed and floating
roof tanks, especially in the cold climates when the product approaches
ambient temperature.

Edwards Engineering Corporation, Pompton Plains, NJ   07444

This organization has developed a wide series of vapor recovery
systems for the collection of hydrocarbon and other condensable gases.
The various designs include collection systems for use in the following
locations:

- Tank truck loading racks
- Gasoline bulk stations

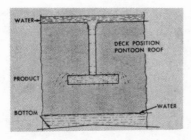

Syphon Drain

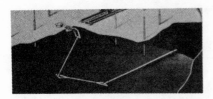

Pipe Drain with Swing Joints   Flexible Hose Drain

FIGURE 4.17. Rainwater Drainage Systems for
Floating Roof Tanks   (Illustration Courtesy of CBI)

- Tanker and barge loading facilities
- Chemical process plants

The vapor recovery units, as can be seen in Figure 4.18 are quite compact.

Figure 4.19 and the cost and technical data sheets that follow are reproduced in part from the Edwards Engineering Corporation Air Pollution Control Manual to clearly define the various vapor recovery units.

FIGURE 4.18. Edwards Engineering Corporation Roof-Mounted Vapor Recovery System

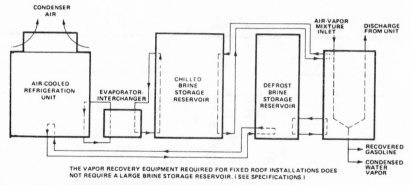

THE VAPOR RECOVERY EQUIPMENT REQUIRED FOR FIXED ROOF INSTALLATIONS DOES NOT REQUIRE A LARGE BRINE STORAGE RESERVOIR. (SEE SPECIFICATIONS)

HOW THE EDWARDS HYDROCARBON VAPOR RECOVERY UNIT WORKS

A cascade refrigeration system follows conventional design, producing temperatures within the evaporator-interchanger in the order of $-90°F$ to $-100°F$. A cold brine pump circulates methylene chloride brine from the brine storage reservoir thru the evaporator-interchanger to obtain the appropriate low temperature fluid (approx $-90°F$) for use in the vapor condenser.

In turn, the low temperature brine coolant is circulated thru the finned tube sections of the vapor condenser. Hydrocarbon vapor and air mixture from the various bulk station filling points is passed over the finned tube sections

of the vapor condenser. Entrained moisture in the entering vapor-air mixture condenses and collects as frost on the cold plate fins. Condensed liquid hydrocarbon is collected at the bottom of the vapor condenser.

At periodic intervals, defrosting of the finned surfaces is accomplished by circulation of warm brine stored in a separate reservoir. The temperature of the warm defrost brine is maintained by heat reclamation from the refrigeration equipment. Defrosting is completed in 10 to 30 minutes, depending upon the amount of frost collected on the finned coil.

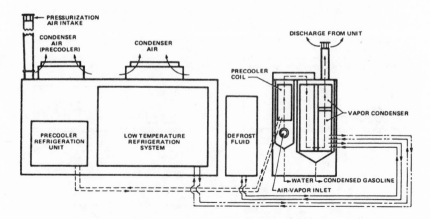

HOW THE EDWARDS HYDROCARBON VAPOR RECOVERY UNIT WORKS

A conventional Edwards refrigeration chiller provides glycol and water at 34°F for precooling the vapors to remove as much water vapor as possible without the formation of hydrates. The effluent vapors leave the precooler at a standardized water vapor dew-point condition of approximately 34°F and 34°F dry bulb.

The vapors, after leaving the precooler, enter the top section of the condensing column where moisture and hydrates are removed. In the next section, heavier molecular weight ends are condensed in the column. The design and use of a direct expansion refrigeration condenser coil heat exchanger permits raising the refrigeration compressor suction pressure and so increasing the capacity of the unit.

At periodic intervals, defrosting of the finned surfaces is accomplished by circulation of warm brine stored in a separate reservoir. The temperature of the warm defrost brine is maintained by heat reclamation from the refrigeration equipment.

Minimal shut-down time is required to accomplish defrosting in the standard DE–unit, since the unit is equipped with a precooler as previously described. The precooler acts to remove most of the water vapor in the entering hydrocarbon vapor-air mixture, thereby reducing the time required for defrost. Defrosting is completed in 30 to 60 minutes, depending upon the amount of frost collected on the finned coil. Dual condenser units with no time lost for defrost are available.

FIGURE 4.19. Schematic Diagrams of Gasoline Bulk Station Systems

EDWARDS HYDROCARBON VAPOR RECOVERY UNITS: PRICE LIST - Domestic

PRESSURIZED UNITS   EXPLOSION–PROOF UNITS

PRESSURIZED MODELS (Note 1) DE	SINGLE CONDENSER (Requires defrost time of ½-to 2 hours every 24 or 48 hours depending upon conditions).	DUAL CONDENSER (Operates continuously through 24 hours.)	EXPLOSION-PROOF MODELS (Note 1) DE	SINGLE CONDENSER (Requires defrost time of ½-to 2 hours every 24 or 48 hours depending upon conditions).	DUAL CONDENSER (Operates continuously through 24 hours)
Single Compressor Units (Note 2)			**Single Compressor Units (Note 2)**		
800P	$115,000	(Consult factory	800X	$144,270	(Consult factory
1200P	$131,670	for price	1200X	$173,000	for price
1600P	$149,000	applicable to unit	1600X	$190,250	applicable to unit
2400P	$184,680	depending	2400X	$225,350	depending
3200P	$218,300	upon extent of	3200X	$260,480	upon extent of
4800P	$291,100	operating time	4800X	$331,850	operating time
Dual Compressor		and loading.)	**Dual Compressor**		and loading.)
6400P	$414,650		6400X	$494,350	
9600P	$577,100		9600X	$657,780	

NOTES:

1. Model numbers apply to DE Series, and to BZ Series by adding suffix DE or BZ to unit model number.

2. Models 2400 thru 4800 also available as Dual Compressor Units; consult factory for pricing.

PRESSURIZED MODELS: provide positive air intake to refrigeration machinery enclosure, and automatic shut-down on positive air failure. Rear deck conforms to explosion-proof requirements.

EXPLOSION-PROOF MODELS: provide for Division II wiring within refrigeration machinery enclosure. Rear deck conforms to explosion-proof requirements.

NOTE: Optional, available thru-out entire unit (consult factory): NEC Class I, Group D, Division I.

NOTE: Prices F.O.B. factory, Pompton Plains, New Jersey

Manufacturer retains the right to change prices, design, and materials without notice.

FORM 8-VRDEP-D-15
Effective Nov. 15, 1979
Replaces 8-VRDEP-D-14

For further information:

Products of

EE *Edwards* **ENGINEERING CORP.**
POMPTON PLAINS, NEW JERSEY 07444
ph. (201) 835-2808   TELEX: 130-131

4.4.3 Other Vapor Containment Systems

One of the simplest techniques for reducing evaporation loss is to keep storage tanks as full as possible to reduce the vapor generation space. However, supply and demand tend to fluctuate and this may not be a practical solution at all times.

Furthermore, evaporation loss from a single, large tank is less than that of several small tanks making up the same total volume or storage capacity. In short, it is preferable to have one large tank rather than a battery of small ones.

The CBI organization has developed a series of designs, which include Vaporspheres, Vaportanks, Hortondome Roofs, and Horton Lifter Roofs, to control vapor releases.

The Vaporsphere and Vaportanks are actually separate tanks that can be installed in a tank field to collect generated vapors through an interconnected piping system. Once the vapor has condensed, it can be pumped back into the main storage tanks on a regular cycle.

Lifter roof and dome designs simply provide additional space to contain the generated vapor on a demand basis. The lifter roof, which is mounted on top of the main storage tank, is liquid sealed and the actual roof of the tank moves upward and downward with vapor volume changes. The dome roofed tanks are mounted on top of the storage vessel and a flexible diaphragm within the hemispherical dome moves in accordance with vapor volume changes.

Another design used to combat vaporization from solar heat evaporation involves a flat roofed tank with the circumferential shell extending above the tank top. A shallow layer of water is then contained within the extended shell to act as a coolant on the tank top.

4.5 INTEGRAL HEATING COILS

In the colder climates it is a standard practice to heat heavier fuel oils through the medium of steam heating coils. Steam generated from a plant boiler is circulated through the steel or black iron coil to transfer heat to the oil in the lower reaches of the tank. This results in the circulation of the warmed oil to the surface of the tank and the colder upper surface oil travels down toward the heating coil.

A number of problems develop from this heating process, one of which has resulted in major spills. When the heating coils are idle during the summer months, the condensate water lying in the coil corrodes the coil metal from the inside, which upon failure (when not under pressure) permits oil to enter the coil. The exhaust from the steam coils on many installations is open to the atmosphere and in some cases to the nearest waterbody. Discharges from coil failure during hours of darkness can go unobserved until daylight.

Some installations return the exhaust steam and condensate back into the boiler water feed system. In installations of this kind, oil can enter the water side of the boiler to act as an insulator and cause burning of the boiler tubes and drums, which in turn can result in metal failure and explosion.

A number of remedies can be applied to prevent spills from steam coil failure. The major procedures are as follows:

- Do not leave moisture standing in steam coils when not in use.
- Install coils fabricated of Admiralty Metal* in lieu of steel or black iron coils. Although the initial cost will be higher, the prolonged life and degree of safety will compensate for the increased purchase cost.

_____

*An alloy of not less than 70 percent copper, about 1 percent in small amounts of other elements, and the balance, zinc; also known as Admiralty brass and Admiralty bronze.

- Do not exhaust condensate from the steam coil to the atmosphere or nearest watercourse. Drain the exhaust to an oil and water separator or to a holding pond.
- Consider the use of external heating coils with outside insulation around the tank.
- Under no circumstances return steam coil condensate back into a boiler water feed system. It is better to lose the condensate than to endanger the plant's steam boiler.

4.6 LIQUID LEVEL MONITORING

For many years industry rejected the use of automatic liquid level alarms, claiming that they were "unreliable," "difficult to maintain," "worked for a month or so and never worked again," and last but not least "the tank gagers place too much reliance on the device." rigid environmental laws and the levying of fines for spills that enter navigable waters is gradually changing the situation. Furthermore, the technology associated with equipment development and manufacture has advanced considerably over the past decade, and there are many more manufacturers and models to select from. Computerized liquid level gaging has developed accuracies to one-half inch, with remote direct digital readout at one or more stations.

Examples of a variety of designs follow. It is stressed, however, that the descriptions* as provided do not in any way endorse any particular product.

Monitor Manufacturing, Elburn, IL 60119
Model CM3A - Material Level Indicator

The CM3A is a highly sensitive measuring device that provides a digital readout on the depth or amount of material contained in a silo, tank, or bin--

---

*When practical, descriptions are taken verbatim from the manufacturer's technical literature.

from 10 to 200 ft in depth. It is claimed to be dependable, easy to use, and accurate--it measures contents to within ±1/10 of 1 foot over the total distance to be measured.

The operating principle of the CM3A unit is simple. A button is pressed on a control panel and the unit is activated. A weighted probe, supported by a coated stainless steel cable, is lowered into a storage facility where it rapidly reaches its reading range. During its downward travel, a counter in the console provides a continuous reading on the amount of material in the bin. When the probe reaches the material level, a slight slack in the cable is automatically sensed in the electrical "head" and triggers the motor to reverse its action--stops the counter, instantly-- and returns the probe to its original position. The counter reading remains frozen until another reading is required (or can be manually "erased").

The controlling console can be located in close proximity to the bin, in remote areas away from the bin, or set up in the production control area with the control devices. A single console is capable of monitoring up to 22 units.

The new CM3A operates efficiently on a variety of materials--chemicals, plastics, cement, coal, paper pulp, vegetable oils, petroleum products, and grains. It is not affected by temperature changes, humidity, or dusty atmospheres. The CM3A is Factory Mutual approved for use in ordinary locations.

Varec Division Emerson Electric Company, Gardena, CA 90247
Omnitrol Liquid Level Control

The concern manufactures a switch actuator control that can shut off a pump control once a desired (predetermined) liquid level is reached within the storage tank.

The Onmitrol switch actuator operates on a unique and proved principle. The inner magnets are constantly engaged with the outer magnet and pivoting action occurs within the magnetic field at all times.

There are four inner magnets, separated by an air space, as shown in Figure 4.20. Magnets 1 and 4 attract the outer magnet. Magnets 2 and 3

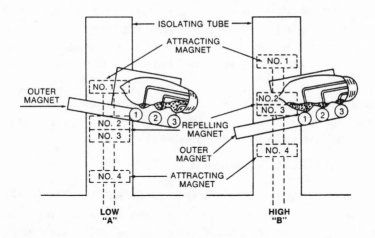

FIGURE 4.20. Principle of "Constantly-Engaged Magnetic Field" (Illustration Courtesy of Varec Division, Emerson Electric Company)

act as a single magnet and repel the outer magnet. All of these inner magnets, as well as the outer magnet, create a separate field and act in such a manner to lock the outer magnet in the position shown in Detail "A." This prevents nuisance switching created by liquid level surges or surface agitation.

As the liquid level rises, the inner magnet stem assembly rises and magnets 2 and 3 will be forced through the field of the outer magnet. The repelling force on the outer magnet will then be acting in a downward direction. Simultaneously, the attracting force of magnet 1 will decrease and that of 4 will increase. The outer magnet or actuator switch assembly will snap into its new locked position as shown in Detail "B." Due to the forces encountered, there is no possibility of the outer magnet finding a null point anywhere between the two locked positions; therefore, a complete and positive snap switching action is assured.

Detail "B" shows the switch and magnet status for high liquid level.
As the level falls, all forces described above are reversed until the inner
repelling magnets have been forced through the magnetic field of the outer
magnet and the actuator is again in the "locked" position shown in Detail
"A."

No springs, bellows, diaphragms, or other mechanical devices are
required. The only movement of the actuator is the slight tilting of
hardened stainless steel shaft in a jeweled pivot bearing.

Automation Products, Inc., Houston, TX 77008
Dynatrol - Level Detector

When used in conjunction with a relay switch receiver, the Dynatrol unit
detects the liquid (or slurry) level and provides an "on-off" switch action.
The control is described as a dynamic-type level switch for application to
either high- or low-point level detection of liquids and slurries. The
design entails a puddle (Figure 4.21) driven into vibration by a driver coil.
A second coil located in the pick-up end produces a voltage proportional to
the paddle vibrational amplitude. When the paddle comes into contact with
the rising liquid, its amplitude of vibration decreases and the output voltage drops to a very low value. The change in output signal operates the
contacts of the relay switch receiver (Figure 4.21). In addition to acting
as a pump shutoff control, the relay switch could be readily wired to actuate a visible/audible alarm signal. The complete package, ready for installation, sells for $384, with a supply time of one week.

Transdata Corporation, Freeport, NY 11520
Liquid Level Instrumentation

The organization manufactures a wide range of liquid level detection
and control equipment, some of which are designed for on-site level reading
while others have a remote readout capability. Figures 4.22 through 4.25
are descriptions reproduced from the manufacturer's sales literature.

GPE Controls, Morton Grove, IL 60053
Telepulse Systems

Telepulse 300 is a simple and accurate system for remote supervision
of storage tank liquid levels and temperatures. It is used mainly for

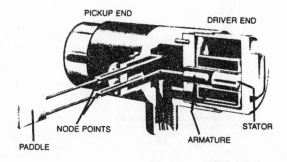

PICKUP END

DRIVER END

NODE POINTS

STATOR

PADDLE

ARMATURE

PATENTS PENDING

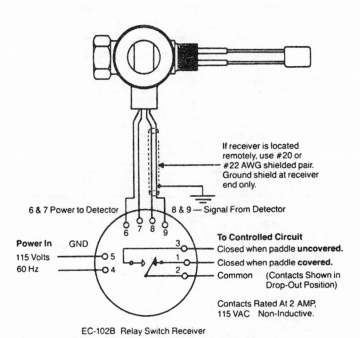

If receiver is located remotely, use #20 or #22 AWG shielded pair. Ground shield at receiver end only.

6 & 7 Power to Detector

8 & 9 — Signal From Detector

Power In   GND

115 Volts

60 Hz

To Controlled Circuit

Closed when paddle **uncovered.**

Closed when paddle **covered.**

Common   (Contacts Shown in Drop-Out Position)

Contacts Rated At 2 AMP, 115 VAC   Non-Inductive.

EC-102B Relay Switch Receiver

FIGURE 4.21. Dynatrol CL-IORH Level Detector and EC-102B Relay Switch Receiver (Illustration Courtesy of Automation Products, Inc.)

TRANSDATA CORPORATION

Fig. I

MODEL 42,000 - AUTOMATIC TANK GAGE

The Protectoseal Model 42000A Automatic Tank Gauge is a float operated liquid level gauge suitable for use on tanks up to 65 ft. high. It provides digital readings in 1/16th inch increments. The gauge head readout can be located at eye level at the bottom of the tank, at the top of the tank, on the outside of a steel firewall or at any convenient location immediately adjacent to the tank. The standard unit is supplied for working pressures up to 15 PSIG. The standard construction consists of aluminum housings with stainless steel trim and a stainless steel float with stainless steel guide wires. The readout can be supplied in the metric system or in tenth of inches or in hundredths of feet.
For pressures beyond the 15 PSIG range, or for corrosive or toxic services, see the Transdata Model 5683 Magna-Float Automatic Tank Gauge.

Fig. II

MODEL 5683 - MAGNA-FLOAT LEVEL GAGE

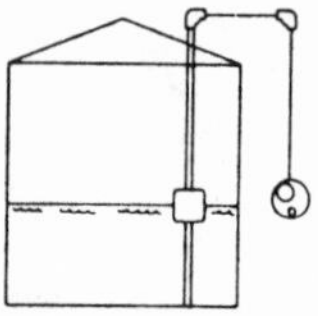

The Transdata Model 5683 Magna-Float Automatic Tank Gauge utilizes the same external components as the Protectoseal 42000A Automatic Tank Gauge except that a non-magnetic floatwell that is sealed at the bottom end extends from the top to the bottom of the tank in order to isolate the automatic tank gauge operating mechanism from the tank contents. The float is center guided by the floatwell and is magnetically coupled to the tape through the wall of the floatwell.
The Model 5683 Magna-Float Automatic Tank Gauge is suitable for use on tanks handling corrosive chemicals, toxic products or liquids being stored at elevated pressures up to 300 PSIG. All of the Transdata accessory equipment that can normally be used with the Protectoseal 42000A Automatic Tank Gauge can also be used with the Transdata Magna-Float Automatic Tank Gauge.

Fig. III

MODEL 42,000-AUTOMATIC TANK GAGE WITH
MODEL 6372-ADJUSTABLE LIMIT SWITCH
(1-6 SWITCHES)

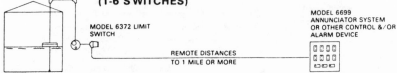

The Transdata Model 6372 Adjustable Limit Switch Assembly can be installed on the gauge head. Up to six adjustable limit switches can be supplied in the explosion proof housing. The standard switches are SPDT, although DPDT are optionally available. They are cam operated and are independently adjustable so that alarm or control circuits can be operated or closed as required at various pre-selected liquid levels. The switches can be used to operate the Transdata 6699 Annunciator/Control System.

FIGURE 4.22. Liquid Level Instrumentation
for Indication and Control

TRANSDATA CORPORATION

Fig. IV

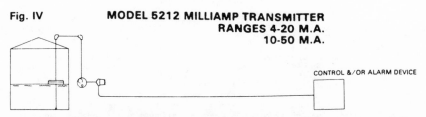

MODEL 5212 MILLIAMP TRANSMITTER
RANGES 4-20 M.A.
10-50 M.A.

CONTROL &/OR ALARM DEVICE

The Transdata Model 5212 Milliamp Transmitter can be supplied for mounting on the 42000A Automatic Tank Gauge to provide a milliamp output as a function of liquid level. The two standard ranges are 4-20 milliamps and 10-50 milliamps.
Adjustable Limit Switches can be included within the explosion proof housing for alarm or control purposes, in addition to the analog milliamp output signal.

Fig. V

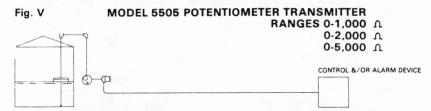

MODEL 5505 POTENTIOMETER TRANSMITTER
RANGES 0-1,000 Ω
0-2,000 Ω
0-5,000 Ω

CONTROL &/OR ALARM DEVICE

The Transdata Model 5505 Potentiometer Transmitter is supplied in an explosion proof housing for mounting on the 42000A Automatic Tank Gauge. This provides a variable resistance as a function of liquid level. Numerous resistance ranges are available, such as 0-1000 ohms, 0-2000 ohms, 0-5000 ohms or other ranges to satisfy a customers specific needs. The Transdata Model 5505 Potentiometer is also available with adjustable limit switches for alarm and control purposes.

Fig. VI

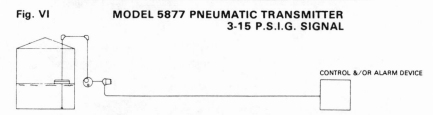

MODEL 5877 PNEUMATIC TRANSMITTER
3-15 P.S.I.G. SIGNAL

CONTROL &/OR ALARM DEVICE

The Transdata Model 5877 Pneumatic Transmitter is designed for mounting on the 42000A Automatic Tank Gauge and is for use in connection with pneumatic instrumentation where a standard pneumatic signal of 3 to 15 PSIG can be used for control or indication. The 3 to 15 PSIG air signal is a function of liquid level but if this proportional analog signal is not desired, the unit can be supplied for on/off control.

FIGURE 4.23. Liquid Level Instrumentation
for Indication and Control.

TRANSDATA CORPORATION

Fig. VII   MODEL 5501/5483 REMOTE READING SYSTEM

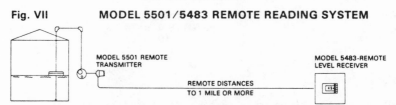

MODEL 5501 REMOTE
TRANSMITTER

MODEL 5483-REMOTE
LEVEL RECEIVER

REMOTE DISTANCES
TO 1 MILE OR MORE

The Transdata Model 5501 Explosion Proof Remote Level Transmitter is designed for mounting on an Automatic Tank Gauge in order to transmit the liquid level readout to a remote location that can be anywhere from a hundred feet to a mile or more away. The Model 5483 Remote Receiver presents the remote reading in a digital display in feet, inches and sixteenths of inches.

The following descriptions illustrate the standard modifications that are available with this basic remote reading system. These options include adjustable limit switches installed in the transmitter for alarm or control purposes.

The Model 5483 Receiver can provide additional output signals in addition to the standard digital display. These output signals consist of adjustable switches, variable resistance, or standard milliamp output signals.

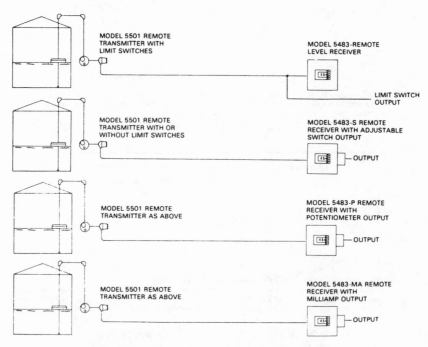

FIGURE 4.24.   Liquid Level Instrumentation
for Indication and Control.

TRANSDATA CORPORATION

Fig. VIII

MODEL 40,000, 41,000 - LIQUID LEVEL INDICATORS

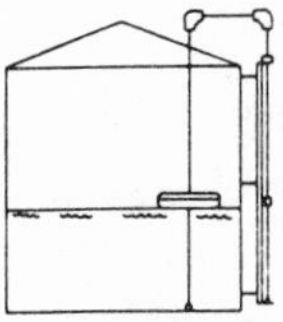

The Protectoseal Model 40000 and the Fig. 41000 Liquid Level Indicators are designed for application to process or storage tanks where a readout to within 1″ of liquid level is satisfactory and provides the additional convenience of being able to be read out at 100 or 200 ft. away from the tank rather than having to approach the gauge head readout device within a few feet. As is the case with the 42000A Automatic Tank Gauge, the float, guide wires and other critical items are of type 316 stainless steel construction. The board on the side of the tank is of heavy gauge aluminum channel construction and is normally calibrated in feet and inches but can be calibrated in the metric system, decimal system or special calibrations as may be required for the application.

Fig. IX

MODEL 6600 - 90° SPROCKET ASSEMBLY

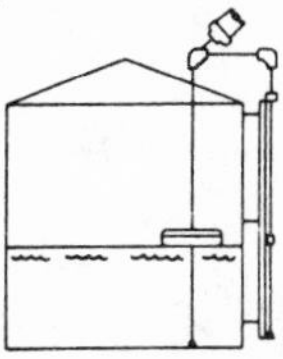

The full range of Transdata accessory equipment that is described for use with the Model 42000A Automatic Tank Gauge can be used in conjunction with the 40000 or 41000 Liquid Level Indicators. This is accomplished by substituting the Model 6600-90 Degree Sprocket Assembly in place of one of the 90 degree pulley assemblies at the top of the tank. The standard cable that is used between the float and the gauge board marker (target) is substituted with a stainless steel bead chain which drives the bead chain sprocket and, in turn, operates the transmitters or switches.

Fig. X

MODEL 6648 - TOP MOUNTED FLOAT SWITCH

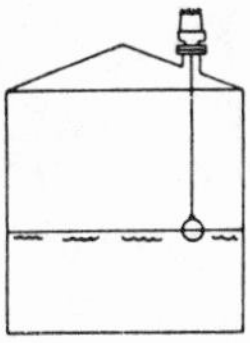

The Model 6648 Float Switch is supplied with a 4″-125# flanged connection for mounting at the top of a tank. As the liquid level rises, the float moves a steel armature into the field of a magnetically operated mercury tilt switch that can be used to either open or close a circuit.

Fig. XI

MODEL 6761 - TOP MOUNTED DISPLACER SWITCH

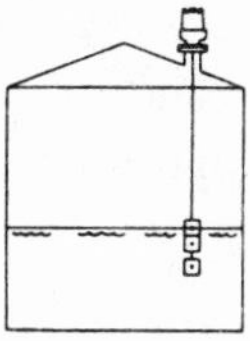

The Model 6761 Displacer Switch is supplied with a 4″-125# flanged connection for mounting at the top of a tank. As the liquid level rises around the displacers a steel armature moves into the field of a magnetically operated tilt switch, which can be used to open or close a circuit. Since the displacers do not float, two points of switch actuation are possible with one switch assembly. The Model 6761 is easily adjusted in the field for level actuation points. These units are especially suited for wide level differential applications, or for surging or agitated liquids.

FIGURE 4.25. Liquid Level Instrumentation for Indication and Control.

inventory control, prevention of overfilling, and loss of pump suction
during discharging. The Telepulse 300 is applicable where there are only
a small number of tanks (from 1 to 20) reporting to a central supervisory
station. Typical installations include storage facilities at marine
terminals, power plants, steel mills, synthetic rubber plants, pulp paper
mills, petroleum product terminals, and chemical processing facilities.

Telepulse 300 displays accurate tank level readings of tanks up to
120 ft high--accuracy is within $\pm$0.25 percent of the level measuring
range. Temperature readings are accurate to within $\pm$1°F. High- and low-
liquid level alarms can be incorporated in the system. Liquid levels are
read in both metric and English measure. A switch on a display console
converts the level display from meters to feet and vice versa.

For single-tank installations, the liquid level is displayed constantly on a digital display meter. In multiple-tank systems, the desired
tank is selected by pushbutton on a central console receiver. The console
can be free standing on a desk, table, or counter, rack mounted, or built
into a custom-designed display.

Also available are the Telepulse 400 Series, which are completely
integrated systems, providing central supervision of tank levels and
temperature gaging in very fine increments. Up to 200 storage tanks
can be remotely supervised, gathering liquid level data within an accuracy
of $\pm$2 millimeters (0.01 ft). Depending on the number of tanks being monitored, Series 400 systems can update tank level readings from four per second
to ten per second. Additional equipment options are available, which makes
it possible to tie in with a computer or provide high-low alarm signals.
The data can also be displayed at more than one location through the use
of satellite units. An additional desirable feature is the ability of the
Series 400 Telepulse systems to "grow" with an installation. By using a
modular concept, they readily adapt to the changing requirements of an
expanding facility.

The cost of the Telepulse actually demands a complete engineering evaluation and cost estimate. The manufacturer could, however, provide cost data on the components of a Telepulse 300:

Level gage	$500.00	(average 50 ft tank)
Transmitter with high-low alarm	664.00	
Telepulse receiver	300.00	(up to 300 ft distance)
Installation kit	100.00	
Intrinsically safe barrier and power supply	660.00	

The manufacturer estimates that the installation costs would equal the part acquisition cost, approximately $2,224. Figure 4.26 graphically displays the receiver for Model TP 400.

Obviously, there are numerous manufacturers and designs of fixed tank level gages available on the market. One additional unit is described laregely because it is portable and of quite unique design.

Metritape, Inc., Concord, MA 01742
Metritape Digilert Level Systems

This unit can be characterized as a portable fill alarm that can be used while liquid cargoes are being transferred from a storage container into a transportation vehicle. Its use would include marine tankers, barges, tank cars, and tank trucks. It can, however, also be used when fixed storage tanks are being filled. The specifications of the unit are as follows:

METRALERT Portable Fill Alarm, Model PFA-80

Dimensions, Overall . . . . Approximately 13"x12"x6"
Total Weight . . . . 11 pounds
Cable Length . . . . 12 ft, standard
Instrument Temperature Range . . . . -20 to +140°F
Sensor Temperature Range . . . . -20°F to 190°F, standard
Sound Output . . . . Rapid-sweep siren, 1,500 to 2,500 Hz.
Sound Intensity . . . . 110 db, max. at 4 ft.

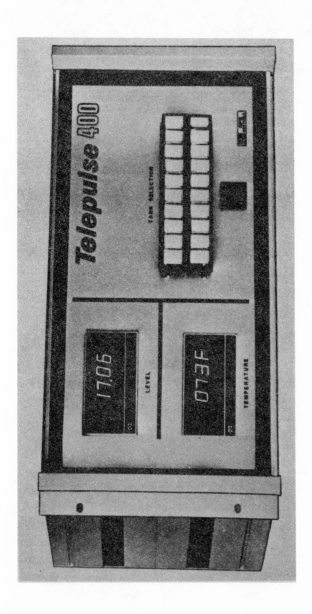

FIGURE 4.26. Telepulse Liquid Level Monitor   Note degree of level accuracy and availability of multiple tank soundings.   (Photograph Courtesy of GPE Controls)

Electrical Safety Rating . . . Intrinsically-safe, Class 1 Groups C and D

Battery Life (alarm sounding). . . . . Approximately 2 hours

Battery Shelf Life (between charges) . . . . 3 to 4 months

Battery Rechargings . . . . Over 100

Exposed Materials . . . . Epoxy fiberglass, phenolic, Teflon,
and neopene, standard

The principle of operation is simple. A pressure sensitive head is first lowered into the tank or cargo compartment to the height of the desired liquid level. The pendant sensor is then activated when the hydrostatic pressure of the liquid cargo activates the internal element of the sensor circuit. The manufacturer designates that a number of parallel contacts are provided to give high switching reliability and that the element has been subjected to 100,000 operating cycles without contact failure or evidence of wear or fatigue. Once activated by hydrostatic pressure, the Teflon-coated sensor actuates a high intensity horn, which produces a high-intensity electronic siren sound sweeping from 1,500 to 2,500 Hz. The sound level reaches 110 db at close range, as loud as can be heard without discomfort.

The concern also manufactures the Digilert level/temperature readout unit, which has a built-in audible fill alarm. The system is battery-powered and provides comprehensive tank readings and alarm functions to the decks of marine transportation vessels, at railroad tank car loading racks, tank truck loading stations, or land-based storage tanks. The packaging is such that it can be fixed mounted or used in a portable mode at any loading station. Figure 4.27 illustrates the compactness and design of the unit, which can be obtained in a variety of models for specific uses.

4.7 LIGHTNING PROTECTION FOR BULK STORAGE TANKS

The National Fire Protection Association (NFPA)[*] Publication No. 78, "Lightning Protection Code 1977," provides detailed information on the topic

---

[*]NFPA, 470 Atlantic Avenue, Boston, MA 02210

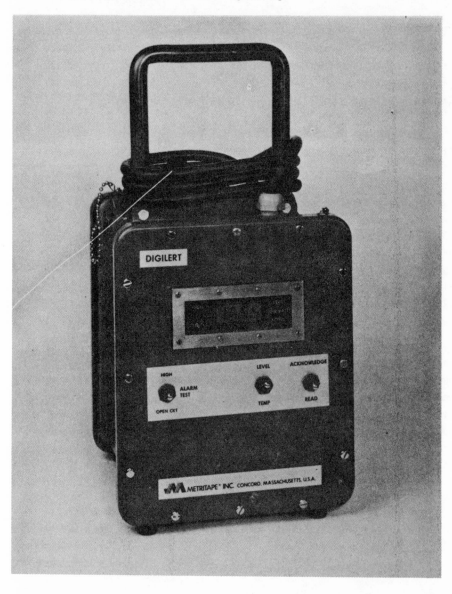

FIGURE 4.27. Digilert Level/Temperature Readout
Unit (Photograph Courtesy of Metritape, Inc.)

under discussion. Additional information is provided in the American
Petroleum Institute (API)[*] Publication RP2003-10/74, "Recommended Protection Against Ignitions Arising Out of Static, Lightning, and Stray Currents."

The voluntary codes contend that the contents of metallic tanks with
steel roofs of riveted, bolted, or welded construction used for the storage
of flammable liquids at atmospheric pressure are protected against lightning
(inherently self-protecting) under the following conditions.

    a. All joints between metallic plates shall be riveted,
        bolted, or welded.

    b. All pipes entering the tanks shall be metallically
        connected to the tank at the point of entrance.

    c. All vapor or gas openings shall be closed or pro-
        vided with flame protection, when the stored stock
        may produce a flammable air-vapor mixture under
        storage conditions.

    d. The metal roof shall have a minimum thickness of
        3/16 in. (4.8 m).

    e. The roof shall be welded, bolted, or riveted
        to the shell.

To gain a maximum degree of protection, however, it has been observed
that most bulk storage facilities still provide grounding protection.

Specialists in the field recommend that there be at least two points
to ground, on opposite sides of the tank. Ground terminals (rods) of not
less than 1/2 in. (12.7 mm) diameter are driven at least 8 ft (2.4 m) into
the earth at a position 2 to 3 ft (0.6 m to 1.02 m) away from the tank.
The rods are either copper clad steel, solid copper, or stainless steel.
The protective systems demand that bonding plates be welded to the tank
at each grounding location and that a No. 6 gage copper wire be attached

---

[*]API, 1801 K Street, N.W., Washington, D.C. 20006.

to each bonding plate and ground rod. The grounding wire should contact
the rod and the ground cable for a distance of 1-1/2 in. (38 mm).

In soil less than 12 in. (0.3 m) deep, the structure should be surrounded with a main-size conductor (a counterpoise) laid in a trench or
in rock crevices. From this counterpoise, a conductor should be run to
pits or hollows where added metal can be deposited. Approximately 9 ft^2
(0.84 m^2) of 0.032 in. thick (0.8 mm) copper plat or equivalent corrosion
resistant metal should be connected to the lateral conductors. The installation should be covered with loose earth for rain absorption.

4.7.1 Floating Roof Tank Lightning Protection

Tanks of this design require special attention. Fires have occurred
when lightning struck the rim of a tank at a time when the roof was high
and the content volatile. Fires have also been started when the roof was
in a low position and, due to a defective floating roof seal, accumulated
vapors above the roof were ignited by a rim strike.

The best defense is to maintain a tight roof-to-tank-shell seal.
Additionally, when the floating roof utilizes hangers within the vapor
space (Figure 4.28), the roof should be electrically bonded to the shoes
(sealing ring) of the seal through the most direct electrical path, at
intervals of not greater than 10 ft (3 m) on the circumference of the tank.
These shunts should consist of flexible Type 302, 28-gage, 1/64-in. x 2-in.
(0.4 mm x 51 mm) wide stainless steel straps, or their equivalent in corrosion resistance and current-carrying capacity.

4.7.2 Ground Resistance Measuring

Following provision of a grounding system, the installation should
be regularly subjected to ground resistance and soil resistivity testing,
on at least an annual basis. A reading of 25 ohms should be maintained.

A typical testing unit is manufactured by the AEMC Corporation of
Boston, Massachusetts. The "Terracontrol" unit manufactured by this concern
is depicted in Figure 4.29. The principle of operation can best be described

For the SR-1 double seal with a mechanical (metallic) shoe type primary seal, CBI recommends that the shunts be located in the same position used for the past 25 years, before the secondary seal was introduced.

For the SR-9 double seal with a resilient (foam) filled fabric type primary seal, and other similar types. CBI recommends external shunts located above and outside the secondary seal.

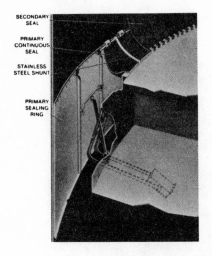

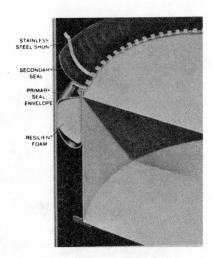

FIGURE 4.28. Floating Roof Grounding
Techniques (Illustration Courtesy of CBI)

by direct reproduction from the manufacturer's Bulletins B130.100 and B132.100, which follow the figure.

The model as illustrated sells complete for $616; a smaller unit, Model B130.100, sells for $406 (July, 1980).

References: (Section 4.7)

1. "Recommended Practice for Protection Against Ignitions Arising out. of Static, Lightning, and Stray Currents," API Bulletin RP2003 Third Edition. October 1974.

2. "Lightning Protection Code 1977," NFPA Bulletin 78.

FIGURE 4.29. Model #B132.100 "Terracontrol"
Electronic Ground Tester (Illustration Courtesy
of AEMC Corporation)

Principle

1° Ground resistance measurement (fig. 1).

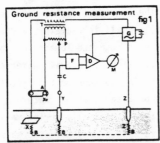

Ground resistance measurement   fig 1

When ground resistance is to be measured, the terminals X and Xv are short circuited and Y, Z are auxiliary electrodes.

A solid state generator G, supplied by six 1.5 V AA dry cells, produces a 225 Hz current sent into the circuit GZBXT. The voltage VX at X between A and B is compared to the voltage VP (produced by the transformer T and its potentiometer P) between A and P.

The resulting voltage VR (VX-VP) is applied to a synchronous amplifier, which supplies the zero galvanometer M.

The balance is obtained by adjusting P so that VR equals zero. The ground resistance is then displayed on the 245 mm (9.64 in) dial 0–10 scale (200 divisions) which is connected to the potentiometer P.

The primary of T has three inputs and gives three ranges 10 - 100 - 1,000 Ω.

2° Soil resistivity measurement (fig. 2)

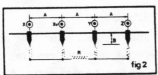

fig 2

The TERRACONTROL can be used for finding the best place to bury ground grids, to place ground mats and ground electrodes, as well as for geophysical prospecting.

In these cases, the strap between X and Xv is disconnected (fig. 2) and 4 electrodes are respectively connected to X, Xv, Y and Z. The basic measuring principle is the same as before, but the resistance R between Xv and Y is measured. The resistivity can then be obtained by using the following formula (Wener formula):

$$\rho = \frac{4\,\pi\,AR}{1 + \dfrac{2\,A}{\sqrt{(A^2 + 4\,B^2)}} - \dfrac{2\,A}{\sqrt{(4\,A^2 + 4\,B^2)}}}$$

Where: A = distance between the electrodes in centimeters.

  B = electrode depth in centimeters.

  if: A > 20 B the formula becomes :

  ρ = 2 π AR (with A in cm).

  ρ = 191.5 AR (with A in feet)

  ρ = Soil resistivity

This value is the average resistivity of the ground at a depth equivalent to the distance A between two electrodes.

3° Non inductive resistance measurement

The TERRACONTROL can also measure a · non inductive resistance or an electrolyte. The set-up is as in fig. 1, the resistance to be measured is placed between X and YZ (Y,Z are connected).

Field Use

Both Terracontrols are designed for field use. They are fitted in rugged leatherette covered plywood carrying cases.

Connections

1) Ground resistance.

Short circuit the terminals X, Xv with the shorting link.

Connect the ground to be measured to terminal X, use the 5m (15 foot) lead. Connect the auxiliary rods to the terminals Y and Z. Use the two 20m (60 foot) leads. (See fig. 3)

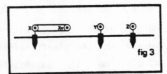

fig 3

2) Soil Resistivity.

Use all four terminals, and separate all four test rods by an equal distance. (See fig. 4)

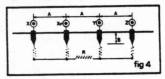

fig 4

NOTE: Reproduced directly from AEMC Corporation sales literature.

4.8 TANK DRAINING

4.8.1 Water Drain Valves

Water drain valves are located in the lowermost section or shell of a bulk storage tank. The valve is usually small in diameter, and an internal drain line goes from the valve down into the lowest possible reach of the tank's lower head. The valve is used to drain water accumulation from the tank. The water collects in the tank bottom from floating roof drains and condensation, or it may be water that has entered and contaminated the product during transportation, especially during marine transport. On a regular frequency, determined by humidity and other atmospheric conditions, the water is drained from the tank through the medium of the water drain valve. It is standard practice to keep these valves in the padlocked-closed position. This is not always the case, however. Padlocks are frequently missing, lying on the ground near the valve, or in some cases, placed back in the valve in an unlocked status.

It is stressed that the water drain valve, after improper opening of a masterflow control valve or tank failure, presents the greates potential for draining and completely emptying the contents of a tank. As recently as September 9, 1980 a 137,000-gallon spill of No. 2 diesel oil occurred in a Baltimore, Maryland bulk storage facility. It was found that the locking chain on a water drain valve had been cut, and the valve opening on a 4-million gallon storage tank. The drain valve in the dike had also been sabotaged, and the oil drained to an oil and water separator having the limited capacity of 250 gallons per hour. The oil overflowed the system and drained into a creek that emptied into Baltimore Harbor.

4.8.1.1 Controlled Drainage Systems and Piping

At many locations the operational practice is to drain the accumulated water directly into the diked enclosure. The operation requires constant attention, since once the tank has been drained free of water, the product in the tank begins to drain. The possibility of this happening is high. Water drainage can be a lengthy process, varying from two hours to two days during some seasons of the year. In such cases, the attention of a worker can be diverted; incidents are recorded whereby workers have actually forgotten they opened the valves and remembered their failure only after leaving work. Product has been lost with massive spills escaping into diked areas and into nearby waterways.

In a serious effort to control such losses, industry has begun to pipe these valves into wastewater treatment ponds, into oil and water separators, and in some instances, into oil retention sumps. The first two receptacles are preferred, largely because they have greater capacities than an oil retention sump, which is normally installed in a plant's storm water drainage system. It is possible, however, to install an audible/visible oil spill detection alarm either in the sump itself or at any convenient location in the drainage system.

4.8.1.2 Dow Imbiber Valves

The Dow Chemical Company, Midland, Michigan has developed an alternative system that permits drainage of a tank's water content into the diked containment area, at the same time ensuring that certain hydrocarbon products do not leave the tank. The "Imbiber Valve," produced by Dow, is designed for attachment on the discharge of the water drain valve. It will restrain the flow of gasoline, benzene, toluene, xylene, ethylbenzene, styrene, polchlorinated biphenyls (PCB's), and 1, 1, 1, trichlorethane. The manufacturer's directions for use are as follows:

1. Attach pipe adapter to water drain valve.
2. Drain any oil entrapped in draw-off into suitable container.
3. Attach Dow Imbiber Valve to water drainoff adapter and tighten (hand tight) against gasket.
4. Open water draw-off valve to allow flow through Imbiber Valve.
5. When flow ceases, indicating contact with water/ oil interface, close water draw-off valve (see note).
6. Remove Imbiber Valve and discard in proper manner.

Note: Occasionally, oil dissolved in the water layer may prematurely activate the Imbiber Valve. If this occurs, close draw-off valve, replace Imbiber Valve with a fresh unit, and resume water draw-off.

The instructions indicate that the unit is disposable. However, Imbiber Valves that utilize a disposable cartridge designed to fit within a sturdy, permanent housing are available. Figure 4.30 depicts the configuration and integral parts of a disposal Imbiber Valve. The cost of these valves is within the following ranges when ordered in varying quantity:

IMBIBER VALVES Shipping Class NMFC #14500:

XFS-43094 Imbiber Valve	up to 10 boxes	$72.00/box
12 units/box	11-75 boxes	66.00/box
Label License for U.S. 3,750,688	over 75 boxes	60.00/box
XFS-43095 Imbiber Valve Adapter	per box	10.00/box
2 units/box; 60 boxes/case	60-120 boxes (1-2 cases)	7.00/box
	over 120 boxes (over 2 cases)	6.50/box
XFS-43096 Imbiber Valve Adapter Gasket	per bag	2.00/bag
12 units/film bag	of 12 units	

IMBIBER CARTRIDGES Shipping Class NMFC #14500:

XFS-43109 Imbiber Cartridge (Diesel)	up to 10 boxes	70.00/box
in boxes of four units	11-50 boxes	63.00/box
Label License for U.S. 3,750,099	over 50 boxes	54.00/box
XFS-43139 Imbiber Cartridge-G (Gasoline)		
in boxes of four units	up to 10 boxes	70.00/box
Label License for U.S. 3,750,088	11-50 boxes	62.00/box
	over 50 boxes	54.00/box
XFS-43110 Imbiber Housing	up to 10 units	97.00/ea.
in boxes of one unit	over 10 units	92.50/ea.
XFS-43111 Imbiber Cartridge Gasket Set	1-10 sets	3.50/ea.
1 set of 2 per envelope	11-49 sets	3.25/ea.
	over 50 sets	3.00/ea.

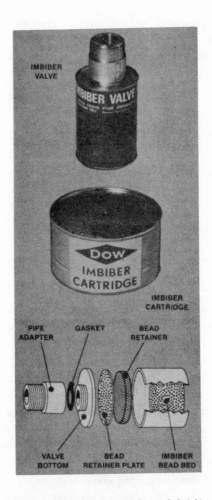

FIGURE 4.30. Configuration of Imbiber
Valve   (Illustration Courtesy of The
Dow Chemical Company)

All prices FOB Midland, MI 48640; prices subject to change without notice.
Send inquiries to:

Imbiber Systems
Functional Products and Systems
2020 Dow Center
Midland, MI 48640

Minimum Order: $50.00

Figures 4.31, 4.32, and 4.33 illustrate the valves and their applications.

4.8.1.3 Security

Water drain valves should be locked in the closed position when not
in service. The opening and closing of the valves should be under strict
authority. Frequent inspections should be made of the valves to see that
the security of the valve has not in any way been violated.

4.9 GENERAL PROCEDURES FOR SPILL CONTAINMENT

4.9.1 Field Observations

Many tank farms are so large that plant personnel travel through them
very infrequently. In some oil fields the tank gager fills the tanks
through the medium of a timer, a telephone, and an intimate knowledge of
the pumping rate and the tank capacity. This fill procedure is of course
aided by computerized liquid level gaging systems. There is a need, however,
for regular inspection of the tanks and their appurtenances. Leakage can
occur from tank seams, valve connections, valve stems, and pipeline joints.
This can be determined only from visual examination, since the quantity of
product lost is too small to be registered by remote gaging systems. For
this reason, each tank should be inspected at least monthly. It is not
sufficient just to walk the dike and examine the tank from a distance. The
inspector should enter the diked area and examine the storage container
from grade level, following which the tank top should be visually examined.
All valves and pipelines within the diked enclosure should be visually sur-
veyed to determine defects or signs of external deterioration. The tank
foundation and metal-to-concrete seal should have special examination, since
a defective seal can permit rainwater to collect under the tank and corrode
the bottom head--sight unseen.

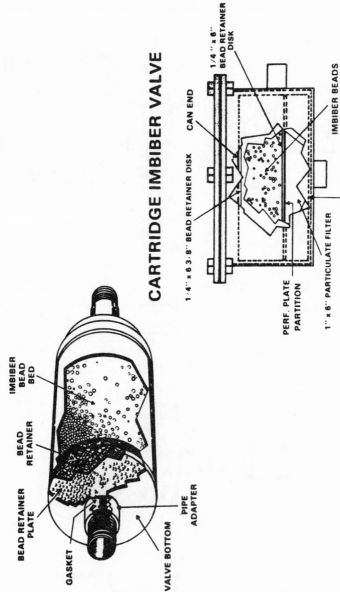

ORIGINAL IMBIBER VALVE
(Screw-on Canister)

CARTRIDGE IMBIBER VALVE

FIGURE 4.31. Originial and Cartridge Imbiber Valves
(Photograph Courtesy of Dow Chemical Company)

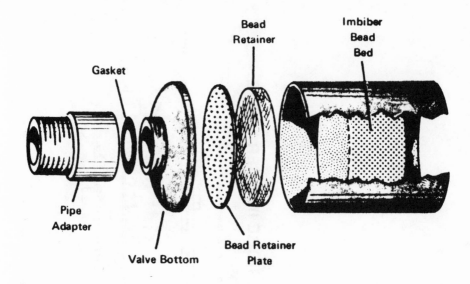

FIGURE 4.32. Dow Imbiber Valve and Method of
Installation   (Photograph Courtesy of Dow
Chemical Company)

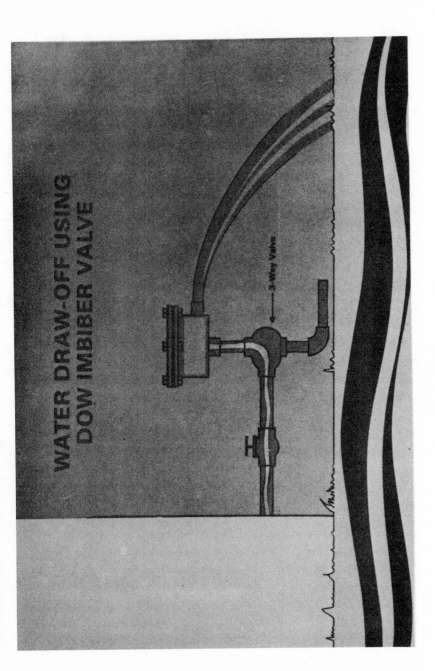

FIGURE 4.33. Water Draw-Off Using Dow Imbiber Valve
(Photograph Courtesy of Dow Chemical Company)

4.9.2 Tell-Tale Signs of Failure

Seams and rivet leakage in older tanks results in tell-tale streaks of product down the tank side. The same can be said of welded seams that have a high degree of porosity (actually, poor welding).

Bolted cleanout openings can also be leak sources through a deteriorated or poorly made gasketed joint.

Valve stem packing can also produce steady leakage due to full compression of the packing or inadequate packing or from physical damage during the opening and closing process.

Settling of pipeline supports can produce a void between the pipe and the support. When this occurs, lack of support can place undue stresses on the unsupported line.

Spalling of the tank's concrete foundation and general deterioration, such as drying out and cracking of the tank's bottom flange bitumastic seal, are two further indications of general deterioration.

Weed overgrowth should be examined and controlled. Excessive weeds within the diked area render it useless as a fire break. Surveys have revealed 10-ft tall trees within a diked enclosure--such growth cannot be tolerated.

As far as practical, the earth within the diked enclosure should be crowned to drain rainwater away from the tank. As stated elsewhere in this manual, conditions have been viewed whereby all rivet heads in the lowermost seam of an older tank were corroded away from standing in water. Only the compression of the rivet body held the tank shell-to-head joint together.

4.9.3 Placarding

For a variety of reasons, including industrial espionage, industry does not like to indicate what product is stored within a particular container. This information can, however, be desperately needed by a fire department when fighting a fire. In the petrochemical field the name alone

does not aquaint the firefighters with the hazards associated with a particular petrochemical. For this reason the federal government through the
U.S. Department of Transportation has adopted a series of symbols to denote
classes of hazards associated with groups of chemicals. The symbols are
internationally recognized for use on hazardous material labels. Persons
dealing with hazardous materials should recognize the symbols and the potential danger they represent regardless of the symbol color. The wording on
the label may be in English or it may be in the language of the country of
origin. Figure 4.34 shown the governmentally accepted symbols.

It should be noted that the Code of Federal Regulations, Title 49,
Part 100-199 contains the regulations for preparation and transportation
of hazardous materials by rail, air, vessel, and public highway.

It is strongly suggested that the symbols be used in oil and petrochemical facilities to denote the chemicals stored in bulk containers,
including process vessels.

4.10 SECONDARY MEANS OF CONTAINMENT AND DIKED ENCLOSURES

Earthen dikes (berms), retaining walls, and curbs fall into the
category of secondary means of containment to impound spill from bulk
storage tanks or processing tanks. The enclosures should effectively contain spillage, overfill, and the entire content of the largest tank within
the enclosure. Earlier in this section of the report an incident was
described in which tank failure followed by a sudden release of the entire
tank content physically washed away a spill retention dike. On the basis
of incidents such as this, dike design and construction becomes a most
important spill retention factor.

Hitherto, dike construction largely consisted of mechanically pushing
fill material into position to provide an earthen barrier around one or
more tanks. For the most part, the permeability of the material was little
considered or was completely ignored. Frequently, erosion would greatly
reduce the effective height of a dike. During construction activities, a
dike would be breached to permit the entry of construction equipment and
the integrity of the dike could be violated for many weeks.

SYMBOLS	UN CLASS NUMBER	DOT HAZARD CLASSES		COLOR OF LABELS
	1	Explosive A Explosive B Explosive C		Orange
	2	Non-Flammable Gas		Green
	2	Flammable Gas		Red
	3	Flammable Liquid		
	4	Flammable Solid		White & Red Stripes
	5	Oxidizer		Yellow
		Organic Peroxide		
	2 or 6	Poisonous Material -	Poison A Extremely Toxic Gas or Liquid Poison B Highly Toxic Liquid or Solid	White
	6	Irritating Material -	The symbol is not required on domestic shipments.	"Irritant" in Red Letters on White for domestic shipments. Black for export shipments.
	6	Etiologic Agent -	Infectious material - Microorganism or Toxin that may cause human disease	Domestic Red and White Combination Export Black and White Combination
	7	Radioactive Materials- (Slight variation in label according to Radiation Levels)	If released, presents radiation and contamination hazards	White I-Red bar Yellow/White II-Red bars Yellow/White III-Red bars
	8	Corrosive Material		Upper Half-White Lower-Black
		Combustible Liquid		No Label Requirement
ORM-		ORM-A		No Label Requirement Refer to Title 49, CFR, Sec. 172.316 for the appropriate ORM designation.
		ORM-B		
		ORM-C		
		ORM-D		

FIGURE 4.34.   U.S. Department of Transportation Hazardous Materials Warning Labels

Some storage containers are surrounded by a concrete block wall. This form of containment has never been satisfactory, for a number of reasons:

1. Settling separates the blocks and, in some cases, crack them, destroying the integrity of the containment wall.
2. Spalling of the mortar between the blocks can destroy the liquid-tightness of the wall.
3. It is difficult to construct a liquid-tight wall from concrete blocks because of their porosity and the vulnerability of mortared joints.

Poured concrete walls are a preferable type of construction.

In older plants limited space, or the confined location of existing tanks, prevents the installation of dikes, barrier walls, or curbs. Such a situation warrants engineered drainage that will direct spilled material to an adequately sized impoundment pond or other temporary storage area.

It is also standard practice to maintain the impoundment area free of water accumulation that would reduce its holding capacity. A pumping system is generally warranted. The exception would be when the spilled material is directed into an operational waste treatment pond that is capable of holding the spilled material in addition to its normal effluent content. The impoundment should also be lined with concrete, impermeable clay, or a synthetic membrane, to prevent percolation through the earth down to ground water. NFPA further recommends that the slope away from the storage tank be from an elevation of 0.5 ft to grade 50 ft away in the direction of the impoundment site.

Returning to diked areas, as with impoundment ponds, the diked enclosure should effectively contain the content of the largest tank in the protected area. To prevent seepage and spill percolation, many industrial plants are installing a veneer of impervious clay or a layer of Gunite® cement over the contained land area. Gunite® is a sand cement mixture that is pressurized through a pneumatic gun onto prepositioned reinforcing steel. The force of application by the use of compressed air develops a piling effect of the larger particles and a dense impervious mass results, having a minimum of entrained air. The proper mixture of sand, cement, and water promotes extreme strength, hardness, and an abrasive resistant surface having a compressive strength that has been tested to 6,000 psi and greater after a setting period of 28 days. The density and liquid-tight qualities of Gunite® allow an application to withstand conditions that ordinarily destroy or damage conventional concrete. It has been successfully used for structures exposed to alkalies, acids, salt water, and certain corrosive fumes. The cost of application varies by state and contractor. It can be closely estimated, however, that a 4-in. thick application pressured into position over #8 reinforcing wire would cost about $3 per square foot. Figure 4.35 illustrates a typical method of Gunite® application.

FIGURE 4.35. Applying Gunite® Cement   Helper lifts reinforcing to ensure that Gunite® gets behind it, leaving the mesh in the center of the lining for strength and freedom from leakage. (Photograph Courtesy of Allentown Pneumatic Gun Company, Allentown, PA)

One of the advanced methods for soil treatment and stabilization has
been developed by the Japanese. Takenaka Komuten Co. Ltd., headquartered
in Osaka, Japan, has developed the Takenaka deep chemical mixing (DCM)
method for ground soil solidification at deep strata. With the DCM technique, soft ground formed by natural deposition is solidified in situ.
The solidifying chemicals with the addition of water are mixed into slurry
form and pumped to the area to be treated by an oil pressure pump. The
slurry is then mixed with soft ground to the depth needed to be solidified.
This technique effectively "locks" existing pollutants in place.

Another treatment known as Volclay as been developed by the American
Colloid Company, Skokie, Illinois. By the application of a veneer of high
swelling Volclay, bentonite pourous soil can be made impermeable. The
Volclay mineral is raked or disked into the top soil and rolled. The
Volclay increase in size on contact with water, jambing itself into voids
in the soil. Volclay is a natural bentonite clay product with the same
constituents as other clays, but with a unique molecular structure that
permits it to absorb many times its weight in water. Volclay swells up
to 15 times its dry bulk when fully wetted. This wetting is reversible;
Volclay can be dried and reswelled an infinte number of times. In extreme
cases where severe ground movement ruptures the soil or a foreign object
penetrates it, Volclay particles migrate with the seepage flow to affect
a resealing action.

4.10.1 Dike Construction and Erosion Control

Dikes are generally raised to a height of around 6 ft and normally
are flattened on the top to provide a 2-ft wide walkway. The slope of the
dike should be consistent with the normal angle of repose of the soil from
which the dike is constructed.

Erosion protection can be gained by covering the dike with a layer
of Gunite® cement at least 4 in. thick, complete with an embedded layer
of reinforcing steel as previously described. The covering also aids in
developing a liquid-tight containment barrier. In recent years, this has
been supported by building dikes that have an impervious base and internal

core. Figure 4.36 denotes how Volclay soil sealant might be used for impermeable dike construction. However, dikes are currently being constructed with a poured concrete base or foundation and a central concrete impervious core. To a certain extent synthetic membranes have been used to develop a liquid-tight barrier. Alternatives to a cement covering over the surface of the dike include steel plates that are positioned and secured permanently over the earthen surface. Some oil-handling facilities have sprayed the external surface of the dike with a bitumastic/ alphatic coating that hardens in position. A concrete dike based on delivered concrete at $30 to $48/yd^3 would cost about $50 to $70/yd^2. A gabion-protected dike would cost around $40 to $70/yd^2. Rip-rapped rock protection would cost $25 to $50/yd^2 of surface.

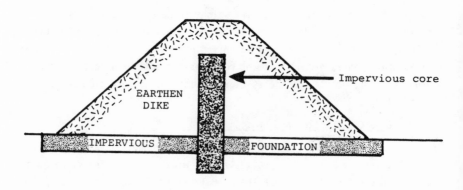

FIGURE 4.36.   Impervious Dike Construction

From a beautification standpoint, erosion can be controlled through
the medium of mulching blankets manufactured by Erosion Control Systems,
Gulf States Paper Corporation, Tuscaloosa, Alabama (205/553-6200, Ext. 427).
The blankets consist of a knitted construction of polypropylene with uniform openings interwoven with strips of biodegradable paper fill. Prior
to positioning the mulching blanket, topsoil should be spread over the
dike surface and grass seed should be spread into the topsoil. The erosion
control blanket is installed immediately after the seeding operation. The
blanket, available in sizes of 5 ft x 360 ft or 10 ft x 360 ft, is then
unrolled over the prepared surface parallel to the direction of flow,
avoiding any wrinkling or stretching that would interfere with continuous
ground contact.

On slopes, each upslope and each downslope end of each piece of
blanket should be placed in a 4-in. trench, stapled on 12-in. centers,
backfilled, and tamped.

On slopes where two or more widths of blanket are applied, the two
edges should be overlapped 4 in. and stapled at 12-in. intervals along the
exposed edge of the lap joint. The body of the fabric should be stapled
in a diagonal grid pattern, with staples 3 ft on center each way.

If any staples become loosened or raised, or if any fabric comes
loose, torn, or undermined, repairs should be made immediately. If seed
is washed out before germination, the area should be refertilized, reseeded,
and otherwise restored.

The staples used to position the blanket are fabricated of 11 gauge
wire, U-shaped with a 1-in. crown and 6-in. legs. They can be manually
pushed into the earth. Figure 4.37 illustrates the principle of the operation and the erosion control techniques. Figure 4.38 depicts the method
of applying the mulching blanket over a steeply sloped surface. The cost
of the described process on a 3.4:1 grade can vary, depending on the type
of seed selected, from 35¢ to 60¢/yd^2. By comparison, the laying of
started sod on a 3:1 slope would be in the vicinity of $2.50 to $3.40/yd^2
and a hay mulch seeding with a jute protection sheet would cost $1.25 to
$1.50/yd^2.

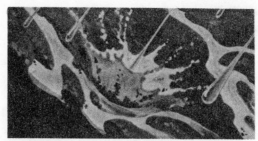

The beginning of erosion: raindrops act as tiny bombs.

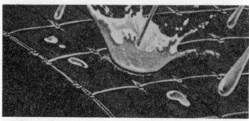

Hold:Gro shields ground from rain's explosive force. Then lets moisture seep through.

Paper acts as mulch to hold moisture in, degrades as new growth begins.

After paper degrades, the yarn temporarily supports the root structure.

FIGURE 4.37. Operation of Mulching Blanket for Dike Erosion Control (Courtesy of Gulf States Paper Corporation)

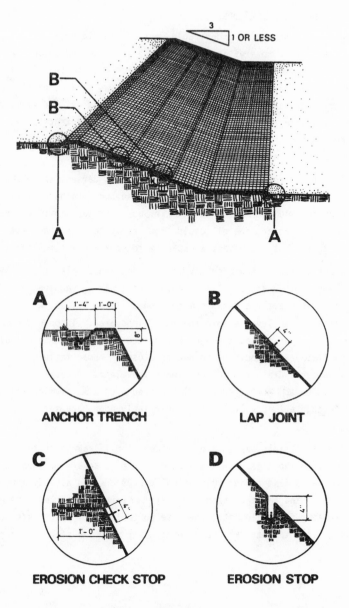

FIGURE 4.38. Mulching Blanket Application Techniques
from Gulf States Paper Corporation

4.10.2 Drainage of Diked Areas

There are a number of <u>undesirable</u> ways of draining rainwater from diked areas. Field observations at one location revealed an interconnected drainage system from one diked area to another. This to the extent that the drainage from an adjoining unowned facility was also connected into the system. The terminal control valve on the system was left in the open position, violating the integrity of the entire spill retention system. Furthermore, an overfill of oil from any tank would have the opportunity to travel over a great area--from one diked area to the other.

At another large oil-handling facility in New Jersey, the plant relied on dropping a large sandbag plug into an open drainage line. This manual plugging would demand that a worker pass through oil to restrain a spill. Until detected, the oil could flow freely to an oil and water separator that might not have the capacity to separate a large spill of oil.

A plant in the Boston area used (note past tense) flap-type valves to drain their diked areas. The procedure of draining required a worker to walk the dikes after a rainstorm to open the flap valves, which were positioned at grade level within the restrained area. Through the medium of a pull cord that passed through an elevated sheave on the crest of the dike, the flap valves could be opened or closed. Human failure to close the valve after a drain action was always prevelant. However, a major spill (200,000 gallons #2 fuel oil) occurred when a massive overfill escaped through a flap valve that was partially open because a stone had lodged under it.

The best possible way to avoid accidental oil leakage is to eliminate all drains from the area. This may appear to be a severe measure, but one of the major (and cleanest) chemical processors has already instigated such action. Management contends that it prefers to first test the accumulated rainwater for pollutants, then use portable pumps to transfer contaminated water into tank trucks and "clean" water into the storm drainage system.

The basic prevention system in the majority of plants is to maintain rigid control over the opening and closing of drain valves from diked areas. The valves are maintained in the locked-closed position. The key for the valve lock and the decision to drain rests with a selected supervisory person such as the maintenance foreman or operational superintendent. Prior to draining, the water quality is assessed for contamination, oil sheen, and the like. Once determined as "clean," draining is completed, the valve is locked by the supervisor and the date of draining is recorded in a time log as a protection against an unjust accusation of causing a sheen on any navigable waterway. Even this action can be improved upon by draining the water through an oil and water separator before final discharge. An added improvement would be to first drain the water through a bed or filter of sorbent material such as Dow Imbiber Beads. The beads provide a sorbent for hydrocarbons (oils), aromatic hydrocarbons (benzene, toluene), and polychlorinated biphenyls (PCP's). The beads are cross linked polymers of alkylstyrene, which has the capacity to absorb a large quantity of fluid without completely dissolving. This holds true even with an excess of the fluid absorbed. As the beads imbibe fluid, they swell in size and fluid is entrapped by the molecular structure of the bead. There is no leaking from the bead which, being oleophilic,* will permit the free flow of water until contacted by an oil material. Figure 4.39 graphically displays the function of an oleophilic sorbent filter.

4.11 TANK CLEANING AND SLUDGE DISPOSAL

4.11.1 General

Most large organizations, being cognizant of the hazards of tank cleaning, maintain a special tank cleaning crew that travels from plant to plant cleaning tanks on a regular frequency.

Unless adequate safety and industrial hygiene precautions are exercised, the accident exposures developed from tank cleaning are extremely high. To

---

*Treated to attract oil

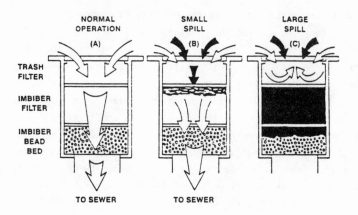

FIGURE 4.39. Dow Imbiber Bead Filter Unit
(Illustration Courtesy of the Dow Chemical
Company)

fortify this statement and to indicate the risks involved, the following
incident is related.

The foreman of a tank cleaning crew arrived at a Tampa, Florida
worksite early; it was in fact, dark. The tank's retention dike had been
broached and the worker entered the diked area in his pickup truck. To
expedite the cleaning process, he decided to open one of the tank's cleanout doors, which entailed removing a large number of nuts that held the
gasketed plate in position. The foreman turned on his truck's headlights
for illumination. As work progressed, the lights drew down the truck
battery and began to dim. At this stage the worker turned on his engine
to brighten the lights and charge the battery. Immediately upon turning
the ignition key, an explosion occurred that demolished the empty tank,
severely injured the worker, and practically demolished his vehicle. The
explosion was later attributed to vapors escaping from the loosened cleanout access door. The vapors hugged the ground and accumulated in the dike
area when they were ignited from a spark from the truck's ignition system.
Luckily, none of the nearby full storage tanks were damaged by the blast,
and a massive spill was avoided.

There are multiple hazards associated with the cleaning operation.
The risks include:

1. Possible explosion and resulting fire.
2. Exposure of tank cleaners to lead when leaded gasoline tanks are cleaned.
3. Lack of oxygen for workers entering the tank.
4. Toxic vapor inhalation.
5. Skin sorbtion, burning, and/or irritation from contact with gasoline, etc.
6. In tanks used for the storage of sour stocks and crude oil having high sulfur content, there is a risk from inhaling hydrogen sulfide (H_2S). H_2S is a colorless, toxic malodorous gas having a threshold limit value of not more than 10 ppm for any 8 hours of work exposure.

4.11.2 Personal Protective Equipment

The use of protective clothing, preferably fabricated of oil resistant rubber or vinyl, is important for personal safety. Positive air pressure respiratory equipment should also be provided for use by tank cleaners. The masks should provide full face protection.

Additionally, there should be no possible source of ignition. Workers should not be permitted to carry matches or lighters into the work area. Shoes with steel nails or tips should be prohibited and all tools should be of non-sparking design. Automotive vehicles should be restricted from entering the area.

4.11.3 Vapor Freeing the Tank

Each tank to be cleaned should be freed of flammable vapors before work is undertaken within the tank. This entails replacing all vapors in the tank with fresh air, which can be done with educators, fans, and blowers designed for use in explosive atmospheres. The American Petroleum Institute advises that testing for entry proceed in the following manner:

"To determine the progress of vapor-freeing operations, the atmosphere in the tank and the surrounding area should be tested frequently throughout the operation with a vapor indicator...

"...The tester should be thoroughly familiar with the reading and handling of the instrument. Before taking readings, he should determine that the instrument is in proper working condition and correctly calibrated. It is important that he adhere to the manufacturer's recommendations for checking and calibrating the instrument...

"...When vapor concentration has been reduced to 50 percent of the lower flammable limit and air is entering the shell manways, the presence of personnel around the tank no longer need be restricted. However, introduction of potential ignition sources within the area should still be subject to rigorous control based on the vapor concentration tests, wind direction and velocity, and other factors.

"When vapor concentration in the mixture leaving the tank is reduced to approximately 20 percent of the lower flammable limit, the first objective (removal of the flammable atmosphere) has been essentially accomplished. However, this condition is not necessarily permanent, and ventilation and vapor testing should be continued. The exact vapor concentration which will be considered safe before proceeding with the next step in the work will depend upon the program set up for sludge removal. This, in turn, will depend on the size of the tank, the facilities available, the amount of sludge and other factors.

"If it has been determined that other toxic substances are not present at levels above TLV,[*] the tank has not contained leaded gasoline since the previous cleaning, the vapor indicator registers a reading not exceeding 20 percent of the lower flammable limit, and oxygen content is at least 19.5 percent, the tester may enter the tank without respiratory equipment to make further tests at various points inside the tank. In most every instance if

---

[*]Threshold limit value.

TLV requirements are met, the LFL* will be well below 20 percent of the LFL. For large diameter tanks it is advisable for the gas tester to wear respiratory equipment for entry even through all of the above requirements have been met. During this testing, ventilation should be continued. For higher concentrations, anyone entering the tank should wear respiratory equipment which provides an independent supply of air in sufficient quantity to produce a positive pressure in the full-facepiece throughout the breathing cycle. Preferably no work should be started within the tank until it is vapor free."

4.11.4 Sludge Removal and Disposal

Initial tank cleaning is normally undertaken by the high pressure hosing of water through the access openings. The sludge and residues are directed to either the water draw or a pump-out connection in this manner. Ventilation of the tank should be continued during this cleaning process, since vapors can be generated and released by the hosing. The disposal of residue mixture warrants strict supervision, and ignition sources should be avoided at all times. API advises retention in a guarded sump, and if the discharge is directed into an open pit or retention pond, all ignition sources should be eliminated. Discharge into a public sewer system should be avoided at all times because of the hazards involved.

Final cleaning is largely a manual effort whereby the sludge water and scale mixture is brushed into piles for manual removal.

API Publication 2015 Second Edition, November 1976, provides considerable detail on personal safety, gas freeing, monitoring, and cleaning of the tank. Unfortunately, it does not describe how to dispose of the sludge removed from the tank. In the past this waste was buried within the confines of the diked retention area. In view of the hazards the buried material presented, some organizations placed a dated burial cross in the center of the site to warn against conducting any excavation in the area. This was the extent of precautions taken, and the area was considered safe a few years after burial.

---

*Lower flammable limit.

This procedure is no longer acceptable. The product contained in the sludge can quickly leach down to ground water and cause contamination. A more environmentally safe method of disposal is demanded.

The US/EPA suggests[*] the following actions as being appropriate for a waste handling strategy.

1. Minimize the quantity of waste generated by modifying the industrial process involved.

2. Concentrate the waste at the source (using evaporation, precipitation, etc.) to reduce handling and transport costs.

3. If possible, transfer the waste "as is," without reprocessing, to another facility that can use it as a feedstock.

4. When a transfer "as is" is not possible, reprocess the waste for material recovery.

5. When material recovery is not possible,

   a. Incinerate the waste for energy recovery and for destruction of hazardous components, or,

   b. If the waste cannot be incinerated, detoxify and neutralize it through chemical treatment.

6. Use carefully controlled land disposal only for what remains.

References:

1. "Cleaning Petroleum Storage Tanks." API Publication 2015, Second Edition, November 1976

2. "Waste Clearinghouses and Exchanges: A Summary. New Ways for Identifying and Transferring Reusable Industrial Process Waste." EPA Report SW-130c.1, a summary of report SW-130c.

---

[*] Federal Register, Vol. 41, No. 161, p 35050-1.

4.12 BURIED AND PARTIALLY BURIED TANKS

4.12.1 General

Although fire codes and regulations specify that certain materials
be stored in buried tanks as far as practical, an underground tank can
rapidly deteriorate when placed in contact with water or moist soil. It
is reasonable to consider that a buried tank can lie in moist soil for at
least 50 percent of its operating life. Under this condition, the metal
of the tank is exposed to a force that demands release. The force causes
the metal to actually dissolve into the surrounding soil or water, after
which it usually combines with oxygen to form oxides. This is the act of
metal returning to its natural state.

The corrosion is a continuous electrochemical process that results
in metal destruction. Underground corrosion of metal surfaces is a direct
result of an electric current that is generated by the reaction between the
metal surfaces and chemicals that exist in the soil or the water. The flow
of current from one portion of the metal to another part of the structure,
through the surrounding soil or water, causes metal ions or particles to
leave the surface of the metal, creating pits. This eventually causes complete destruction of the metal. The rate of current flow determines the life
span of the metal. One ampere of current discharge from iron can remove 22
pounds of metal in one year. Harco Corporation has developed cumulative leak
curves that indicate that after several years of trouble-free operation,
leakage is possible between the fifth and sixth year of unprotected burial.
Leakage is then a positive indication of increasing future trouble as three
additional leaks can be anticipated in the next year.

With the increasing cost of material and labor for tank burial and
removal, the provision of cathodic protection becomes cost beneficial if
installed at the time of tank burial. Figure 4.40 depicts a typical buried
tank cathodic protection system and placement of protective annodes and D.C.
rectifier. Section 7.3.2, External Corrosion, provides additional details
on cathodic protection.

FIGURE 4.40. Galvanic Protection of
Buried Tanks   (Illustration Courtesy
of Harco Corporation, Medina, Ohio)

4.12.2 Protective Coatings and Wrappings

In addition to cathodic protection, buried tanks should be cleaned
by abrasive blasting and coated with a protective coating before burial.
Basically, the principles of protecting a buried tank duplicate those
for protecting buried pipeline. A variety of protective coatings are
available. They include coal tar, somastic, and polyethylene tape.

The Tapecoat Company, Evanson, Illinois has specialized in hot-applied
coal tar coatings, cold-applied tape coatings, and mastic coal tar coatings
that can be sprayed on a tank. Figure 4.41 shows a mastic coal tar coating
being applied prior to tank burial.

4.12.3 Hydrostatic Testing

As with pipelines, buried tanks should be regularly subjected to
hydrostatic testing. Piping arrangements should be available to isolate
the tank for hydrostatic testing purposes. A five-year testing frequency
is advisable, using a water test pressure not over 1-1/2 times maximum
working pressures. The pressure should be maintained for a period of at
least an hour, preferably more, using a test pressure gage to detect any
pressure loss during the test. The water used to fill and pressure-test
the tank should be at room temperature, but not below 70°F. If desired,

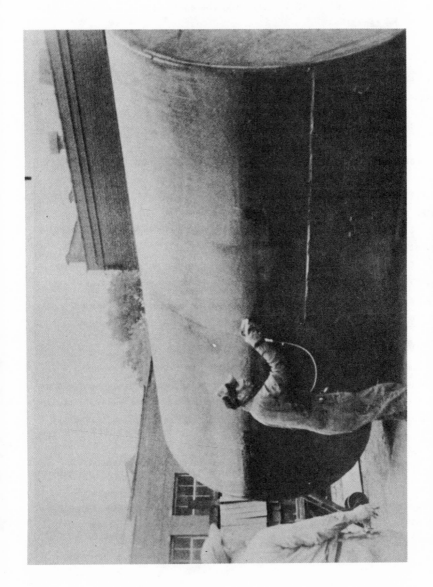

FIGURE 4.41. Applying Mastic Coal Tar Coating Using Spray Techique (Photograph Courtesy of the Tapecoat Company)

a dye can be mixed with the water; should a leak occur, the dye in the water will more readily aid in detecting the approximate location of the leak in the tank.

A number of publications provide guidance on hydrostatic testing of tanks. The main document is the American Society of Mechanical Engineers' "Unfired Pressure Vessel Code." The National Safety Council's "Accident Prevention Manual for Industrial Operations" also provides a well-composed and condensed section on hydrostatic testing.

4.12.4 Partially Buried Tanks

Partially burying tanks or resting them directly on the earth should be avoided. Metal deterioration is rapid, for obvious reasons, and the deterioration cannot be detected until failure occurs. Tanks should be raised above the surrounding earth and supported by adequately designed support structures installed on a sound foundation. There appears to be no benefit from partially burying or leaving a tank directly on the earth other than eliminating the cost of properly engineered supports. This cost saving is quickly cancelled by the rapid deterioration of the tank metal and short life span of the vessel.

4.13 BULK STORAGE OF BARRELLED AND DRUMMED MATERIALS

4.13.1 General

Numerous commodities are stored in bulk in drums, barrels, and portable refillable tanks. Materials such as solvents, greases, dry chemicals, and in some cases, fuels and other flammable or combustible liquids are transported and stored in 55-gallon capacity and even smaller drums. Spill prevention surveys have indicated a degree of sloppiness, carelessness, or an outright lack of knowledge in the safe way to store and handle drums. In a like manner, the fueling of vehicles directly from barrels of fuel presents a high risk of spillage and possible explosion and fire. Additionally, hazardous waste materials are frequently stored in drums prior to shipment to a waste disposal facility. Many plants do not use good drums for this purpose, since the drum will eventually be buried. A reconditioned 55-gallon steel drum can cost around $12, while new drums range between $15 to $16, so there is a tendency to use old drums, many of which have corroded and leaking bottoms or bottom seams.

It is normal to wait until the number of drums makes up a full truck load to reduce transportation costs. Leakage is common during both the storage and the transportation period because of the deteriorated condition of drums used for disposal; this to the extent that the haulers urge their truck drivers to refuse to transport "leakers."

Drums are often stored incorrectly and are left where they are exposed to being hit by vehicles.

Drum handling can also result in spillage. Heavy drums can be dropped while being manhandled into position, without properly designed lifting and carrying equipment. Accidents, with resulting spillage, can also be attributed to the use of gravity flow to draw materials from drums. Pumping with a hand pump is much better than gravity flow transfers. Racks of drums also warrant an effective grounding* system to protect against static electricity generation and lightning strikes. Bonding* is also warranted between the container and the receptacle to which a volatile liquid is being transferred.

4.13.2 Drum Storage

Preferably, drums should be stored in a multiple-tier rack constructed of substantial angle iron that is designed to hold the maximum weight in full drums that can be supported in the rack. The rack should be positioned within a curbed area that will effectively contain the liquid content of at least one drum plus 10 percent of other drums in the enclosure in the event of drum failure. When spigots extend beyond the curbing, a catch tray or cut-off drum should be positioned beneath the spigot to contain drippings and spillage.

_____

*Grounding and bonding are two separate systems of static protection. Grounding eliminates a difference in potential between a drum, tank, or object and the ground. Bonding eliminates a difference in potential between two objects such as a drum and another container.

When stored in an up-ended position without a rack, the drums should be contained within a curbed area as defined above. Pallets, or other structures, should be used to raise the drums off the grade within the curbed area so that the lower head and seam do not stand in collected water, where they can rust. Many plants spread a sorbent material within the curbed area. This can, however, become waterlogged with rainwater, which reduces its effectiveness. Both curbed-rack and plain-curbed storage areas should be protected with substantially constructed yellow- and red-striped pipe barriers that can draw vehicle operator's attention to the presence of the rack. The structural strength of the barrier should be such that it can absorb the initial impact of a backing vehicle that could otherwise hit and damage the stored drums.

4.13.3 Grounding and Bonding

Storage racks should be effectively grounded by using a clamp on an unpainted section of the rack with at least a No. 6 gage copper wire attached to a 1/2-in. (12.7 mm) diameter copper-clad steel, solid copper, or stainless steel rod. The rod should be driven at least 8 ft (2.4 m) into the earth at a position 2-3 ft (0.6 m-1.02 m) away from the rack.

Bonding does not eliminate a static charge, but it does equalize the potential between two receptacles so that a spark does not occur during a flammable liquid transfer. The bonding wire is basically a conductive wire attached to the storage container and to the container being filled by the liquid transfer. Alligator-type clamps can be used for physical attachment to each container. An alternative procedure that eliminates human failure to attach the wire clamps involves the use of a conductive fill line. Many transfer hoses have a built-in bonding line or have a woven metal outer covering that serves the purpose of a bonding wire.

4.13.4 Drum Handling

Considerable spillage of drummed products has been caused by a single individual attempting to handle a drum. A typical 55-gallon drum of oils, minerals, or lubricants could weigh 321 pounds (145.5 kg). This is too

great a weight to be handled by one person. Attempts to roll drums on the
bottom edge or rim frequently result in the drum being dropped on its side,
which releases the upper head (if clamped with a circular ring clamp) or
bursts the drum and discharges its contents. The worker is frequently
injured in the process.

It requires at least two men to up-end a drum unless a specially
designed lifting aid that provides a lever-type action is used. Barrels
are easier to handle with lifting aids than are drums, since barrels are
designed with two protruding bulges that encircle the container. The
bulges greatly aid the use of mechanical lifting devices. Circular clamps
with a levered closure action are also available. Once positioned and
clamped under the bulge of a drum, they provide handles on each side that
facilitate a two-man lift. These devices are generally used only on small
drums and barrels; fork lift trucks and pallets provide the safest means
of raising and lowering heavy 55-gallon barrels or drums.

American Petroleum Institute Publication 2008, "Safe Operation of
Inland Bulk Plants," provides the following comments on drum storage:

> Material stored outside should be safely and neatly
> arranged. Drums should not be stored under or around
> tanks. Drum storage areas should be protected from
> access by unauthorized persons. All empty drums
> should have their bungs in place. Leakers should
> be disposed of to prevent oil creating a potential
> slipping or fire hazard. Lighting should be ade-
> quate to ensure safety for personnel working at
> night.

In addition to maintaining good housekeeping in the drum/barrel storage
area, at least one portable fire extinguisher of a type suitable to combat
flammable or combustible liquid fires should be located not less than 25 ft
(7.62 m) from any outside flammable liquid storage area.

Additional safety factors are as follows:

- Petroleum products stored in drums should be clearly marked. Unidentified petroleum products can result in improper use, a fire hazard, an operating failure, or possible damage to equipment.

- Smoking or the use of open flames within 50 ft (15.24 m) of where flammables are being used or transferred or where equipment is being fueled should be prohibited.

- Areas in which flammable or combustible liquids are transferred at one time, in quantities greater than 5 gallons (0.019 m^3) from one tank or container to another tank or container should be separated from other operations by 25 ft (7.62 m) or by construction having a fire resistance of at least one hour. Drainage or other means should be provided to control spills. Adequate natural or mechanical ventilation should be provided to maintain the concentration of flammable vapor at or below 10 percent of the lower flammable limit.

- Workmen should be required to guard carefully against any part of their clothing becoming contaminated with flammable or combustible fluids. They should not be allowed to continue work when their clothing becomes so contaminated.

- Storage of containers, not more than 60 gallons (0.227 m^3) each, should not exceed 1,100 gallons (4.16 m^3) in any one pile. Piles should be separated by at least 5 ft (1.5 m) and should be at least 20 ft (6.1 m) from buildings.

- Electrical lighting should be the only means used for artificial illumination in areas where flammable liquids, vapors, fumes, dust, or gases

are present. All electrical equipment and installations shall be in accordance with provisions of
the National Electrical Code for hazardous locations.
Globes or lamps should not be removed or replaced
nor should repairs be made on the electrical circuit
until the circuit has been de-energized.

- Flashlights and electric lanterns used in connection
with the handling of flammable liquids should be
the type listed by the Underwriters' Laboratories,
Inc., or another nationally recognized testing laboratory, for use in such hazardous areas.

- Equipment using flammable liquid fuel should be
shut down during refueling, servicing, or maintenance
operations. This requirement may be waived for diesel-
fueled equipment that is serviced by a closed system
with attachments designed to prevent spillage.

4.13.5 Chlorine Cylinders

Although not considered to be fuel, chlorine warrants special comment
by virtue of the energy needed to produce it and to respond to a chlorine
spill. Replaceable and refillable cylinders or tanks are used to transport
and store chlorine. The vapor is poisonous if inhaled and will burn the
eyes. The liquid will burn the skin and eyes and cause frostbite and is
harmful to aquatic life even in very low concentrations.

For many years, The Chlorine Institute (TCI) of New York City, has
organized and coordinated mutual aid and spill response services. CHEMTREC,
the spill response service system developed by the Chemical Manufacturer's
Association, can rapidly alert specially trained personnel in the appropriate
geographic location, who will go to a spill site, evaluate the needs and,
if necessary, alert a mutual aid response team trained to stem leakage. The
teams have emergency kits that are designed to cap a variety of possible
leak sources. TCI reports that 6,000 leak control kits have been distributed
in the North American Continent. Three types of kits are available from TCI,
at costs of $625, $735, and $742 (September 1980) (Figure 4.42).

FIGURE 4.42. Chlorine Institute Emergency Kit "B"
(Photograph Courtesy of the Chlorine Institute)

From a spill response training standpoint, TCI conducts secondary or classroom training seminars (not hands-on training).

Information on training seminars or kits is available from:

Mr. Michael E. Lyden
Senior Engineer
The Chlorine Institute, Inc.
342 Madison Avenue
New York, NY 10017
Telephone: 212/682-4324

A similar service can be obtained in Canada through the Canadian Chemical Producers Association (CCPA), Suite 803, 350 Sparks Street, Ottawa, Ontario K1R7S8, Mr. Jean Belanger, 613/237-6215. The CCPA operates CANUTEC, which to some extent duplicates CMA's CHEMTREC.

References:

1.   "NACE Surface Preparation Handbook." (1977)

2.   Manufacturers' literature as listed in text.

3.   "Evaporation Loss Prevention." Chicago Bridge and Iron Co., Bulletin 533.

4.   "Oil Storage Tanks with Fixed Roofs." Chicago Bridge and Iron Co., Bulletin 3310.

5.   "Horton Floating Roof Tanks." Chicago Bridge and Iron Co., Bulletin 3200.

6.   "Lightning Protection Code 1977." National Fire Protection Association, Publication 78.

7.   "Recommended Practice for Protection Against Ignitions Arising Out of Static, Lightning and Stray Currents." American Petroleum Institute, Bulletin RP2003, Third Edition (1974).

8.   "Safety--General Safety Requirements." Department of the Army Corps of Engineers Publication EM385-1-1, June 1977.

9.   "Manual of Accident Prevention in Construction." The Association General Contractors of America, Inc. 1971.

10. "General Industry Safety and Health Standards." U.S./DOL Occupational Safety and Health Administration OSHA 2206 (CFR1910 Revised January 1976).

5. Electrical Transformer Dielectric Spill Control

5.1 GENERAL

This section deals with the prevention of spills of cooling liquids from electrical transformers and capacitor banks. Large transformer banks can contain as much as 250 gallons per phase and the dielectric can be either mineral oil or polychlorinated biphenyl (PCB). This material is on the US/EPA Toxic Chemicals List. A typical PCB insulation coolant mixture could be 70 percent PCB (AROCLOR 1254) and 30 percent trichlorobenzene. Federal investigations[*] have revealed that PCB's released into a waterbody pose an unacceptable risk to the health of man and the environment.[**] Even though the acute toxicity of PCB's is not too severe, their extreme persistency and bioaccumulation in the environment make spills of PCB's of particular significance. Cleanup actions for spills of this nature have been costly.

Normally, PCB's used as dielectrics in transformers and large capacitor have a life equal to that of the equipment, and with proper design, leakage does not occur. When the equipment is scrapped, the quantity of dielectric normally justifies regeneration.

Askarel is the generic name given to the PCB that is used in electrical equipment. Each electrical manufacturer has given a trade name to the particular grade of askarel used in its products. Pyranol, Inerteen, NoFlamol, and Saf-T-Kuhl are among the better known trade names.

Askarel has a specific gravity of 1.4 to 1.5, depending on its particular formulation. When it is spilled on land, it rapidly permeates the soil, finding its way to a water table; when it is spilled on water, it sinks to

---

[*] EPA Federal Water Pollution Control Act as Amended, particularly those portions that deal with water quality criteria and standards, effluent standards, and spill prevention.

[**] EPA Toxic Substance Control Act; EPA Resource Conservation and Recovery Act; OSHA, HEW/NIOSH, and FDA.

the bottom and remains there to be ingested by all forms of aquatic life, initiating its passage through the food chain.

5.2 SPILL CONTAINMENT

At most installations, should a transformer or a capacitor casing fail, the PCB content would flow to the nearest storm drain, then to a drainage ditch, and ultimately to a waterbody. Oil retention sumps in storm sewer lines have little value in containing PCB's, since they do not fall into the category of a floatable material.

To contain a spill, transformers and capacitor banks should be enclosed within poured concrete curbs or walls. The wall need not be high as long as its configuration will contain the dielectric content of the largest transformer within the enclosure. With capacitor banks, consideration should be given to "flashover," where more than one capacitor could be damaged by the failure of one unit. The enclosure should be equipped with a floor drain that can be filled with Dow Imbiber Beads (see Section 4.8, Tank Draining). With this type of drainage system, it is possible to leave the drain valve from the restrained area open to drain away rainwater. Should PCB/s make contact with the system, the beads will abosrb the PCB's and swell to 27 times their original volume in the process. This absorbtion is in the nature of the chemical rather than a physical process; the absorbed fluids cannot be squeezed from the beads, even if the bead is cut with a knife. The beads are unaffected by water, which can flow through the floor drain, but they will absorb the contaminating fluid from the water. They are of special value in containing askarel, since the pollution threat is removed from the environment once the askarel has been imbibed. Following a spill, the contained material and the askarel-saturated beads can be shipped to a US/EPA authorized PCB disposal facility.

Table 5.1 provides source data for the sorbent materials.

Figure 5.1, provided by Dow Chemical U.S.A., details the construction of a typical retention and drainage system.

This section is not intended as an endorsement of a single manufacturer's product; Dow is the only known manufacturer of the imbiber beads.

TABLE 5.1.

Imbiber Bead Product Distributors

MANUFACTURER AND TRADE NAME		TECHNICAL DETAILS
Dow Chemical U.S.A. Midland, MI 48640 Imbiber Systems Functional Products & Systems 2020 Dow Center	DESCRIPTION:	cross-linked solid organic polymer with imbibitive (true absorbent) properties for most organic fluids. Composition is mainly tertiary-alkyl styrenes with other monomers to imbibe (or absorb) organic solvents without dissolving.
	SPECIFICATIONS:	beads average 150 or 350 microns in diameter
	-	specific gravity 0.96-0.97
	-	absorb (or imbibe) up to 27 times its own weight (depends on solvent and its viscosity).
	-	bulk density of imbiber beads ∿38 lbs/cu ft (626 kg/cu M)
U.S.A. DISTRIBUTORS	PRODUCTS:	Imbiber Beads
	-	Soil Sealing (drainage devices)
Callahan Chemical Co. Box 53 Palmyra, N.J. 08057 (609-665-6640)	-	Packets and Blankets (spill control)
	-	Valves and Cartridges (water drainage)
Illinois Chemical Co. Box E Highland Park, IL 60035 (312-433-1145)	-	Drainage Devices for diked areas
	-	Shipping Containers (safety)
	-	Disposal of hazardous and nuclear contaminated oils to prevent liquid flow
R. G. Metz Co. 3914 Miami Road Cincinnati, OH 44227 (513-271-2468)	PACKAGING:	Imbiber Beads 40# (18.2 kg)/fiber drum
	-	Imbiber Packets (17.5 cm x 17.5 cm) 100/box
	-	Small Blankets (35 cm x 53 cm) 15/box
	-	Large Blankets (53 cm x 106 cm) 5/box
Licensed for use in The State of California for "Oil Spill Cleanup"	-	Imbiber Valve (8.89 cm diameter x 11.43 cm long) 12/box
	-	Imbiber Cartridge (15.24 cm diameter x 7.62 cm high) 4/box
	-	Imbiber Shipper Kit 1/2 liter 6/box
	-	Imbiber Shipper Kit 1 liter 4/box

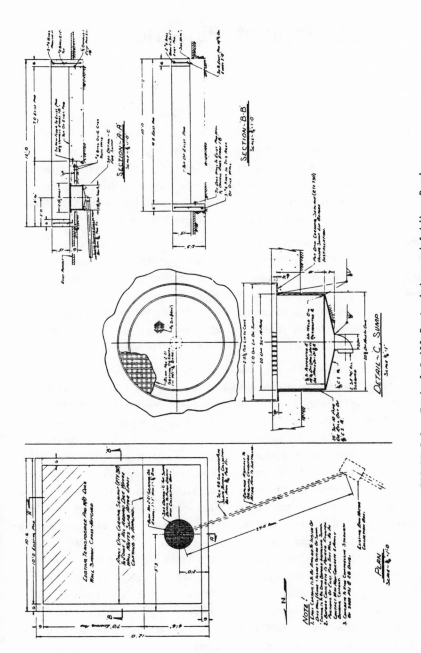

FIGURE 5.1. Typical Spill Restraining and Imbiber Bead Drainage System   (Illustration Courtesy of the Dow Chemical Company)

6. Inplant Pipelines

6.1 GENERAL

Inplant pipelines have historically been a major source of spillage. At many plants, lines have been buried as long as 30 years. This was a time when coatings were not as good as they are today and cathodic protection was rarely used. This latter deficiency is very predominant in the Northern states. However, the Southern states, especially in the Gulf Coast area, have been more inclined to provide cathodic protection.

The first sign of leakage generally occurs when the product comes bubbling up to the grade surface. In many cases, leakage could continue for many months before detection and line replacement could be undertaken. Very few facilities expose and examine inplant buried pipe sections, and it is even less customary to conduct hydrostatic testing of the lines. There is a noticeable trend, however, toward installing new lines above ground and in some cases raising lines up out of burial trenches.

Hydrostatic testing of lines, when conducted, should be at 1-1/2 times the lines' maximum designed working pressures. This test pressure is specifically designated by the U.S. Coast Guard regulations for cargo hose and pipeing (33CFR Part 126, Section 126.15). The regulation does not, however, designate the liquid that should be used for test purposes. The Military Standardization Handbook - Petroleum Operations (MIL-HDB-201B) also specifies testing at 1-1/2 times the normal operating pressure but not less than 100 psi. Pressure testing in this publication is recommended on an annual basis while the lines remain filled with the same product used in normal operations. Using this approach it would appear that pressure testing in excess of normal pressure levels in badly corroded or poorly welded pipelines could result in massive spills. If the lines contained their usual product, pollution would result and cleanup would be necessary. With low flash fuels, such spills could be of extreme hazard to life and property. This is recognized to some degree in the Federal Safety Regulations for Liquid Petroleum Pipelines (49CFR195.306), which specifies the following test medium for newly constructed steel pipeline:

195.306 Test medium

a. Except as provided in paragraph (b) of this section,
    water must be used as the test medium.

b. Except for offshore pipelines, liquid petroleum
    that does not vaporize rapidly may be used as the
    test medium if --

    1. The entire pipeline section under test
        is outside of cities and populated areas;

    2. Each building within 300 feet of the
        test section is unoccupied while the
        test pressure is equal to or greater
        than a pressure which produces a hoop
        stress of 50 percent of specified mini-
        mum yield strength;

    3. The test section is kept under surveil-
        lance by regular patrols during the test;
        and

    4. Continuous communication is maintained
        along entire test section.

The regulation does not, however, specifically designate the test
pressure that should be used.

6.2 COLOR CODING OF LINES

There has been no definite trend toward color coding or the labeling
of lines to indicate the product contained therein or pumped through the
line. The advantage of color coding and labeling warrants the expense
involved, since human error in opening the wrong valve is greatly reduced
once workers have been indoctrinated to the color code.

Color coding currently varies by plant. There has been no leadership
in assigning a certain color to a certain petroleum product or petrochemical.

6.3 OVERHEAD LINES

The height of inplant lines is an important factor. One known spill was caused by a visiting truck that tore down a bridge of lines passing over a plant roadway. Overhead lines should be positioned at heights well above that of any motor vehicle, and plant security should be alerted to obtain height information on visiting vehicles should they give an indication of excessive height (vehicle or load). The maximum height of pipelines should be posted at the entrance security office.

6.4 PIPELINE SUPPORTS

Until recent years, inplant pipeline supports were rarely an engineered item. During the course of nationwide surveys a wide variety of support structure has been viewed and critiqued. One New Jersey facility was satisfied to support a 6-in diameter line with a piece of 2 in x 4 in. scantling wood and an empty grease can. At other locations dock lines are frequently supported by sections of creosoted lumber of railroad tie dimensions. Much to their remiss a Massachusetts' plant experienced a major spill from this practice. The creosote had dried out of the wood until it retained rainwater, which in turn corroded the underside of the pipeline. The corrosion, in combination with abrasion caused by the pulsating action of the line during a pumping period, literally ate away a section of the pipe the size of a man's hand. During an after dark offload action, the leak went undetected for an undetermined period of time. The entire harbor, however, was covered by a fairly thick oil slick.

Pipe-supporting elements should not in any way restrict thermal expansion and contraction of the line. The spacing of supports should be close enough to eliminate sag. Naturally, the supports should be designed for maximum loading conditions. This design factor is considered important since lines running from loading/unloading docks can be used to transfer a variety of products that will have a wide weight variance and the dynamic loading on the line will be further affected by product changes. In short, the supports should not be considered as a simple means of propping up a

line: the design criteria should meet the maximum weight and dynamic
loading associated with the transportation system.

Basic steps to prevent pipeline leakage can be listed as follows:

- Using nondestructive, ultrasonic testing techniques,
  make at least an annual check of pipeline thickness
  to determine the extent of internal corrosion. Comparative records should be made and retained to gage the
  rate of internal deterioration and to estimate the line's
  anticipated lifespan.

- On a suggested three-year basis, partially rotate the
  lines to extend the life of the line from support contact wear and rainfall exposure of the upper half of
  the pipe.

- Provide supports that make a minimum of physical contact
  with the pipe metal. Rubber roller type supports are
  suggested since they rotate with the pulsing action of
  the pipe, are resilient, and do not interfere with the
  thermal expansion and contraction of the pipe. The
  dimension and strength of the internal spindle in the
  roller should be adequate to support the full load of
  the pipe section it supports. Cast iron should not be
  used except for roller bases, rollers,[*] or anchor
  bases, and so on; under mainly compression loading.

- Maintain a good protective coating on the lines at
  all times.

- On at least a five-year basis, subject the inplant
  lines and attached fittings to a hydrostatic test at
  a pressure 1-1/2 times the maximum designed working

---

[*]Some facilities provide V-type support spools that are of rotational
design.

pressure for the system. This pressure should be
maintained for at least 24 hours. Hydrostatic testing of new or modified lines is mandatory under
Federal law (33CFR126.15).

● Whenever practical, "dry" lines should be maintained.
The lines can be cleared of contained product by
injecting an inert gas into them. This action is
generally conducted from the plant's loading/unloading
dock to clear the entire length of the line back into
the storage container.

6.5 VALVES

Naturally, valves should be designed to meet the pressure demands of
the entire pipeline system. Valves should also be compatible with the pipe
or fittings to which they are attached and the liquid they transport.

Unfortunately, few plants have been found that have developed a safety
and/or security practice of locking valves in the closed position when not
in active service. It is most important that master flow control valves
be locked-closed to avoid unauthorized tampering or accidental opening of
the wrong valve. The time delay to unlock valves will give the attendant
time to clearly think over what he is doing. Basically, the opening and
closing of pipeline valves should not be a rapid operation. Sudden opening
of a valve can "shock" the downstream section of the line unless prepared
in readiness for the liquid flow. On this basis, time spent to unlock and
lock a valve can be time well spent.

Locking devices can consist of an intertwined chain and a padlock, but
most valves secured in this manner can be "cracked" open enough to gain
liquid flow. A more secure method is to hinge finger pieces to the valve
body, which when raised, encircle either the outer rim of the valve wheel
or one of the valve wheel spokes. A padlock is then used for positive closure.

A comparatively new method for detecting leakage of acids and caustics
as handled in a refinery is to lace covers over the valves or over the flange

couplings (Figure 6.1). The covers are a standard color, but the color
changes upon contact with a leaking product. The color change is so striking that it can be immediately perceived by a passing worker or an aerial
survey team. The manufacturer of these products, Slickbar, Inc., Southport,
Connecticut can provide samples of the material for testing with any product
to determine the extent of color change that can be expected. Table 6.1
defines the reaction of the covers to a variety of materials. Figure 6.2
illustrates a flange cover distributed by Slickbar, Inc.

6.6 ONE-WAY FLOW CHECK VALVES

The use of one-way flow check valves can provide a high degree of
protection against valve failure and product backflow. The installation
of check valves on fill lines that travel to a storage tank are exceptionally
important if the line connects into the bottom of the tank. Without a check
valve, the unintentional opening of a valve can drain the content of a line;
should the master flow control valve at the tank be open, considerable backflow can be anticipated. For this reason fill lines should have a one-way
flow check valve positioned at the termination of the line immediately
adjacent to the master flow valve on the receiving end of the pipe. Although
not a common practice, a check valve installed on the line between the tank
and the tank's flow control valve can provide an added degree of safety. The
position of the failsafe check valve at the base of the tank provides a high
degree of protection against line valve failure. Unfortunately, dual purpose
lines used both for fill and discharge purposes cannot be protected in this
manner, although a number of designs now incorporate two lines--a fill and
a suction. Fill lines that are positioned into the tank above the liquid
level do not require check valves unless provided as protection against backflow of the product remaining in the fill line (Figure 6.3).

On manufacturer, Wheatley/Geosource, Inc., Tulsa, Oklahoma produces
a wafer check valve that provides a compact, light-weight, and inexpensive
means to control backflow in close-quarter applications. It is available
in carbon steel and 316 stainless steel in two basic designs.

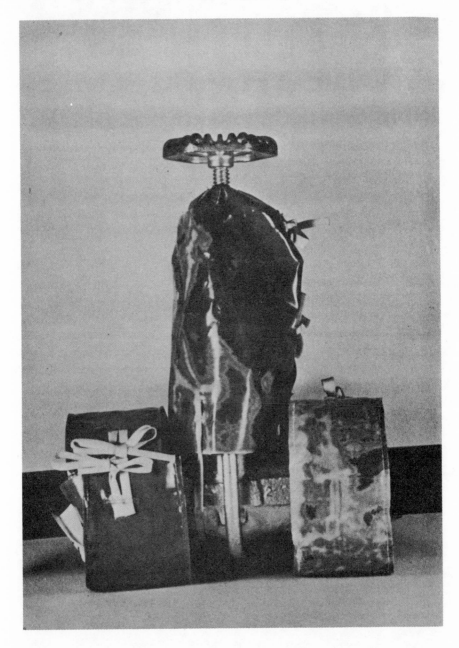

FIGURE 6.1.   Color Changing Valve Cover
(Illustration Courtesy of Slickbar, Inc.)

TABLE 6.1.

MILSHEFF SPRAY-STOP
VALVE & FLANGE COVERS
INDICATING COVER REACTION LIST

Indicating type covers will change color in those areas in contact with "triggering" reagents. Below is a list of some reagents and their effect on the covers. The color change occurs on the inside of the cover but will make a mark on the outside. Reagent concentration will affect the time and color change. The color of the untriggered cover is international orange. For those reagents not listed we will be pleased to furnish a sample of indicating material for your test and evaluation.

TYPE REAGENT	CONCENTRATE	REACTION/COLOR/TIME	OUTSIDE MARK
Acetic	100	None	None
Adipic	1.4	None	None
Ammonia	28.	None	None
Calc. Chloride		None	None
Chromic	50	None	None
Chromic	25	None	None
Ferric Chloride	70	Bleached Gray	Slightly Dark
Ferric Chloride	25	None	None
Formaldehyde	40	None	None
Hydrobromic	48	Bright Yellow Immediately	Bright Yellow
Hydrochloric	38	Bright Yellow - 5 min. To White	Colorless Area
Hydrofluric	100	None	None
Hydrofluric	50	Yellow - 10 min.	Slight Stain
Hyarofluric	25	Yellow - 10 min.	Slight Stain
Hydrofluric	15	Slight Yellow - 10 min.	Slight Stain
Hydrofluric	10	White Salty Deposit	None
Nitric	70	Color Removed - 5 min. White	Color Removed
Nitric	30	Brilliant Yellow 2 or 3 minutes	Brilliant Yellow
Nitric	20	Brilliant Yellow 2 or 3 minutes	Brilliant Yellow
Nitric	5	Slight Oil Mark	Slight Oil Stain
Oleic	100	Oily Mark	Oily Mark
Oxalic	51	Pink Stain - 20 min.	Pink Stain
Perchloroethylene	100	Oily Mark	Oily Mark
Phenol	5	Oily Mark	Oily Mark
Phosphoric	85	Dark Yellow 3 min.	Dark Yellow
Silver Nitrate	50	Black 15 min.	Dark Stain
Sodium Bichromate	50	Dark Stain - 3 min.	Dark Stain
Sodium Hydroxide	50	Light Yellow - 3 min.	Light Yellow
Sodium Hydroxide	35	Light Yellow Immed.	Light Yellow
Sulphuric	96	Dark Yellow - 25 min.	Med. Yellow
Sulphuric	70	Light Yellow - 15 min.	Light Yellow
Sulphuric	30	Light Yellow - 5 min.	Light Yellow
Sulphuric	10	Light Yellow - 3 min.	Light Yellow
Sulphuric	5	Light Yellow Immed.	Almost White
Hydrogen peroxide		None	None
Carbon tetrachloride		None	None

SLICKBAR Inc. 250 Pequot Ave. Southport, Ct 06490 U.S.A. 203-255-2601

Prices and specifications subject to change without notice.   All prices F.O.B. Southport, Ct. U.S.A.

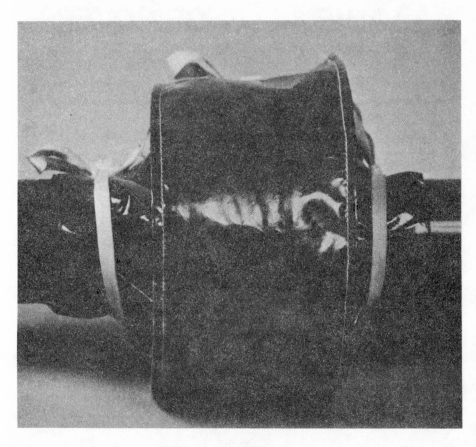

FIGURE 6.2. Drip Control Cover
(Illustration Courtesy of Slickbar, Inc.)

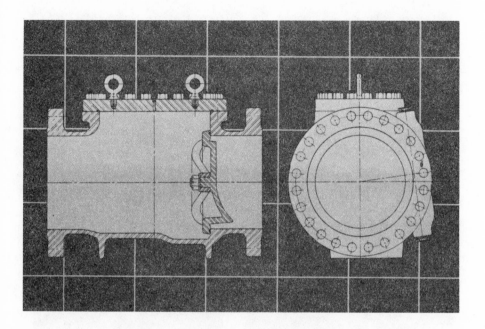

FIGURE 6.3. Check Valve Configuration
(Illustration Courtesy of Wheatley/Geosource, Inc.)

Sizes 2 inches through 12 inches are furnished with either a drop-in clapper
or a through-pin clapper. Sizes 14 inches through 48 inches are constructed
with a through-pin clapper.

The valve's materials of construction, face-to-face dimensions, and
performance conform to API-6D; the API monogram can be applied upon request.
The clapper seat O-ring is located in the valve body to eliminate washout.
Nitrile rubber is standard with fluorocarbon and butyl rubber and Viton A
are optional; the latter material being the best suited for hydrocarbon
products. Optional body facings (serrated, ring joint, or plain face smooth)
are also available. Working pressures range from 275 to 3,600 psi. The
valves can be readily inserted between two flanges if the line has play

enough to be separated by crowbar.   Once installed, a bubble-tight closure
is claimed by the manufacturer.

    The major change involves the replacement of the flange bolts to
accommodate the wafer check valve.   A 2-in. diameter valve is 3/4 in. wide
and costs about $90.   A 12-in. valve varies between 1-1/2 in. to 2 in. in
width and costs between $470 and $700, depending on the working pressure
requirement (Figure 6.4).

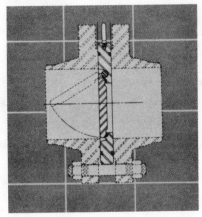

FIGURE 6.4.   Wafer Check Valve and Method of Installation
(Illustration Courtesy of Wheatley/Geosource, Inc.)

6.7 STEMMING A PIPELINE LEAK

One of the simplest methods for reducing and possibly fully containing a leak in a line is to drive a number of soft wooden plugs into the rupture or corroded opening. A good maintenance crew should have wooden plugs available for such an emergency (Figure 6.5).

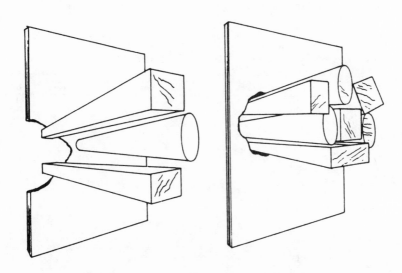

FIGURE 6.5. Plugging a Leak with Wedges
Note: Although a flat surface is depicted, the technique can also be used on pipelines.
(Illustration Courtesy of U.S. Navy Damage Control)

Another remedy is to have available within the plant a range of clamps that can be used to encircle the defective line and be tightened into position. The range of pipe diameters of inplant pipelines is normally not too broad, and clamps of varying length and diameter can be stored as a response measure. This is, however, a technique generally ignored by plant personnel. In 34 years of plant surveying by one of the authors, this leak containment technique has never been viewed.

7. Transportation Pipelines

7.1 GENERAL

The transportation of oil in overland pipelines is expanding annually. Some typical statistics on existing lines is contained in Table 7.1.

TABLE 7.1.
Some Typical U.S. Pipeline Installations

Products Line	System Mileage	System Pipe Diameters, in.	Capacity, b/d	Pump Stations	Installed Horsepower	Delivery Terminals	Tankage Capacity, bbl
Colonial Pipeline Co.	2845	6 to 36	800,000	27	296,250	164	18,819,000
Great Lakes Pipe Line Co.	6228	4 to 12		53	101,515	41	13,242,200
Buckeye Pipe Line Co.	2670	6 to 16	629,000	34	49,580	51	3,155,300
Plantation Pipe Line Co.	3213	4 to 18	354,000	39	121,230	169	5,829,400
Texas Eastern Transmission Corp., Little Big Inch Div.	3222	4 to 20	334,000	59	76,830	16	20,976,600 (includes underground storage)
Magnolia Pipe Line Co.	3225	4 to 12	319,000	34	19,300	49	925,000
American Oil Co.	2360	3 to 12		28	27,075	27	
Southern Pacific Pipe Lines	1923	3 to 16	217,000	19	22,365	18	4,836,791
Laurel Pipe Line Co.	447	10 to 24	200,000	4	12,800	22	3,555,440
Atlantic Pipe Line Co.	983	4 to 14	138,000	10	12,500	56	465,000
Phillips Pipe Line Co.	2633	6 to 12	122,000	28	58,675	9	7,713,450
Mid-America Pipeline Co.	2538	4 to 10	107,000	27	39,700		2,582,000
Dixie Pipeline Co.	1090	6 to 12	75,000	14	14,550	7	47,000
Continental Pipe Line Co.	746.46	4 to 8	71,900	9	5,365	10	297,000
Salt Lake Pipe Line Co.	1483	6 to 8	70,700	10	12,500	14	684,000
Cherokee Pipe Line Co.	1481	3 to 36	60,000	11	10,600	11	2,358,900
Champlin Petroleum Co.	710	2 to 6	54,600	9	3,725	7	1,095,742
Kaneb Pipe Line Co.	693	6 to 10	50,000	13	6,850	6	1,604,220
Yellowstone Pipe Line Co.	744	3 to 10	40,180	5	5,200	14	106,000
Gulf Refining Co.	474	8 to 12	18,000	3	930	4	801,500
Okan Pipeline Co.	447	3 to 6	14,600	2	2,860	5	

Colonial Pipeline Company Table

Transportation rates have reached 1,584,000 barrels per day. Probably one of the greatest pipeline construction projects completed within the last decade is the Alaskan pipeline. Built and operated in a very hostile environment, this line transports crude oil from oil-producing wells on the Alaskan North Slope 800 miles south to an ice-free deepwater loading terminal at Valdez, Alaska. The line crosses three mountain ranges, 800 streams, and miles of permafrost on its journey. It is a unique engineering feat and the largest privately funded construction project in history.

148

The oil industries have calculated that for overland movement, pipelines are the least expensive transportation medium. This to the extent that it costs 18 cents to send a one-ounce letter from Houston to New York, but a gallon of gasoline travels the same route by pipeline for less than 1-1/2 cents. Excluding utility gas distribution lines, there are some 550,000 miles of land and sea lines in the United States, transporting 7 billion barrels of oil and 20 trillion cubic feet of gas annually.[*] The low transportation cost has to a large degree curtailed the need for marine terminals and has reduced the quantity of refined products transported by coastal tankers. It has been found more economical to locate a marketing terminal on a stub line from an overland main line. By this action it is possible to maintain full tanks of product without the need to locate on a navigable waterway, which under any circumstance can be categorized as a potential trouble spot.

7.2 PIPELINE SAFETY

Modern pipelines are vastly improved over those constructed not many years ago. This resulted from advanced technologies, improved materials and welding techniques, more accurate testing procedures, computer programming, the advent of versatile pipelaying equipment, and remotely actuated and automated flow control systems. A small staff of people located at remote control centers, assisted by technicians and maintenance personnel, can keep an entire pipeline system running and maintain a continuous stream of oil or gas from source to destination.

They are, however, supported by sophisticated equipment--measuring and monitoring devices, computers, controls, telecommunication and other safety systems.

Turbine metering systems are available to provide a flexible means for reliable and accurate measurement of oil transported through pipelines. Digital telemetering of all controls, indications, and data between two stations permits accurate, rapid comparison with flow measurements at a receiving terminal. The comparison technique permits continuous leak detection monitoring and immediate remote shutdown of any or all pipelines.

---

[*]"Energy Lifelines Under The Sea." API Publication

It is possible to provide safety shutdowns and/or alarm devices for parameters such as flow, liquid level, temperature, pressure, overspeed, and vibration. The accuracy of available systems is in the vicinity of $\pm$1.0 percent of the total line flow. If the pumping station's flow rate should be greater than that of the receiving terminal by more than 1 percent, an alarm can be actuated, indicating a possible loss of oil. Normally, two alarms within a period of five minutes result in a shutdown of the pipeline system.

An alternative or additional monitoring system consists of flow-pressure relationship testing. Should a decrease in discharge pressure and an increase in flow rate occur at the same time, the pumping unit on the pipeline shuts down and an alarm activates.

A third safety system that can be installed on a line to detect any variance is a low pressure shutdown on the pumping unit. Upon detection of abnormally low discharge pressure the pump is immediately shut down.

The use of block and check valves on line sections that cross waterways are also vital in minimizing a spill in a watercourse. The block valve can be either manually or remotely closed to restrict natural flow in the line after pumping has ceased. Naturally a remotely closing valve is preferable, since the time factor for a worker to get to and close the block valve is eliminated. By positioning a one-way flow check valve on the line on the opposite side of a waterway, backflow is restricted and the only oil released is that contained in the line between the block and check valves.

In short, in addition to control of operations at manned locations, it is possible to maintain constant and complete control of operations for the full length of the line. Through the medium of a computer the entire system can be scanned as frequently as every 10 seconds. Any change from the previous 10 seconds will be indicated to the line operator on his control panel. The operator can rapidly assess the change and in the shortest possible time, instigate action to stop or start units, close valves, and so forth, to mitigate any deviation from normal operation.

7.2.1 Planning the Line

Although the markets determine the general location of the line, the most desirable route to a given terminal can be rapidly evaluated from detailed topographic maps and by aerial survey, using plain and infrared or multispectral photography. From these surveys, a ground survey of the right of way can be instigated. Wherever possible, the crossing of waterways and wetlands should be held to a minimum, since they can definitely be considered as potential trouble spots in the event of an accidental leakage.

Aerial photographs can be used to advantage in the development of a federally required Spill Prevention, Control, and Countermeasure Plan (40CFR112). The photography can help to determine where spill response equipment can best be stored for fast deployment. The maps and aerial survey will indicate heavily industrialized and populated areas which, as far as practical, should be avoided. Environmentally sensitive areas should be bypassed; this would include spawning areas such as wetlands, marshes, and so on. Attention to this factor should aid greatly in gaining a permit to construct the line.

The selection of materials is of major importance. High-yield and tensile-strength pipe of seamless or electric-weld pipe offers many advantages over lap- or butt-weld pipe. Pipelines also benefit greatly from economies of scale--the larger the pipe diameter, the lower the transportation cost will be. An initial high cost will be returned by the ability to transport more product, with an energy saving over the life of the line. Engineering computer programs can be run to determine the correct size of pipe to gain maximum throughput. Finally, the pipe selected should meet the maximum stress requirements that will be placed on the system.

The pipeline industry has determined from years of experience that pipeline joints need special attention. At least 15 percent of all girth welds should be x-rayed for welding defects (this exceeds federal requirements by a factor of 5 percent), and 100 percent of road, railroad, and river crossing welds should be x-rayed (although federal law requires only 90 percent testing.

7.3 CORROSION PREVENTION

7.3.1 Internal Corrosion

Maximum protection can be gained by keeping both the product and the line free of water. The products can now be treated with rust inhibitors. It is advisable that even the water used for hydrostatic testing of the line be treated with a rust inhibitor.

7.3.2 External Corrosion

External corrosion is one of the principal factors in line failure, especially when the pipe is buried and hidden from view. Bernard Husock, P.E., Vice President, Harco Corporation, Medina, Ohio[*], stresses four factors that relate to underground corrosion:

1. Corrosion of iron and steel is a natural process.
2. All ferrous metals corrode at essentially the same rate.
3. Corrosion of iron and steel underground or underwater results in selective and concentrated attack.
4. Once leaks start to occur on an iron or steel pipe, they continue to occur at an exponentially rising rate.

Husock defines the two basic mechanisms responsible for underground corrosion as follows:

- Electrolytic Corrosion: Electrolytic corrosion is a result of direct currents from outside sources. These direct currents are introduced into the soil and are picked up by an underground pipe. The locations where the current is picked up are not affected

---

[*]"Causes of Underground Corrosion." Paper HC-36, 21st Annual Applachian Underground Corrosion Short Course. West Virginia University, May 1976.

or are provided some degree of corrosion protection.
But the locations along the pipe where this current
leaves the pipe to enter the soil, those locations
are driven anodic and corrosion will result. The
corrosion of an iron or steel pipe under this influence will be at the rate of 20 pounds per ampere per
year. This type of corrosion is often referred to
as stray current corrosion. If the outside source
of current is a cathodic protection rectifier on
a pipeline belonging to others, the corrosion problem is referred to as an interference problem.

● Galvanic Corrosion: Galvanic corrosion is the self-
generated corrosion activity which results when the
pipe is placed in the soil. Differences in potential
develop along the pipe or between different pipes.
These differences can result from dissimilar metals
placed in the soil or they can develop on the same
metal as a result of differences within the soil.
These potentials generate corrosion currents which
leave the metal to enter the soil at anodic areas
and return to the metal at cathodic areas. Corrosion
occurs at the anodic areas where current leaves the
metal to enter the soil.

Based on this information (with certain restrictions that follow),
resistivity testing of the soil on a pipeline route may be a beneficial
factor. Table 7.2 was also drawn from Mr. Bernard Husock's technical
paper.

The paper continues with the following precautions.

This classification is not intended to be considered
as an exact guide for classifying soils according
to corrosiveness. It merely serves to indicate that

TABLE 7.2.

Soil Resistivity Classification

Resistivity--ohm-cm	Category
0-2000	Very corrosive
2000-5000	Corrosive
5000-10,000	Moderately corrosive
10,000-25,000	Mildy corrosive
Over 25,000	Progressively less corrosive

the corrosion rate is lower in soils of higher resistivity. However, it must be appreciated that if a pipe is in the ground long enough, it will develop leaks even in soils of more than 25,000 ohm-cm.

In addition, there is a further complication encountered in the higher resistivity soils. Large variations in resistivity within relatively short distances are often seen in higher resistivity areas. It is not unusual to have variations in resistivity from less than 10,000 to more than 100,000 within 500 feet of pipeline right-of-way. These variations in themselves indicate variations in soil composition which can be responsible for the promotion of galvanic corrosion activity.

There are some positive ways of retarding the deterioration process of a buried pipe as follows.

● The pipe should be abrasive blast cleaned and painted with a protective coating such as asphalt, coal tar, somastic, and polyethelyne tape, or any proven synthetic coating developed for pipeline protection. The coating should be the one determined to provide the best protection for the soil condition at the trenching location (Figures 7.1, 7.2, and 7.3).

FIGURE 7.1. Hot Applied Coal Tar Enamel Installed Prior to Burial of Line (Figure Courtesy of Koppers Co., Inc.)

FIGURE 7.2. 36" Pipeline Tapecoat 20 Application Process   36" pipeline being
laid in a residential area that has been coated with Tapecoat 20 (Hot Applied Coal
Tar Coating) in a bend which goes down in the ground more than 14 feet. The men
in the ditch are making a tie in weld and are making sure that the pipeline is
properly lined up prior to making the weld. (Photograph Courtesy of the Tapecoat
Company, Evanston, Illinois)

FIGURE 7.3.   Hot Application of Tapecoat 20 Coal Tar Coating (This particular installation was applied 20 years ago.) (Photograph Courtesy of the Tapecoat Company)

- Under federal regulation it is also mandatory that a
  buried or submerged line be protected by a system of
  cathodic protection. A typical protection system
  involves connecting the pipeline and a sacrificial
  anode, generally graphite or magnesium, to a direct
  current rectifier as a source of electrical current.
  A controlled quantity of current is then passed
  through the soil and to the pipeline. In this manner
  the anode is corroded instead of the pipeline metal.
  Regular testing of the system and renewal of the
  anodes is obviously warranted (Figure 7.4).

Specific details on federal requirements for operation and maintenance
of overland pipelines can be found in CFR 49, Part 195 as revised October 1,
1979.

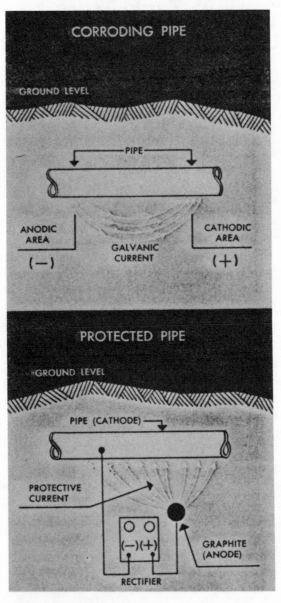

FIGURE 7.4. Corroding and Protected Pipe
(Illustration Courtesy of Harco Corporation,
Medina, Ohio)

8. Tank Truck Loading/Offloading Racks

8.1 GENERAL

Considering the vast amounts of flammable petroleum liquids handled daily through truck loading/offloading racks, the losses due to spills, explosion, and fire have been relatively small in number. However, a single incident can be catastrophic. As the trucking industry continues to grow, the risk increases in proportion. An upward trend in losses in recent years prompts the need for better loading-rack design and enhanced training for those who operate the equipment. This would include company-employed loaders and common carrier personnel who conduct their own loading operation. The following factors (as researched by Industrial Risk Insurers Company, Hartford, Connecticut) contribute to losses at loading racks.

- Faster loading rates, which run between 500 and 1,200 GPM (0.03 to 0.07 m^3/s).
- Extensive use of product filters, which are prolific producers of static electricity.
- Drier, cleaner products, especially in the distillate range, which retain their static electric charges for longer periods of time.
- Increased handling of more hazardous products such as jet fuel (JP4), benzene, toluene, and xylene, which, because of their vapor pressures, present longer exposure to vapor conditions inside the tank compartment within the flammable range.
- Extensive use of large-capacity aluminum tank trailers, which are prone to rapid structural failure when exposed to intensive ground fires.
- Larger and more complicated loading racks, designed to handle many large trucks at one time. Some up-to-date racks are equipped with extensive automated data processing equipment for recording and billing deliveries.

160

- Limited availability of real estate at some locations, which not only creates congestion and complicated truck traffic patterns, but also exposes adjacent tankage and facilities to fire.
- Key-stop (unattended) operations are gaining more attention, with the result that there is less supervision over loading practices and less possibility of controlling fires once they have started.
- Labor turnover at loading racks is higher today. Loaders are not as experienced as in the past.

To minimize losses and to attain a safe, productive record, it is necessary for those concerned with loading rack operations to fully understand the inherent hazards, especially those related with the hazardous practice of "switch loading," which involves loading a vehicle with a product other than the one previously carried.

In general, industry needs to design better loading racks and to renovate existing racks so as to minimze operational hazards, to constantly educate loaders and truck drivers in the safest loading/offloading practices, and to implement the best possible engineering protection to control spills and/or fires.

8.2 CONSTRUCTION AND LOCATION OF LOADING RACKS

Proper, adequate spacing of loading racks is of primary importance both against exposure fires and for control of fire emergencies and spills on the racks themselves. Table 8.1 illustrates the recommended spacing for loading racks as developed by Industrial Risk Insurers Company.

Loading racks should be constructed entirely of noncombustible materials. In no case should materials such as wood decking, wood stairs, wooden ramps, and frame enclosures for the protection of loading personnel be used. Roof sheeting should also be noncombustible, with no tar or asphalt used as coatings or sealing materials. Racks should be limited in size to accommodate no more than six trucks at one time. Larger installations should consist of no less than 50 ft (15.2 m) of open space between structures.

TABLE 8.1.

Recommended Spacing for Truck Loading Racks

MINIMUM DISTANCE IN FEET	PRODUCT TANKS	LPG TANKS	LOADING RACK	PUMP HOUSES	SERVICE BUILDINGS	DOCKS	FIRE PUMPS	FIRE HYDRANTS	COMPRESSORS	PROCESS UNITS
LOADING RACK	250	100	50	200	200	200	150	100	200	200
	76.2 m	30.5 m	15.2 m	61 m	61 m	61 m	45.7 m	30.5 m	61 m	61 m

Proper drainage of loading rack areas is extremely important in loss prevention and control in the event of a fire. Rack areas should be fully paved, curbed, and drained (Figures 8.1 and 8.2), so that any spill from trucks or equipment would quickly flow to trapped drains that are capable of removing the hydrocarbon liquid without any backup of liquids at the drain.

Drains should be located out from under the trucks on the outboard side of the rack and should be frequently checked to see that they are clear of debris and blockage.

In situations where pumps, additive storage tanks, or undiked product storage tanks are located uphill from the loading rack, diversion dikes should be installed to prevent any hydrocarbon leaks from entering the rack area. Conversely, if the loading rack is located uphill from other hazardous or valuable property, diversion dikes should be provided to stop possible spread of fire from the rack to those areas. Any oil storage in drums or tanks should be stored within a confined (walled or curbed) area, with sorbent materials spread within the contained area to contain spillage and drippings. They should be kept as far as possible from the rack.

FIGURE 8.1. Truck Loading Rack Drainage
This depicts a well-drained and curbed loading
area; however, center drain exposes the underside of truck to possible fire and/or explosion.
Water spray heads have been installed in the
metal grating for added protection. (Photograph
Courtesy of IRI Company)

FIGURE 8.2. Preferable Truck Loading Rack Drainage
System This illustrates a better drain arrangement,
where spills would quickly pass into catch basins out
from under the tanker trailer. Note good area curbing.

Curbing or guards of sufficient strength and height should be installed around the loading islands to prevent trucks from striking and damaging the rack. Enough space should be provided so that even the largest trucks have ample turning radius when entering or leaving the rack. Spacing should be provided for trucks to park 100 to 200 ft (30.5 to 61.0 m) from the rack while awaiting a turn to load or to make minor repairs. Loading racks should be located as far as possible from tankage and other hazardous structures, such that trucks would not have to pass through any other potentially hazardous areas between the street and the rack. Separate entrance and exit gates should be implemented to prevent possible truck collisions and jam-ups. Finally, safety signs should be properly posted at conspicuous locations to constantly remind rack personnel of concern for safe loading practices.

Closed circuit television monitoring systems have been found useful for the supervision of loading/unloading racks. A well-positioned monitor in the plant office can provide a means of viewing the operational procedures. These devices present an opportunity to observe unsafe acts or conditions during daily operations. A typical installation would cost $5,300, plus an installation cost of around $1,000.

8.3 STATIC GENERATION

Electrostatic generation in hydrocarbons passing through piping is not critical, since turbulence is usually low and any charges generated quickly dissipate to the metal piping. However, hydrocarbons passing through produce filters that utilize cotton, paper, or felt elements generate strong static charges, requiring at least 30 seconds of relaxation time to dissipate in the flow between the outlet of the filter and the tip of loading tube. (Longer lengths of pipe and special relaxation chambers yield additional relaxation time if needed.)

Currently, industry is placing more emphasis on higher purity products, particularly as produced by hydrodesulfurization. The 30-second relaxation time formerly thought to be adequate may now be subject to serious questioning. Surveys of test results indicate a significant increase in relaxation times, from 100 to 500 seconds (1.6 to 8.3 min.) or more under some conditions.

Electrostatic charges are also generated as the oil enters the truck tank compartment. The intensity is proportional to the amount of flow turbulence, which itself is dependent upon two factors: the rate at which the product enters the compartment and the design of the loading tip. Several oil companies, in the hope of reducing spills, have investigated various loading tip designs. The straight tips (formed by cutting a loading tube at 90-degree angle to its centerline), seem to cause turbulence and tend to be thrown upward when loading is initiated. A loading tube cut at a 45-degree angle to its centerline, however, has satisfactory turbulence characteristics at a moderate flow rate and is the less costly of the two. All loading tubes should be constructed of metal and of sufficient length to rest on the bottom of the tank compartment.

The rates for loading jet fuels (JP4), benzene, toluene, xylene, or distillate class materials after handling gasoline (switch loading) should not exceed a velocity of 3 ft/second (0.9 m/s) until the outlet is completely submerged. The full loading rate should then not exceed 15 to 20 ft/second (4.6 to 6.1 m/s). This shifting in rate can be accomplished by utilizing a special loading regulator tip, shown in Figure 8.3.

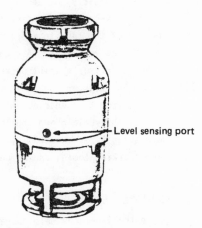
Level sensing port

FIGURE 8.3. Loading Regulator Tip

The special loading tip automatically shifts to the full loading rate when submerged to a safe depth. The circular deflector tip design minimizes spraying and splashing and prevents the loading tube from being jolted out of the tank compartment when loading has begun.

Instances of disasters caused by static electricity are plentiful. In one example, case history No. 30,081 from Industrial Risk Insurers (IRI) Company, "a diesel-powered tractor pulling an 8,600-gallon (32.5 m3) aluminum trailer was being loaded with #2 fuel oil when there was an explosion and subsequent fire. The driver had pulled in at the rack and had started to load. The meter showed 6,379 (24.1 m3) gallons had been loaded when there was a terrific explosion followed by fire. The yard man shut down the nozzle. The foreman called the fire department, threw the pump disconnect switch, and actuated the emergency valves. The fire chief and foreman checked the truck and found the ground wire was in use and the ignition key was in the "off" position. Static electricity caused by fast loading of the fuel oil appears to have been the source of ignition."

Causes other than static electricity can readily ignite explosive vapors. In another example a gasoline transportation vehicle had crossed town in an empty status; the travel greatly increased the vapor content in the cargo compartment. Upon arrival at the loading terminal, the maintenance supervisor climbed on top of the tank to inspect the loading dome, which was reported as having a defective gasket. A cigarette lighter (found in the aftermath) fell from the supervisor's shirt pocket into the vapor-filled tank. The fall activated the lighter, which in turn caused an explosion. The worker was blown 60 ft from the vehicle. Although badly injured, he survived the ordeal. However, the vehicle and the loading rack were a total loss as a result of an ensuing fire. Naturally, sources of ignition such as lighters, matches, and sparking tools, and so on, should be prohibited in loading/unloading areas.

8.3.1 Neutralizing Static Charges

"In-line" static hcarge neutralizers can help to reduce static by producing charge gradients. Such devices are constructed of a plastic insulator

tube, enclosed in a metal housing that has sharp grounding points. They
should be installed downstream from filters, pumps, and other equipment
generating static charges. These neutralizers remove the static electrical
hazard from the oil before it reaches the loading tip, but they have no
effect on charges caused by excessive turbulence at the loading arm tip.

Specially formulated additives known as fuel conductivity or anti-
static additives are most often used in petroleum distillates such as
kerosene, furnace oil, and diesel fuel, since these fuels have a tendency
for higher charging, lower conductivity, and lower vapor pressures. Even
proper grounding of low conductivity fuels does not guarantee that charges
will not accumulate in a receiver and result in significant surface voltages. If this surface voltage exceeds the critical set value, an incendiary discharge between the fuel surface and the receiver interior may occur
in the available vapor space, which can ignite a flammable fuel/air mixture.
These additives increase fuel conductivity and increase the dissipation of
electrostatic charges through the fuel. Low concentration of these additives
will increase fuel conductivity to 50 picosiemens/meter. (Studies indicate
that a fuel with a conductivity of 50 picosiemens/meter will satisfactorily
dissipate electrostatic charges.) There are several models of meters that
are designed to measure fuel conductivity in picosiemens/meter directly and
are simple to operate.

Although these additives serve to protect against accidents caused by
equipment failure and human error, they should not be used in lieu of well-
designed equipment and proper operational practices.

8.3.2 Bonding

Bonding involves the joining of two pieces of metal with a wire to
eliminate the probability of a spark. Good bonding is essential between the
loading rack and the tank truck to eliminate sparking at the hatch (Figure
8.4). The bonding or grounding cable should be attached before the loading
pipe is inserted. Bonding is usually accomplished by means of a heavy
flexible copper or stranded steel cable that is securely attached to a product pipe riser at one end and clamped to the compartment shell being loaded

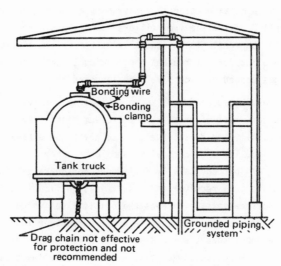

FIGURE 8.4. Grounding Tank Trucks Immediately After the Truck is Spotted and before the Loading Spout is Inserted   (API Illustration)

at the other end with a heavy-duty alligator clip or pressure clamp. In most cases, trucks will have permanently attached special bolts made of noncorrosive material with spherical heads. The bonding cable would then be equipped with a pressure clip to match the spherical bolt head. The condition of alligator clips, especially the joint between the bonding wire and the clip, should be periodically inspected by persons other than rack personnel to ensure effective bonding circuits. This is very important, since vehicles are frequently driven away from the rack without disconnecting the bonding wire. Some organizations place a barrier over the truck's windshield as a reminder to the driver that all connections, fill line and grounding wire, be disconnected before pulling away from the rack. As long as the resistance in the circuit does not exceed 1×10^6 ohms, the electrostatic bonding is considered adequate. In some designs, loading rack ramps drop down and the vehicles back into the ramp to offload. The rear of the truck can then make physical contact with a bumper stop or concrete wall. This is not an effective electrical bonding for loading/offloading operations.

In the concept of grounding (Figure 8.4), the bonding system is electrically connected to the earth so that any static charges or stray electrical currents from the truck are led to a common ground. The testing of proper rack grounding is accomplished by using a common volt/ohm meter, attaching one end to the rack and the other end to an earth ground (for example, a fire hydrant or water pipe). Readings in excess of 10,000 ohms show extremely poor grounding and correction is recommended. Resistance is usually less than 5 ohms on good grounding.

Electrostatic drain and ground indicators can be used on loading racks. These indicators have signal lights to indicate safe grounding conditions. They can also be arranged so that loading pumps cannot be started unless proper grounding has been established. This feature proves very effective, especially where loading racks are left unattended and loading is conducted by individuals other than company employees.

8.4 CONTROL OF THE ATMOSPHERE ABOVE THE OIL

The control of the atmosphere above an oil cargo involves a reduction in oxygen level in the compartment being loaded to a point where combustion cannot take place. Ordinarily, with gasoline and other products having a Reid vapor pressure above 4.5 psi (0.3 kg/cm^2), the vapor in the compartment is "too rich" to burn and need only be properly bonded. Intermediate products having Reid vapor pressures below 4.5 psi (0.3 kg/cm^2) and a flash point below 100°F (37.7^{o}C) produce concentrations within the flammable range, which can easily be ignited by static charges on the surface on the oil. To achieve maximum safety in loading these products, oxygen levels in the compartment can be reduced below combustion levels by injecting sufficient amounts of CO_2, nitrogen, or inert gas into the vapor space before loading. Equally effective is the practice of gas blanketing the compartment with fuel gas or heating the oil sufficiently to maintain a "rich" vapor condition over the oil.

Currently, these procedures are not generally followed because of the time consumed and the extra expense involved. The only alternative, which does carry some degree of risk, is to minimize static potential as discussed in the preceding sections.

8.5 SWITCH LOADING

According to the Risk Insurers Association, the practice of switch loading accounts for 70 to 80 percent of the serious losses reported at loading racks; as a result, switch loading merits the special attention of all those connected with loading rack safety.

Switch loading is the loading of low vapor pressure products into a compartment that previously contained a high vapor pressure product (for example, loading kerosene or diesel fuel into a compartment previously containing gasoline). A flammable vapor space is present over the lower vapor pressure product because the compartment has been partially ventilated when unloading the previous load of gasoline, thus reducing the "over rich" condition to a flammable one. The lower vapor pressure product usually retains static charges long enough to develop a potential ignition sources. Switch loading losses seem to occur most often when compartments are one-quarter to one-third full and temperatures are close to $30^\circ F$ ($-1^\circ C$). Proper bonding and grounding has no effect in preventing such losses, because the spark discharge in switch loading ignitions occurs between the oil surface and the metal of the fill pipe or truck compartment. As previously discussed, the switch loading hazard can be eliminated by injecting sufficient quantities of an inert gas (that is, CO_2 or nitrogen) into the vapor space prior to loading, but such action is generally not taken because of the added time and expense involved. The principal means of preventing switch loading fires would be to eliminate the practice altogether. In all practicality, this cannot be done but much can be accomplished by proper dispatching to reduce switch loading to a minimum.

Another instance (Figure 8.5) illustrates disregard for safety precautions. Case History No. 18,576, IRI Company, reads as follows: "A tank truck which had formerly carried gasoline was being switch loaded with a distillate. The rack was equipped with positive ground indicators which indicated that the truck was properly bonded. Wired, brass coupled rubber hose extended approximately 2 ft (0.6 m) into the compartment. The loading rate through this hose was approximately 18 fps (5.5 m/s); the product was also being filtered. Several minutes after loading began, an explosion occurred. Apparently the difference in electrostatic potential between the liquid surface and the rubber loading hose caused an incendiary spark which ignited the flammable atmosphere."

FIGURE 8.5. Switch Loading
This presents the most serious exposure of loading rack disasters. (Photograph Courtesy of IRI)

By using a proper combination of the following precautions, relative safety in switch loading can be accomplished and instances like Case No. 18,576 eliminated.

1. Reduce filling rates. Initial filling rates should be 3 fps (0.09 m/s), with a final rate not to exceed 15 fps (4.6 m/s).

2. The installation of static neutralizing devices on pipe risers handling intermediate oils having Reid vapor pressures below 4.5 psi (0.3 kg/cm^2).

3. There should be at least 30 seconds relaxation time for product flows between filters and loading tips.

With the current trend toward higher purity products,
a relaxation time of 100 to 500 seconds (1.6 to 8.3
min.) or more may be necessary to achieve a safe
static charge level if products are handled without
some type of charge-neutralizing device or additive.

4. Special loading arm tips should be installed that
    produce a minimum of flow turbulence.

5. Loading tip should touch the bottom of the tank.

8.6 ELECTRICAL IGNITION

Statistics indicate that electrical energy (other than static electrical charges) is not a frequent cause of fire loss at loading racks;
however, it does present an ever-present source of ignition that requires
definite precaution. The hazard from ignition where there is a sufficient
voltage difference between the truck frame and the loading rack piping can
be eliminated by making sure that the truck's electrical system is turned
off and that the truck is properly grounded before hatch covers are opened.
Also ground wires should not be removed until all hatches have been secured
after loading is completed.

The Industrial Risk Insurers Company recommends that all loading racks
handling low flash materials be wired to conform to National Electrical
Code requirements for Class I, Division II, Group D locations, except within
3 ft (0.9 m) of possible dome cover openings, where Class I, Division I,
Group D electrical equipment is required. (See Figure 8.6, "Electrical
Classification.")

The Division II area should extend to all points within 25 ft (7.6 m),
to an elevation of 25 ft (7.6 m), and to all points that are not over 4 ft
(1.2 m) above grade between 25 ft (7.6 m) and 50 ft (15.2 m) from the rack.
It should be noted that overfilling and possible equipment leaks or failures
can rapidly create vapor concentrations within the flammable range throughout
the entire loading rack area. Any ordinary equipment (office calculators,
data recording equipment, and so on) needed on the rack should be located
in vapor-tight rooms that are provided with positive air pressure taking

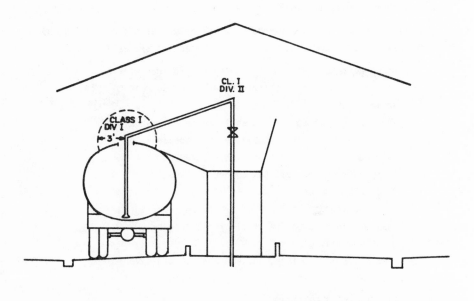

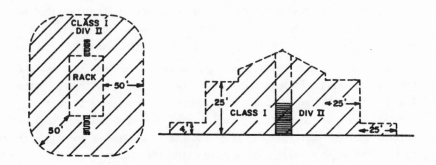

FIGURE 8.6. Industrial Risk Insurers' Recommended
Loading Rack Electrical Classification

suction from a clean air source, preferably a stack terminating not less than 10 ft (3.0 m) above the roof of the rack or 25 ft (7.6 m) above grade. This type of installation should incorporate an alarm to indicate sustained loss of proper air pressurization to the enclosure.

All electrical equipment should be inspected periodically to ensure electrical safety. Some items requiring inspection are as follows:

1. Make certain that conduit seals are provided within a 1.5 ft (0.5 m) of all arcing devices, such as relays, etc.
2. Proper sealing compound (mastic materials) should be installed in all conduit seals, especially on new additions to the electrical system.
3. All junction and pull box covers must be properly secured.
4. Light fixtures should be of explosion-proof design and checked frequently for broken or missing glass globes.
5. Radios, electric fans, and so forth should never· be used in the area.

8.7 BOTTOM FILLING OPERATIONS

New installations of bottom loading racks are no longer favored in certain sections of the country because of the increasing necessity of vapor recovery-type racks in compliance with air pollution authority requirements.

There are some definite advantages to bottom loading. Besides the foremost advantages of being less expensive and less complicated, bottom loading allows faster loading rates, reduced splashing the turbulence, reduced chance of hatch fires (since loading tubes are not inserted), and reduced need for personnel on top of the tank truck during loading operations. Bottom loading is easier, cleaner, and eliminates cumbersome counterbalance systems and supporting members. Additionally, the metallic loading arms eliminate weak joints such as flexible hoses, bellows, and ball joints. They also eliminate a leakage problem common to sliding arm or telescopic fill lines. In use, the sliding section becomes damaged from constant impact with the edge of the fill dome. The uneven section of the sliding sleeve then tears the soft packing installed between the sleeve and the fixed section of line and product leakage results. Figure 8.7 illustrates a typical bottom loading installation.

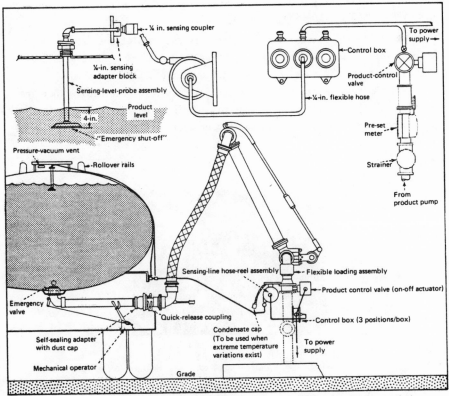

BOTTOM-LOADING SYSTEM minimizes liquid spillage. Sensing device trips foot valve when tank capacity is reached

FIGURE 8.7. Bottom Loading Installation
From: "Transporting, Loading and Unloading Hazardous Materials," William S. Wood, Chemical Engineering, June 25, 1975.

Conversion to bottom loading can, however, be a costly procedure. In addition to the cost of modifying the loading rack, all of the tank trucks in the fleet may warrant replacement or costly modification for bottom loading. One organization found vehicle replacement to be the least costly approach to the conversion problem, since the cost of modifying the older tank trucks was prohibitive.

8.8 VAPOR RECOVERY LOADING EQUIPMENT

Air pollution authorities' requirements and regulations are setting a trend for the use of vapor recovery-type loading arms. These arms produce a gas-tight fit over the dome hatch and have a separate vapor line that extends to a remote compressor and gas holder (Figure 8.8). They are fitted with float mechanisms to shut off loading when the compartment is full, and they have short pipes so that products are splash loaded (Figures 8.9 and 8.10).

FIGURE 8.8. Loading Rack Vapor
Recovery Compressors and Gas Holder

The fill connections provide a double purpose--the load of the fluid while simultaneously recovering the vapors. A single operator can readily handle a bank of these arms. Since the loading arm is automatic when used with set-stop meters, the operator can proceed to the next loading location

FIGURE 8.9. Hydraulically Operated
Loading Arm   Note short pipe, which
necessitates splash filling. (Photographs Courtesy of Industrial Risk
Insurers Company)

FIGURE 8.10. Loading Arm Inserted
into Truck Hatch   A float mechanism
inside loading pipe shuts off when
flow compartment is full.

immediately after each hookup. The systems meet the most stringent local
and federal vapor containment regulations. According to one of the major
manufacturers, FMC Corporation Fluid Control Division, Brea, California, all
components that require periodic cleaning inspection and lubrication are
readily accessible with ordinary hand tools. Maintenance consists primarily
of routine inspection of pneumatic controls, debris removal from the control
valve, and light lubrication of swivel joints.

Manufacturers claim benefits that include protection against tank over-
pressurization, overfilling, loss of line pressure, and loss of seal at the
loading dome.

8.9 GENERAL SAFETY AND FAILSAFE DEVICES

The following pictorial review of safety devices should serve as a general guideline to state-of-the-art safety systems. Although they do not
include every safety device, use of these failsafe engineering systems will
increase plant safety and drastically reduce the possibility of spills.

To substantially reduce the incidence of overfilling compartments, it
is recommended that all manually operated filling valves be of the spring-
loaded self-closing type (Figure 8.11). This is actually a form of "dead man"

FIGURE 8.11. Spring-Loaded Loading Valve
Note that the valve is arranged so that it
cannot possibly be oeprated without the
loader's hand on the valve. (Photograph
Courtesy of Industrial Rick Insurers Company)

control; the loader must physically control the valve or the liquid flow
automatically stops.

Attended loading racks should have switches capable of killing power
to all loading pumps. These switches should be clearly marked to indicate
their specific use and should be remotely loacted from the filling station,
preferably at the bottom of the stairs (Figure 8.12). Actually, stop
switches should be available at each end of the loading rack, with at least
one additional switch located not less than 100 ft (30.5 m) from the rack,
along in the route of easiest passage (Figure 8.13).

Unattended (key-stop) loading racks should be equipped with automatic
fire detection devices that actuate a water spray or deluge system and shut
down electrical circuits. The detection devices should be placed over the
truck bays and close to the ground, where they can quickly react to either
a dome fire or a ground fire beneath the truck. There should also be a
tie-in between the fire detection circuit and the alarm circuit. The system should provide early warning of the problem to the plant guard house,
a central station alarm headquarters, or a local fire department.

FIGURE 8.12. Emergency Loading Pump
Stop Switch Located at Bottom of Rack
Stairs   (Photographs Courtesy of In-
dustrial Risk Insurers Company)

FIGURE 8.13. Emergency Loading
Pump Stop Switch Located Remote
from Rack at Property Entrance

In cases where only a moderate hazard exists and mobile water spray
cannot be justified, several fixed 500-GPM (0.03 m^3/s) capacity water nozzles
should be provided (Figure 8.14 and 8.15). These units should be equipped

FIGURE 8.14.   FIGURE 8.15.

Fixed Fire Monitor Nozzles with Quick Opening Valves that
can Provide Protection between 50 and 100 ft (Photographs
Courtesy of Industrial Risk Insurers Company)

with a combination straight-stream/fog nozzle tip and should be set back at least 50 ft (15.2 m) from the rack. (Actual placement will depend on the area of rack strucutre, truck traffic patterns, and location of tanks and building.)

Dry chemical hand-held and actuated fire extinguishers should be available for use in the event of a loading rack dome fire. At least two 30-lb (13.6 kg) dry chemical units should be positioned at each loading island, one at platform level (Figure 8.16) and the other at ground level (Figure 8.17). To supplement the hand-held exginguishers, especially in areas lacking special protection systems, 150- and 350- lb (68 and 158.8 kg)

FIGURE 8.16. Dry Chemical Hand Extinguisher at Platform Level (Photographs Courtesy of Industrial Risk Insurers Company)

F.IGURE 8.17. Dry Chemical Hand Extinguisher at Ground Level

wheeled extinguishers should be used. The larger capacity stationary dry chemical units [i.e., 100 lb (463.6 kg) or more] can be used with fixed piping where no fire water systems are available. Also, strong wash-down lines should be available at the base of the rack to dissipate spills before trucks can be started and moved. The wash-down line can also be used as a first aid supplement in fire emergencies.

8.10 LOADING DIRECTLY FROM THE STORAGE TANK--NO RACK PROVIDED

8.10.1 General

There are some facilities that load tank trucks directly from storage tanks, without the benefits and safety features associated with loading racks. This is fairly standard among many domestic fuel oil distributors.

The vehicles are loaded through the medium of the gravity head pressure of the oil within the tank or from gravity gained by elevating the storage tank on cradled supports. Some installations operate with badly damaged fill hoses. Return reels are not provided, and the hoses left on the ground are run over and flattened by vehicles. Static eliminators and grounding systems are either nonexistent or ignored. The earth in the loading area is frequently blackened and oil saturated from leakage and overfill. More frequently than not, the bulk storage tanks lack a secondary means of containment. The storage area is unfenced in many cases and flow control valves are rarely locked, thereby developing a situation ripe for malicious mischief. Clearly, there is a need for revision and revamping installations as described.

8.10.2 Design Criteria

One installation viewed in the Baltimore, Maryland area has, through engineering design, greatly reduced the possibility of accidental spillage when loading directly from the tank.

The vehicle backs down into the graded concrete ramp that is fully curbed at the back and on both sides. Once a bumper stop has been reached, the vehicle is in position for loading. At the bottom of the ramp a sump is provided, with enough volume to contain a fairly sizeable overfill. Should an overfill occur, the spilled oil would be pumped back into the storage tank or, if found to be contaminated, disposed of in a nonpolluting manner (Figure 8.18).

Return reels are provided for hoses and the bulk storage tanks are surrounded by earthen retainment dikes. The entire installation was concrete

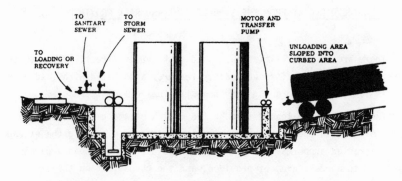

FIGURE 8.18. Containment Curb-Type Spill Catchment System, Depressed Area Form From: G. N. McDermott, Journal Water Pollution Control Federation, August

poured and equipped at a cost of $7,500, using plant labor. One improvement that might have been added is the provision of a vapor return hose system; aside from this, the loading facility is well-equpped for spill prevention.

9. Railroad Tank Car Loading/Offloading Racks

9.1 GENERAL

Spills continually occur at tank car loading areas. This statement can be substantiated by examining the earth around the loading area; at most facilities, it is saturated with oil from overfills. Also common is the practice of draining the residue from a tank car directly onto the area between the tracks. The effect of many years of waste drainage has resulted in contaminated ground water, leaching, and seepage. At some plants the problem is so acute that during periods of heavy rain the ground water physically pumps the oil out of the earth. This to the extent that on the East Coast two plants cannot confine the oil onto their own property. It bubbles out of the earth and flows out of the plants into the drainage ditches of a nearby public highway.

9.2 LOADING AND UNLOADING

Basically, loading racks should be of metal construction. Wooden racks are a fire hazard, and the timbers must be frequently replaced because of rot. The Industrial Risk Insurers recommend that racks be positioned at least 50 ft apart from each other and preferably 200 ft away from other machinery, service buildings, or bulk storage tanks.

The loading area should be hard surfaced, with engineering drainage to spill-retention sumps or holding ponds or surface drainage to oil and water separators.

Prior to a loading operation, the wheels of the line of cars should be securely blocked in position.

The rack and the truck should be effectively grounded (Figure 9.1), because static charges of sufficient voltage to cause sparking can be generated from the turbulent flow of hydrocarbon products in the fill lines. It is also important that each section of track that is separated by expansion joints be connected by welding or bolting a bonding cable to the separated track section. Grounding rods for track use could duplicate those described in Section 4.7, Lightning Protection for Bulk Storage Tanks.

183

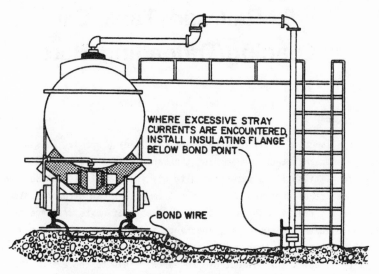

FIGURE 9.1. Bond Rails of Tank Car Loading
Spur to Piping (API Illustration)

Prior to commencing a product flow, the tank car should be inspected
for defects, since once filled it can go into immediate shipment and unobserved
defects could result in an en route spill. Typical areas warranting inspection
are air brake hoses, couplers, wheels and bearings, and all sections of the
under carriage. Once on the route, it becomes difficult for the railroad
or an individual tank car owner to maintain regular examination of the car.

The bottom drain cap on the tank car should be examined to ensure that
it has no defects, and it should be properly tightened in position. Leaking
drain caps can result in a continuous spill en route with considerable loss
of product, possibly the entire content of the tank car. Unfortunately,
drain caps are never locked in the closed position. Sitting unattended
on a side track, marshalling yard, or open track area, the cap is a prime
target for sabotage or malicious mischief even though a large cumbersome
wrench is needed to loosen and tighten the cap.

The type of product being loaded determines the extent of fire protection warranted at a loading rack. Heavy oil loading would require less
fire protection than a rack handling gasoline and doing switch loading.[*]
The hazards associated with loading gasoline or other high vapor pressure
products are greatly reduced when good bonding and grounding is introduced
before loading hatches are opened. At least one minute should be allowed
for static charge discharge before either spouts are withdrawn or load
samples taken. When light products are loaded, the tank car soon becomes
"too vapor rich" for ignition from static electrical charges. In any event,
a check should be made to see that an adequate number of fire extinguishers
are provided--at least one every 100 ft, with one at grade level at each end
of the rack. In addition, an adequate supply of water should be available
with a length of hose sufficient to reach all of the cars positioned at the
loading rack. No part of the rack should be more than 100 ft away from a
water supply source.

The Industrial Risk Insurers do in fact indicate that one or more of
the following special protection systems, which employ fixed piping, would
be desirable at the loading rack.

   a. Automatically actuated water spray system designed
       to immediately cover the entire rack area at a min-
       imum density of 0.25 GPM/ft^2 (10.2 liters/m^2/min.).

   b. Automatic foam-water system. This system would
       provide the fire-blanketing effect of foam, and
       when the foam supply has been exhausted, it can
       continue to provide the cooling effect of water
       spray.

   c. Automatic fixed dry chemical system. This type
       of special system can be installed where a fire
       water system is not available. This system

---

[*]Changing from one product to another.

would provide the extinguishing effect of dry
chemical, but does not provide the cooling
effect of water spray. Furthermore, when its
charge of extinguishing medium is exhausted,
it cannot be readily recharged to provide for
continued operation in the event that a given
fire is not extinguished by the initial charge.

All automatic systems should be provided with manual actuation stations, preferably two stations at widely separated locations and a minimum
of 100 ft (30.5 m) from the rack. It is also advisable to provide inter-
connections to shut down loading pumps and sound alarm systems upon operation of the special protection systems. When special protection systems
are being planned, drainage and curbing in the loading rack area should
receive special consideration.

It should be remembered that where overhead structures extend out over
racks, many railroads demand a fixed clearance between the top of the car
and the structure. Should the automatic fire extinguishing system project
over the railroad cars, as it probably should, a check should be made with
the servicing railroad to determine the desired clearance.

The actual filling procedure is frequently a manual operation whereby
the loader may be filling more than one tank car at one time. The fill
spouts are inserted in the domes and the pumps started to fill the car.
This is still a typical operation and considerable vapors are released
to the atmosphere during the entire fill period. Meanwhile, the loader
or fill rack attendent must pass back and forth on the rack, especially
during the "topping" period. Past evidence of overfill spillage can
provide proof of the inefficiency of such an operation. These installations
can be materially improved by vapor recovery loading arms, computerized fill
control devices that measure the desired cargo load and automatically shut
off the pumping system.

Another predominant spill source that duplicates leakage problems associated with tank truck telescopic loading arms is damaged trombone slide

type arms. Loading rack attendents are prone to slap the loading arm into the fill dome. The action damages the ferrous metal arms and can eventually dent and distort their shape. This in turn tears the packing in the sliding seal, and extensive product leakage can occur. Repair will not correct the situation--replacement of the expensive arm section is required. A rack viewed in one of the larger New Jersey refineries had leakage from six loading arms that were in service. The rack operator completely ignored the situation and the plant engineer accompanying one of the authors on the survey also turned a "blind-eye" to the multiple leak sources. The situation was definitely of long duration; the earth under the rack was oil saturated, and the pumps and piping, located under the rack were oil covered in a manner not conducive to good pump operation. The spillage definitely presented a fire hazard, since a refined product rack was involved.

It can be stated that in view of the conditions described, strict supervision is warranted to maintain desirable conditions in the loading/unloading area.

References:

1. "Truck Loading Rack Safety." Industrial Risk Insurers Publication.
2. "Gasoline Plants Recommendations and Guidelines." Industrial Risk Insurers Publication.
3. "Flammable and Combustible Liquids Code 1977." National Fire Protection Association Bulletin #30.
4. "Transporting, Loading, and Unloading Hazardous Materials." W. S. Wood, Chemical Engineering, June 25, 1973.
5. "Guidelines for Chemical Plants in the Prevention, Control, and Reporting of Spills." Manufacturing Chemists Association, Inc., 1972.
6. "Preventing and Dealing with Hazardous Spills." G. F. Wirth, Chemical Engineering, August 18, 1975.
7. "Spill Prevention, Control and Countermeasure Practices at Small Petroleum Facilities." U.S. EPA 600/7-80-004 1980.

10. Loading/Unloading Piers (Waterfront Facilities)

10.1 PIER DESIGN

Piers designed over 15 years ago have become rapidly outdated. Wooden structures with spaces between the decking that permit drippings and minor spills to enter a waterway have little place in plants working toward spill-free operation. Additionally, piers with wooden pilings have already been subjected to investigation by the US/EPA as water polluting sources themselves. The pilings have as much as 22 lb of creosote pressured into each cubic foot of wood. When subjected to summer sun, the pilings exposed above water can reach temperatures as high as 190°F. The pressure built up within the piling can result in creosote leakage that violates the Federal regulation prohibiting generation and release of a sheen on the watersurface. Although an excellent protector of wood from marine worm attack, creosote has been characterized as an aromatic and marine pollutant and a pesticide. The use of creosoted pilings has not been banned, however, and to the knowledge of the writers, no enforcement actions have resulted from sheens developed by creosoted piling. The Federal government (USCG, USN, and USA) still uses creosoted pilings for pier construction. Most alternative wood preservatives contain arsenic (between 22 percent to 45 percent), which has presented manufacturing problems that are being investigated by US/HEW National Institute of Occupational Safety and Health.

Newer loading/unloading piers are of concrete pile and deck construction. The actual pier is comparatively small in physical dimension, servicing only the vessel's loading/unloading manifold and the access and egress companion-ladder. The vessels themselves, especially the large tankers, are moored to a spread of dolphins (clusters of piles) that are connected to the pier structure by metal catwalks.

The pier surfaces are curbed with controlled drainage. When desired, rainwater can be drained by opening the restraining valve to gain flow. Once contained by the curb, contaminated water or oil spillage can be drained to an onland sump tank, which is later emptied into a tank truck for reclamation or nonpolluting disposal. To some extent, this has eliminated the need for catch trays to be positioned under pipeline connections. However, some facilities have retained catch trays that drain into previously described sump tank. In this manner, spillage from manhandling catch trays during the emptying process is eliminated.

10.2 HOSE HANDLING

It is possible for hose handling to be engineered into a fingertip-controlled process. Lifting and positioning of the hose can be accurately handled through automated lifting and manipulating systems. The same system can be used for cleaning lines (pigging) and opening and closing valves and for emptying lines by the inert gas method.

The dockman can also be provided with direct voice communication with the tank field and the deck crew of the vessel being serviced.

USCG regulations (CFR 33 Part 154.500) require that each hose section be designed to have a minimum bursting pressure of 600 psi or more and a working pressure of 150 psi or more.

Each hose section must also be designated and marked to indicate the following:

1. The products for which the hose may be used or the words "oil service."
2. Date of manufacture.
3. Bursting pressure.
4. Manufacturer's designed working pressure.
5. Date of last hydrostatic test.
   (Note: The Federal regulation pertaining to hose test is very vague on the recommended frequency of testing for hose sections. The hydrostatic test pressure should, however, be 1.5 times the maximum working pressure; as a minimum this would be 1.5 x 150 = 225 psi. Many plants, however, maintain their own

blank flanges and hydrostatic testing pumps
and examine and pressure test hose semi-
annually.)
6. The pressure used for that test.

The exception to marking the hose involves maintaining test records
that can be associated with each hose--specifically, marking hoses so that
each section can be identified with the maintenance record.

Chaffing gear should be used with each hose section to which a lifting
sling is attached. This will protect the hose from being cut by the steel
sling and from abrasion during the movement of the hose during the pumping
action. The hose should have sufficient slack in its length to compensate
for ship movement. If the hose is rested on a ship's stanchion or bulwark,
it should also be protected from chaffing action.

10.2.1 Making and Breaking Hose Connections

Hose connecting is a prime time for minor spills. It is difficult
for the hose handlers to raise the end of the hose rapidly enough to contain
the product that remains in the hose after a loading or unloading procedure.
Generally, such spills are caught in catch trays, usually the lower half of
a cut-off 55-gallon drum.

Some industrial locations have overcome the spillage problem by attaching butterfly valves. Hereto industry was reluctant to use this positive
type of closure valve on flexible hose lines because of the added weight.
Complaints were made that "they stick" and "are hard to open" after closing.
To a large extent modern engineering has eliminated most of the problems
associated with their use.

Butterfly valves are available in most sizes up to 48-in. diameter.
Figure 10.1 depicts the external and internal workings of the valve. Industrial practice is to install a butterfly valve on the final flange of the
flexible hose line, then to add an additional spool piece to the most offshore
side of the valve. The spool is then connected to the shipboard loading/
offloading manifold at deck level. From the spool is attached a drain line,

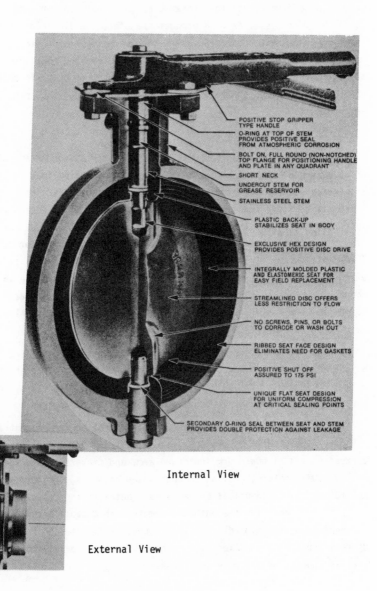

POSITIVE STOP GRIPPER
TYPE HANDLE

O-RING AT TOP OF STEM
PROVIDES POSITIVE SEAL
FROM ATMOSPHERIC CORROSION

BOLT ON, FULL ROUND (NON-NOTCHED)
TOP FLANGE FOR POSITIONING HANDLE
AND PLATE IN ANY QUADRANT

SHORT NECK

UNDERCUT STEM FOR
GREASE RESERVOIR

STAINLESS STEEL STEM

PLASTIC BACK-UP
STABILIZES SEAT IN BODY

EXCLUSIVE HEX DESIGN
PROVIDES POSITIVE DISC DRIVE

INTEGRALLY MOLDED PLASTIC
AND ELASTOMERIC SEAT FOR
EASY FIELD REPLACEMENT

STREAMLINED DISC OFFERS
LESS RESTRICTION TO FLOW

NO SCREWS, PINS, OR BOLTS
TO CORRODE OR WASH OUT

RIBBED SEAT FACE DESIGN
ELIMINATES NEED FOR GASKETS

POSITIVE SHUT OFF
ASSURED TO 175 PSI

UNIQUE FLAT SEAT DESIGN
FOR UNIFORM COMPRESSION
AT CRITICAL SEALING POINTS

SECONDARY O-RING SEAL BETWEEN SEAT AND STEM
PROVIDES DOUBLE PROTECTION AGAINST LEAKAGE

Internal View

External View

FIGURE 10.1. WECO Butterfly Valve
(Illustration Courtesy of FMC Corporation)

which can be directed to the tanker's cargo compartment or into a suitable container. At cessation of a product transfer, the flow control valve on the ship's manifold is closed, as is the butterfly valve. The limited quantity of oil left in the spool piece can be drained back into the tanker or into the previously described waste oil container prior to breaking the shipboard connection. Figure 10.2 shows the valving, spool piece, and drain arrangement.

10.3 MARINE LOADING ARMS

A number of organizations, such as FMC Corporation, Brea, California and Continental Emsco, Dallas, Texas manufacture marine loading systems that eliminate the need for flexible hose lines. These hydraulically powered, all-metal arms for loading and unloading liquid cargo have been increasing in number since they were introduced in 1956. They do reduce the possibility of massive spills that have occurred in the past due to ruptured hose lines. The manufacturers claim that they reduce labor costs and dock clutter and expedite loading/unloading operations. The arms are available in models ranging from small, single-arm, manually operated units for barge handling to multiple bank, 24-in., hydraulically operated systems capable of transferring 90,000 barrels of liquid per hour per arm.

Once they are connected to the ship's flanges, the arms are self-adjusting and will follow all movements of the ship as it rises and falls with the tide, wave action, and draft changes. Alarms are provided to warn operators when the maximum drift limits are being approached.

The Continental-Emsco loading arm is controlled from a central control console. The console is equipped with a selector valve that assigns the three control valves (inboard arm, outboard arm, and traverse) to a specific station. Two hydraulic motors, driving through speed reducers, control and boom's vertical movement. An identical drive package actuates the rotating counterbalance, which in turn actuates the outboard arm through a cable belt. Traverse motion is supplied by two hydraulic cylinders that are mounted on the trunion assembly. Hydraulic power is supplied by an explosion-proof electric motor-driven pump that is mounted on the reservoir tank.

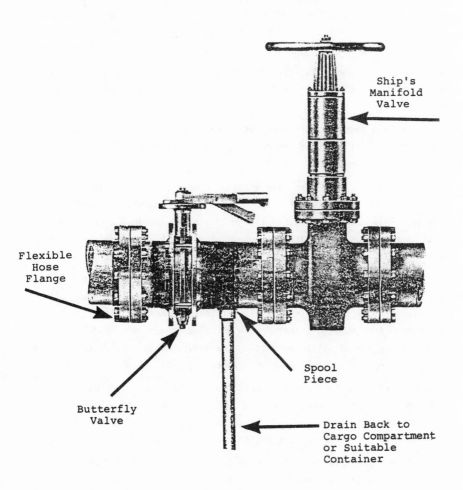

Ship's
Manifold
Valve

Flexible
Hose
Flange

Spool
Piece

Butterfly
Valve

Drain Back to
Cargo Compartment
or Suitable
Container

FIGURE 10.2. Typical Butterfly
Valve Installation

By operation of the three directional control valves, the arm unit can be made to extend, retract, or traverse. The operator maintains positive control at all times, as the arm is maneuvered into position.

Once the hookup is made, the hydraulic control is switched into neutral and the arm can then follow the movements of the ship. Any sudden surge of the ship before the arm is in free-wheeling will cause the arm to override the controls.

Should a power failure occur while positioning the arm for hookup, the counterbalance of the unit prevents violent motion of the inboard and outboard pipe sections.

Hydraulically actuated jaws engage the back face of the tanker's manifold flanges and clamp the face of the flange against the loading arm flange. This option eliminates flange bolting at the ship's manifold. In case of emergency, the quick disconnect coupler can immediately be released from the deck manifold to assist in the disconnection.

Continental-Emsco offers a marine loading arm that is remotely controlled by radio. This system is operated entirely from the tanker's deck, where the operator has full control of all vertical and horizontal movements of the arms. Arms move on signal from a portable transmitter held by the operator. There are no cables connecting the radio transmitter to the loading arm, so that the operator can move anywhere on deck, without restriction.

With this combination it is possible to connect or disconnect marine loading arms by radio control without using any hand tools. The arms take up a minimum of dock space, and installation is simple, requiring only the bolting down of the base plate to the dock manifold.

Another safety factor at some locations is to divert the passage of tanker crews clear of the operating area, by providing a separate, fenced passageway for crew members who do not need to enter the loading/unloading zone.

10.4 OPERATIONS MANUALS

Under Federal law, each waterfront facility must develop and submit
to the U.S. Coast Guard an operation manual that describes:

1. The equipment and procedures used to meet
   operating rules.
2. An outline of the duties and responsibilities
   of personnel involved in oil transfer operations.
3. A map and description of the facility's geographical location and a physical description,
   including a plan that shows mooring areas,
   transfer locations, control stations, and
   storage locations of safety equipment.
4. Operational hours, sizes, types, and number
   of vessels that can transfer oil simultaneously.
5. A list of other products that may be handled
   at the facility that may be incompatible with
   oil.
6. The minimum number of persons on duty during
   transfer operations and a description of their
   duties.
7. Names and telephone number of Federal and
   industrial personnel who will be called by
   the facility in the event of an emergency.
8. The duties of watchmen required by law to
   guard or protect unmanned vessels in the
   facility.
9. A description of each communication system
   at the facility.
10. The location and description of personnel
    shelters on the property.
11. A description of drip and discharge collection
    and vessel slop reception facilities if any.

12. A description of emergency shutdown systems
    and their location.
13. Quantity and type, location, and use tech-
    niques of spill response containment
    equipment.
14. Maximum relief valve setting or, if relief
    valves are not provided, the maximum system
    pressure of each oil transfer system.
15. Procedures for:
    ● Loading arm operation and limitations of same
    ● Oil transfer
    ● Completion of pumping
    ● Emergencies
    ● Contingency plan for reporting and containing
      oil discharges
16. A brief summary of Federal, state, and local
    oil pollution laws and regulations.
17. A description of training and qualifying persons
    in charge of oil transfer operations.

Many organizations have introduced a system of supernumerary loading
and offloading. A specially assigned person, preferably a retired deck
officer, supervises the transfer procedure. This individual inspects (check-
lists are desirable) all equipment aboard the tanker and on the loading dock
prior to transfer line connection. Having had marine experience, he also
approves the mooring lines and the method of mooring. The supernumerary also
checks and approves the ship's spill response equipment and cargo papers. He
verifies readiness for cargo discharge with the ship's captain and the shore-
side personnel and authorizes the start of pumping operations.

At the end of the pumping operation the supernumerary supervises the
hose disconnection and storage procedures. By this failsafe type of operation,
a number of plants that had developed a reputation of "chronic pollutors" have
begun to operate as a spill-free operation.

11. Spill Detection Systems on Water

11.1 GENERAL

National oil spill statistics indicate that accidental spills frequently occur during hours of darkness when it is difficult to detect oil on water and/or during weekends when facilities have a limited number of plant personnel on duty. The spillage and leakage of oil and oil products occurs predominantly at marine terminals during oil loading/unloading operations when workers and tanker personnel are involved with making and breaking flange connections, raising and lowering flexible hose lines, opening and closing valves, metering flow rates, checking tank or cargo compartment oil levels, or handling mooring lines.

To reduce loss of products, costly response actions, expensive fines, and resulting pollution, spill detection systems can be cost beneficial.

Spill detection systems placed in effluent discharge areas or in a waterbody in proximity to a loading/unloading dock can alert plant operators to spillage that could otherwise go undetected for a lengthy period of time.

Most systems utilize visual and/or audible alarms that facilitate an early response action to contain and otherwise mitigate a spill.

A number of major petroleum refineries and manufacturing plants in the United States, Canada, and the Carribbean have installed electro-optical monitors, which permit automatic, continuous remote oil spill detection without water contact or sampling by sensing only a single point on the surface of a moving body of water. This type of spill detection system, the infrared oil film monitor, was developed in 1972 and has been extensively tested by the U.S. Coast Guard in both laboratory and field environments.

The infrared oil monitor is sensitive to oil concentrations of less than one part per million (ppm) in turbulent wastewater streams and it can detect even smaller concentrations of oil in calm water. The exact concentration in parts per million is somewhat meaningless because of the uneven distribution of oil within the water column. Tests have shown that one

197

milliliter of oil spread evenly over 10 m^2 (109 ft^2) of water results in a 0.1-micrometer film. Even with extreme turbulence and entrainment, sufficient oil always rises to the surface, hence triggering an alarm.

Water surface monitoring is responsive to the Federal Water Pollution Control Act, wherein prohibitions are described in terms of appearance: oil discharges must not "cause a film or sheen upon or discoloration of the surface of the water." The effluent regulations for the National Pollutant Discharge Elimination System (NPDES) limits oil concentrations to 15 ppm on water. On this basis, the correlation of surface appearance and film thickness with volume measurements becomes extremely difficult.

Spill detection monitors are currently employed in a number of commercial applications. The following installations, extracted verbatim from the 1977 Oil Spill Conference Proceedings, in New Orleans, Louisiana, illustrate applications of oil detection based on monitoring a single point at the surface of the water column.

- At a large east coast petroleum refinery. Here remote oil monitoring instruments are mounted over five separate cooling water streams. Several of the monitored streams are in open concrete troughs while others are in buried sewer pipes. Explosion-proof infrared oil film monitors are installed on catwalks and in manhole covers from 5 to 15 feet above the streams. Average flow rates range from 0.5 to 5 feet per second. Instrument locations were chosen so that the oil spill detector could pinpoint condenser coil leaks by isolating the section of the refinery served by a particular cooling water stream. Maximum sensitivity is assured with special short time constant response circuits. The instruments respond to oil slicks approximately 6 inches in diameter passing at a 5 foot per second flow rate.

  Monitor outputs are recorded by a paper chart recorder. Additional alarm circuitry causes a master alarm to be

activated when the oil alarm rate exceeds a certain predetermined value (e.g., 10 alarms per hour) or when the
duration of a single oil alarm exceeds a certain time
limit (e.g., 20 seconds). Process water may be diverted
to a holding pond or, if necessary, a section of the
refinery operation can be shut down when a major leak is
detected. In this application, the instruments function
essentially as an early warning system by locating small
leaks and thus preventing costly ruptures and product
losses...

- Monitoring refinery effluent in the Caribbean. Other
remote oil detection instruments are used in the Caribbean
to monitor refinery effluent immediately prior to discharge
into a harbor. In this application, the effluent is a combination of treated ballast water from oil tankers and
refinery process water which has been routed through a
series of separators and filters. An infrared oil film
monitor is mounted over a culvert where discharge water
is channeled under a road and then into the sea. The
instrument monitors an 8-inch spot in the middle of a
20-foot wide stream. Operating height is approximately
10 feet above the water surface and the stream velocity
is roughly 2 feet per second. Experience in this application has shown that the single point monitored in the
center of the stream was representative of the surface of
the entire stream from bank to bank. Scanning across the
stream originally was considered by refinery personnel but
found to be unnecessary. One monitor has operated continuously at this tropical installation for more than six
months with no maintenance or cleaning, despite average
daily temperatures in excess of $85^{\circ}F$ and an extremely
dusty atmosphere. The instrument's signals are transmitted
by telephone wires to a control room about 1,000 feet from

the outflow. Here an oil spill alarm is displayed by
a red light and buzzer. In the event of a major oil
spill, a containment boom can be released where the
outflow enters the harbor. In this application, the
monitors function as a last defense against oiled
beaches in a highly popular tourist area...

● At a manufacturing plant in Pennsylvania. A remote oil
detection instrument has been installed inside an underground cavern where there is a series of chambers that
once served as sand filters for a municipal water supply.
These chambers have been emptied and now are used as
holding tanks for both process water and storm water
runoff. The sensing instrument, because of its non-
critical alignment and automatic gain control, is able
to adjust to variations in water level below the instrument, and the flow rate is approximately 0.1 foot per
second.

The underground chambers function as gravity separators
allowing the water to flow through while any oil
present floats on the top and triggers the infrared oil
film monitor. The instrument's output is continuously
monitored and, when a constant oil alarm has been recorded
for a preset time, a vacuum pump is used to remove the oil
accumulated on the water surface. If the caverns were not
monitored, the capacity of the separators could be exceeded,
resulting in oil spilling into a river which supplies drinking water for several communities downstream...

● At a Canadian refinery. An infrared reflectance monitor was
installed in a manhole and used to detect oil spills in an
ostensibly clean sewer line which empties into the Saint Lawrence River. The manhole cover was removed and the instrument
installed at ground level. The sewer pipe is four feet in

diameter and the flow rate is approximately 5 feet per second; a ladder in the manhole creates an extremely turbulent water surface 9 feet below the instrument.

In tests conducted by instrumentation engineers at the refinery various types types of oil were introduced into the monitored sewer line at carefully calibrated rates from 1,000 to 5,000 feet upstream of the instrument. Both heavy and light hydrocarbons in concentrations ranging from 1 ppm to 15 ppm were reliably detected and recorded by the infrared oil film monitor. In this application, the instrument is used as a continuous effluent monitor to demonstrate compliance with discharge regulations. The outfall is periodically checked by Environment Canada.

There are many more applications of spill detection systems throughout industry. The following section is devoted to the current manufacturers, suppliers, and operational characteristics of oil monitoring systems.

11.2 MANUFACTURERS AND GENERAL SUPPLIERS OF OIL MONITORING SYSTEMS

AQUALERT MODEL-240

Manufacturer

    Bull and Roberts, Inc.
    785 Central Avenue
    Murray Hill, NJ   07974
    (201)464-6500

Operational Characteristics

Range - Heavy Oil 0-50 ppm
        Light Oil 0-100 ppm
Accuracy - 5% full scale (standard application)
Repeatability - 2%
Operates with less than 1% light transmission
Ambient Temperature - 32°F to 140°F
Lamp Life - 5,000 hours

The Aqualert Model-240 is an in-line, continuous, oil detector, pollution monitor. Instrument operation is based on the measurement of light transmission through the fluid being monitored and its potential contaminant. The measured value of the transmitted light is used as an index of sample stream contamination. Two segments of light spectrum are monitored. The first is an ultraviolet wavelength, which is selectively absorbed by organic materials. The second is the visible light portion of the lamp output, used as a reference that compensates for the presence of light absorbent materials that are not specific absorbers in the UV wavelength. Absorption depends on the absorption coefficient of the material. The reading is compared with a curve for a specific contaminant and indicates the contaminant concentration. A sketch of this device is shown in Figure 11.1.

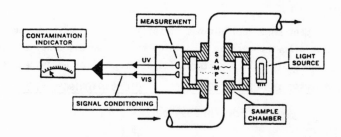

FIGURE 11.1.   AQUALERT Model-240

Cost:

Aqualert Oil Detection System consisting of the following:

    Aqualert Model 240 Control Box

    Sample Assembly consisting of sensor enclosure, flow chamber
    (1/2" NPT x 2" viewing depth), and ultraviolet lamp enclosure

    10-ft sensor wire or cable (2SWA-43)

    Instruction manuals (2)

    BASIC SYSTEM TOTAL $4,160.00

Flow chamber, 1/2" NPT x 6" viewing depth for increased
sensitivity (in addition to above price). . . . . $ 300.00
The following accessories depend upon specifications:

Two 1/4" solenoid valves for stem cleaning 154.70

1.2" sample flow indicator 137.90

Cooling coil (if sample temp. exceeds 125°F) 305.00

Audible alarm 60.00

Recommended spare parts 129.80

Additional sensor wire 60/ft

Additional sensor cable 2.30/ft

Additional instruction manuals (standard). 20.00/each

Set vendor drawings (standard) 10.00/set

Set vendor drawings (standard) 20.00/set

Services by engineers available as per service policy
#Po-79-51

Explosion proof, NEMA class, and specially designed
systems available upon request

OIL-IN-WATER ANALYSIS SYSTEM

Manufacturer

Teledyne Analytical Instruments
333 West Mission Drive
San Gabriel, CA 91776
(213)283-7181

Operational Characteristics

The TAI oil-in-water analysis system is a dual-wavelength ultraviolet
analyzer combined with an essential sample conditioning system. The dual-
beam analyzer employs a reference signal at a wavelength of 4,000 Angstroms,
at which oil does not absorb. Although the reference signal is affected by
turbidity, as is the measuring signal (2,540 Angstroms), electronic circuitry
within the unit compensates by comparing the two signals and ratioing turbidity to zero. The unit functions as the sample is fed into both sides of a

conditioning system for sample preparation. A high-speed, high-sheer homogenizer disperses any oil in the sample, including small and large oil droplets, and oil adsorbed onto foreign matter. A portion of the stream is conditioned to remove all oil (dissolved and undissolved) without altering the background (organic or inorganic non-oil compounds). The analyzer then subtracts the background from the total and reads the contribution of oil only. The analyzer is calibrated with a known standard one a one-time basis. Figure 11.2 is a sketch of the system.

The unit can be used continuously or intermittently to gain an accurate analysis ($\pm$1 percent) of oil in effluent stream discharges. The samples are drawn from both sides. A continuous record of oil content is obtained, and relays are provided that would permit the use of an alarm system such as a light, bell, or siren. The cubicle provides shielding against cold weather and human interference.

Cost:

The base price of a standard unit Model 661C is $10,530; an explosion-proof unit, Model L660C, adaptable for industries that handle flammable liquids is $12,030.

INFRARED OIL FILM MONITOR

Manufacturer

Wright and Wright, Inc.
P.O. Box 1728
Oak Bluffs, MA   02557
(617)693-2608

Operational Characteristics

This monitor was developed under USCG and U.S. DOE sponsorship. It utilizes infrared reflectance at the 3-micron water molecule resonance band to detect floating oil slicks. The instrument consists of a transmitter and a receiver in rugged, cast aluminum housings that are mounted above the water to be monitored. The transmitter projects a light beam to the water surface and the reflected infrared light is analyzed by the receiver.

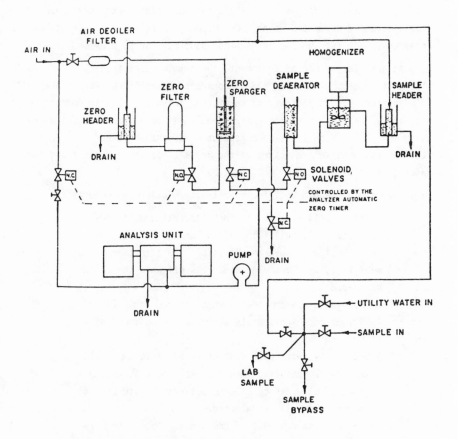

FIGURE 11.2. Teledyne's Oil in Water Analysis System

An alarm is activated when oil is detected. If desired, instrument response time can be adjusted to ignore small slicks.

Once installed, the system can provide continuous spill detection under any weather conditions. Since the unit, or any of its components, does not make physical contact with the water, it is not susceptible to marine fouling or impact damage. It can be mounted on a piling, pier, bridge, or small craft bow, with the sheen detection instruments mounted from 6 to 30 ft above the water. One model is available for use at heights in excess of 30 ft, and explosion-proof housings can be provided.

An auxiliary piece of equipment, the remote readout unit, is available. This unit is cabinet mounted and is connected by cable to the infrared oil film monitor. It is designed to permit the infrared oil film monitor to activate automatically remote control equipment (e.g., motors, solenoids, skimmer, etc.). With the remote readout unit, a technician at a remote locaiton can observe the status of the infrared oil film monitor and the water surface below it.

The remote readout unit incorporates the following features:

1. A red lamp (OIL ALARM) to indicate that the instrument has detected an oil film. This lamp is normally off and lights only when an oil slick is within the instrument field of view. The lamp goes off when the slick has passed.

2. A loud audible alarm sounds when the red lamp (OIL ALARM) is on. The audible alarm can be switched off, if desired.

3. A DPDT relay (with each contact rated at 10 amps , 115 volts AC) is activated when the red lamp (OIL ALARM) is on. This relay can be used to activate remote control equipment when oil is detected.

4. An analog meter readout of the signal processing circuitry of the infrared oil film monitor.

Low meter reading - clean water (no oil)
High meter reading - oil on water

5. A green lamp (INSTRUMENT STATUS) to indicate that the
infrared oil film monitor is operating properly. This lamp
is normally on and goes off only when a component failure
has occurred within the instrument. The infrared oil film
monitor has been designed for at least six months of continuous unattended operation; no maintenance is required
during this period.

6. Built-in logic to prevent OIL ALARM from activating when
INSTRUMENT STATUS lamp is not on.

The following comments and Figure 11.3 describe and depict a typical
Wright and Wright, Inc. installation.

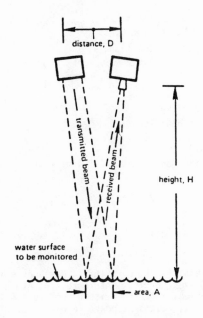

FIGURE 11.3. Infrared Oil Film Monitor Installation

The infrared oil film monitor is mounted 6 to 30 ft (H) above the
water, directly above the water surface to be monitored (A), normally one
square foot of water surface. The transmitter and receiver are less than
1 ft apart (D), and the tilt angles are within ±1 percent off vertical.
Above-average tidal ranges of 8 to 10 ft will have no effect on this
instrument, since it is equipped with an automatic gain control that is
activated by any weakening of the signal.

Cost: (July, 1980)

Infrared Oil Film Monitor, in explosion-proof housing

    Model D500 . . . . $9,500.00[*]
    Approved by Factory Mutual for Class I,
    Div. 2, Group D hazardous locations,
    Temp. Class T1

    Model D250 . . . . 9,500.00
    Approved by Factory Mutual for Class I,
    Div. 1, Group D hazardous locations,
    Temp. Class T2D

    Model D150 . . . . 9,500.00
    Approved by Factory Mutual for Class I,
    Div. 1, Group D hazardous locations,
    Temp. Class T3C

Additional transmitters for Models D500, D250, or D150. . 1,609.00
Infrared Oil Film Monitor, Model E250 . . . . 4,900.00
Additional transmitter, Model FL250 in
floodlight housing 331.00
Film Thickness Discriminator Option 285.00

---

[*]Prices are in U.S. dollars, F.O.B. Oak Bluffs, Massachusetts, and do
not include applicable taxes, duties, insurance, and other shipping
charges. Prices are subject to change without notice. All sales are
subject to Wright & Wright, Inc.'s Standard Terms and Conditions of
Sale.

Selectable Criteria Alarm Option:

when specified with new instrument . . . .$ 705.00

when purchased separately. 1,230.00

Remote Readout Unit:

Model R2, for rack mounting 1,140.00

Model R3, for rack mounting, with paper chart recorder . . . 1,620.00

Custom Housings and Systems . . . . Contact Factory

For Delivery Information . . . . Contact Factory

D-O-W (DISPERSED OIL-IN-WATER) MONITOR

Manufacturer

C-E Invalco

Division of Combustion-Engineering, Inc.

P.O. Box 556

Tulsa, OK

(918)932-5671

Operational Characteristics

The dispersed oil-in-water (D-O-W) monitor provides continuous on-line monitoring of effluent streams, detecting oil in water normally in a range of 0 to 150 ppm, with accuracy of ± 5.0 percent. The monitor operates effectively at temperatures up to 175°F and working pressures of 60 psig.

The monitor utilizes a dual-beam light measuring system, which continuously probes the rate of ultraviolet absorption in the discharge stream. Clear water transmits ultraviolet (UV) with very little absorption, whereas most petroleum oils and their derivatives absorb UV either partially or completely. Thus, variations in absorption provide a sensitive and accurate means for the determination of oil contained in water.

Output from the system is displayed on an indicating meter or an operational chart recorder. Audible and visible alarms could be incorporated in the system.

Power requirements: 95-130 VAC, single phase 60 Hz, 60 watts
Input range (crude, fuel and lube oils):   0-20 or 0-150 ppm
Maximum Fluid Temperature: $175^\circ F$
Ambient Temperature: $40^\circ F$ to $150^\circ F$
Maximum Pressure of Detector Cell: 60 psig working, 100 psig test
Nominal Sample Flow Rate: 1 gpm
Pressure Drop at 1 gpm: 10 psig
Pressure Drop at 1 gpm with static mixer: 20 psig
Detector Cell Connections: 1/4" NPT inlet and outlet
Connections Required in Main Stream Piping: 1" NPT

For use at a loading/unloading pier, the incorporation of a small
pump would be required to draw a continuous water sample from the dock area
and feed the sample into the monitoring system.

Cost:

The base price of this unit with no housing is $6,273.72; with weather-
proof housing, $6,611.76; and with explosion-proof housing, $8,314.92.
About two weeks lead time can be anticipated following an order.

OIL SENSOR MONITORS

Manufacturer

Rambie, Inc.
Irving Business Park
1100 E. Airport Fwy.
Irving, TX 75061
(214)438-6909

Operational Characteristics

These active infrared sensors illuminate the water, then spectrally
analyze the reflected infrared energy for the presence of floating hydrocarbons. When a hydrocarbon is indicated, the alarm system is activated
with a relay closure. The user-selected time delays discriminate between
persistent large spills and short period smaller spills.

The sensors are mounted remote from the water and operate in the extremes of tropic and arctic marine environments. They are seldom affected by waves, water level changes, or floating debris. False alarms from other infrared sources such as the sun, lights, and emission from the water are completely eliminated. They continuously test themselves for faults.

The Oil Spotter OS20 is housed in a single 11-lb cylindrical container; and it is mounted directly above a point on the water that is to be monitored. The unit is simple to install, operate, and test (without spilling oil).

Should an explosion hazard be present, the oil spotter can be housed in an explosion-proof junction box. In this housing the sensor becomes known as the Oil Spotter OS 25.

The sensor uses a transmitter mounted on a tower and a receiver located across a body of water on a second tower. The receiver and transmitter can be scanned to monitor an extended line between the two towers. The power supply, mounted near ground level, is housed in an explosion-proof box. The receiver and transmitter are normally mounted 20 or more feet above the ground, which according to the manufacturer precludes the need for explosion proofing at hazardous locations.

Although the device was originally designed for monitoring a water surface, it is equally useful in monitoring a solid surface such as soil or concrete. It will warn when either water or liquid hydrocarbons cover the monitored solid surface.

Cost:

The oil spotter is priced at $4,780. (When housed in an explosion-proof case, the cost increases to $6,190.) The scanning oil sensor costs $19,500. These prices are FOB Irving, Texas and the terms for payment are net 30 days after delivery.

OIL SPILL DETECTION SYSTEM

Manufacturer

Versatile Environmental Products
60 Riverside Drive
N. Vancouver, BC
Canada V7H1T4
(604)929-5451

Operational Characteristics

Detector/Transmitter

The oil detection element is constructed in the form of a loop from a porous oleophilic material. Two pieces of the material are butt joined using an oil-soluble, water-insoluble adhesive. In the presence of hydrocarbons, the join parts separate, thus permitting the sensing tube to drop and initiate the transmission of the detector signal. The detector element will part within a few seconds of oil contact in the case of #2 or light oil and within one minute in the case of #5 oil or equivalent. Oils of higher viscosity have difficulty in penetrating the element and as a result, the unit is not recommended for such oils. The element is replaced by the use of two ball lock pins. Adjustment of the joint line to just above the water level is easily performed.

Each positioned buoy is identified with a three-digit number, with the first digit identifying its group. Prior to shipping, this number is set into the telemetry system of the transmitter, permitting its identification by the receiver. Once the sensing tube has tripped the magnetic switch, the telemetry system and transmitter are automatically turned on. At that time, a double series of coded, two-tone frequency shift-keyed signals are transmitted. To ensure reception of the signal, it is repeated after 20 seconds and every minute therafter. The system provides a high degree of security and permits reception even in the presence of other transmissions.

Because the security of the system is considered to be extremely important, a battery-level check is provided. Should the batter voltage drop below

the required operating level, the transmitter will initiate a separate
signal. This results in a flashing visual display and the lighting of the
'Low Battery' indicator on the receiver.

To ensure water tightness, an air valve is installed, which permits
pressurizing of the transmitter capsule and checking for leakage. To minimize damage to seals, the unit has been designed to be charged through the
antenna. Similarly, an 'On-Off' switch is provided, which is activated by
a magnet mounted in a plastic holder, screwed into a blind hole on the top
of the transmitter capsule.

Detector/Receiver

Each receiver is coded to accept signals from a group of monitoring
buoys. Upon receiving a signal from a buoy, the receiver automatically
scans the message for accuracy and coding. Should the incoming signal
match the group coding of the receiver, the oil spill indicator light is
illuminated, and the external alarm contacts close. Subsequent signals are
displayed as they arrive, with the preceding signals being moved into the
storage register. Messages stored are shown by indicator lights, which are
numbered from one to four. Pushing the 'Cancel' button permits the recall
of the preceding messages. This memory feature allows more than one buoy
to signal into the receiver without loss of a message and permits determination of the size or direction of travel of a spill when a number of buoys
are being used.

When incoming signals are not coded to match the receiver coding, the
signal is ignored. When the signal approximates the receiver coding but
contains a data error (which is determined by the receiver logic), the buoy
number is displayed in a flashing mode and the 'Data Erro' indicator is
illuminated, but the external alarm contact will not close.

The receiver is not fitted with an 'On-Off' switch, in order to prevent
the inadvertent shutting down of the receiving system. A small light appearing in the upper display window indicates 'Power On.' To ensure that all
lights are operational, a 'Lamp Test' button is provided. Should it be

desired to listen to the channel of the system in operation, the 'Audio Test' buttom may be pressed and locked into position. When the 'Audio Test' button is released, all background noise is filtered out and the distinctive noise of the frequency shift keyed (FSK) signal is heard. This may be turned off by pressing the 'Audio Off' button.

An antenna is provided for use with the receiver. In some instances, it may be preferred to use a remote antenna, which has proven to be satisfactory.

The power input is normally set for 110 volts AC with a normal draw of 1/10 ampere. When desired, the receiver can be provided with a 200-volt AC connection. One receiver can run up to 99 transmitters if necessary.

Cost:

The base price of the unit is as follows:
   Oil spill detection transmitter - $2,000.00
   Oil spill detection receiver   - 2,000.00
F.O.B. Vancouver, B.C.
Note: 15 percent import duty applicable for deliveries within USA.

OIL DETECTION SYSTEM

Manufacturer

   Spectrogram Corporation
   385 State Street
   North Haven, CT   06473
   (203)281-0122

Operational Characteristics

The system operates on the principle of petroleum products exhibiting a fluorescent characteristic when subjected to high energy activation. When an oil sample is irradiated with high energy emission such as short wavelength ultraviolet or x-ray energy, the sample will absorb a portion of the excitation energy and reradiate lower energy of a longer wavelength such as visible light. Since both the wavelength of maximum energy absorption and the wavelength of reradiated energy are a function of the molecular composition of the oil, the oil detection buoy provides an alarm signal upon the detection and identification of a specific oil type.

The basic system consists of a land station, three buoys (two simultaneously operational, one standby), and the interconnecting cables. The land station or main console contains the power supplies, strip chart recorders, and the alert/alarm logic circuitry. Each buoy contains an excitation energy source, a multichannel optical detection system, solid state detectors, integrated circuit photometric amplifiers and logic circuitry, and various local power supplies. The buoys derive operating power from the main console via the interconnecting cable. This waterproof cable also carries the necessary data signals from the buoys to the console recorders and the alert/alarm network, thus providing final contact closures for external and remote indications such as lights, audible alarms, or the "shut-down" of the transfer pumping system.

The units have had a 12-month test period under varying weather conditions. One system was installed at a tidal river location and a second at a barge loading/unloading dock.

The basic system typically includes three oil detection buoys, one land-based recorder/power console, and the required interconnecting waterproof cable.

Spectrogram Corporation has recently stated that under contract with the United States Coast Guard, the land console used with a five-wire waterproof cable has been updated with a free floating buoy system using solar cell battery recharge panels. This modification requires no cable to land and all data communication is done by radio frequency telemetry.

Cost:

These new buoys are priced at $4,500 to $5,000 each. The computerized land-based station is priced at $15,000. A system consisting of three buoys and a computerized base station has a selling prive of $30,000. Delivery is nominally 90 days and the costs are FOB North Haven, Connecticut.

MULTISPECTRAL ACTIVE/PASSIVE SCANNER
(Fluorescence Oil Spill Detector)

Manufacturer

Baird-Atomic, Inc.
125 Middlesex Turnpike
Bedford, MA 01730
(617)276-6140

Operating Characteristics

The oil sensor projects a beam of ultraviolet radiation and simultaneously observes the fluorescence emitted by oil on water. To overcome
the effects of varying range, natural sea fluorescence, atmospheric conditions, and lamp aging, the ratio (rather than the amplitude) of the observed
fluorescence in two spectral regions is used to determine the presence and
type of oil.

Recent tests indicate a maximum useful range of 600 ft at night and
30 ft in daylight with a 6-month lamp life. A range of 900 ft has been
achieve, but with a lamp life of only 4 hours continuous use.

Dimensions and Operating Demands

Head:	24" W x 30" D x 12" H, 95 lbs
Base:	24" W x 30" D x 24" H, 110 lbs
Power:	110 volts AC, 60 Hz, 1,400 watts
Operating Temperature:	-30°F to 120°F
Environment:	Waterproof, oilproof, dustproof, explosion proof (internally pressurized with dry inert nitrogen, pressure switch power interlock)
Lamp Life:	6 months (12 hours per night)

Because light is mostly reflected at shallow illumination angles,
fluorescence cannot be effectively stimulated within 6 degrees of the horizon.
Thus, to achieve a range of 600 ft, the detector must be at least 60 ft above
sea level to provide an illumination angle of more than 6 degrees. This maximum

ratio (10:1) of range to height represents one of the major installation considerations in some applications. The following specifications would fall into a preliminary category.

Range:	600 ft maximum, less in fog
Hours of Operation:	45 minutes after sunset to 45 minutes before sunrise
Elevation Scan:	6 to 30 degrees below horizon (adjustable) at 0.5 degrees per second
Azimuth Scan:	$\pm$150 degrees (adjustable) at 2.5 degrees per second
Scan Pattern:	Programmable
Alarm Criteria:	Detection on two consecutive scans
Beam Size:	0.5 degrees elevation x 2.5 degrees azimuth (projects to maximum of 50 x 25 ft on sea surface at 600 ft and 6 degress below the horizon)
Minimum Spill Size:	75 percent of beam area
Minimum Spill Thickness:	1 micron (at maximum range
Display:	Remote via RF or wire
Indicators:	Power On
	Equipment Active
	Equipment Failure
	Oil Alarm
	Oil Type
	Position

Cost:

The base price of a unit has been established at $75,000. Motorola AR.81 FM transmitter alarms are available at $2,500, which greatly reduce the expense of a travel line.

SPILL SENTRY

Manufacturer

Sentry Systems, Inc.
5304 Allum Road
Houston, TX 77045
(713)721-0200

Operational Characteristics

Rather than being classified as an oil sensing system, this product would fall under the category of a spill prevention unit. It comprises a system of power actuators that operate loading valves and controls associated with tanker or barge loading operations. A master panel incorporates controls for opening, closing, or throttling loading valves. It is also equipped with an "emergency shut down knob" (ESD knob); the actuation of which immediately shuts down the entire loading operation. A lightweight hand-carried "emergency station" equipped with one ESD knob on either the dock or the on-loading vessel can shut down the entire loading process once a potential spill has been sensed or is imminent.

Normally the manufacturer would send a company representative to survey and measure the dock, pipelines, and valving to determine where the various mountings would be located and to determine the engineering requirements for either cylinder or rotary actuators for adaption of existing valves. This action can be eliminated if the plant can provide a detailed engineering description of the dock and its loading equipment.

Figure 11.4 indicates a typical installation of the system.

The control valves (F) in the power circuit, located on the master panel, are positioned manually to open, close, or throttle the loading valves (J). Speed controls (H) govern the opening and closing rates. For gate valves, the closing thrust, supplied by the actuator (I), is limited by regulators (C2), to ensure that the force for opening the valves on the dock (J) will be greater.

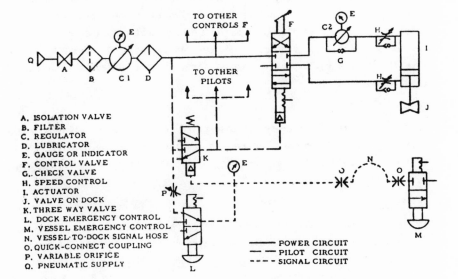

A. ISOLATION VALVE
B. FILTER
C. REGULATOR
D. LUBRICATOR
E. GAUGE OR INDICATOR
F. CONTROL VALVE
G.. CHECK VALVE
H. SPEED CONTROL
I. ACTUATOR
J. VALVE ON DOCK
K. THREE WAY VALVE
L. DOCK EMERGENCY CONTROL
M. VESSEL EMERGENCY CONTROL
N. VESSEL-TO-DOCK SIGNAL HOSE
O. QUICK-CONNECT COUPLING
P. VARIABLE ORIFICE
Q. PNEUMATIC SUPPLY

——— POWER CIRCUIT
— — PILOT CIRCUIT
- - - - SIGNAL CIRCUIT

FIGURE 11.4.   Spill Sentry System

During normal operation, pressure is maintained in the signal circuit, holding valve K in the position shown. Actuation of valves L or M,
or breakage of hose N, as by excessive motion of the vessel, results in
the loss of signal pressure. This causes valve K to shift, thereby initiating Emergency Shut Down.

When valve k shifts, pilot pressure is applied at the control valves
(F), shifting them to the extreme out position. This closes (or opens, if
so intended) Valves J, shutting down the loading operation. Valves J can
be reset only from the master panel.

Other devices that can be incorporated in the system are not shown.
Among these are sensors to detect level of liquid in the tanks, automatically
initiate EDS, sound an alarm, or perform other functions, such as shutting
off pumps. Similarly, protective devices can be included to respond to excessive flow, pressure, explosive vapors, fire, or other undesirable or unsafe
conditions.

Cost:

It is difficult to present a base price because the installation would
differ by terminal; however, an average loading dock having 8 to 16 flow control valves would cost in the vicinity of $1,600 a valve to modify into an
automated system.

References:

1. Manufacturers' Sales LIterature (as per each unit described).
2. Meeting between J. L. Goodier and Baird-Atomic representatives.
3. Little, Arthur D., 1976. A Systems' Study of Oil Pollution Abatement
   and Control for Portland Inner and Outer Harbor, Casco Bay, Maine.
   Prepared for the State of Maine Department of Environmental Protection
   State House, Augusta, Maine, March, 1976.

12. Wastewater Treatment Ponds

12.1 GENERAL

At many production facilities the wastewater treatment ponds occupy
more acreage than the production buildings. At one large installation,
it is normal for the flow of process wastewater to run in the vicinity of 8
to 9 million gallons per day (gpd). Until a few years ago, up to 14 million gpd was treated. The reduction was obtained by tighter controls that
were introduced by plant operators who concentrated on reducing the daily
quantities of process water used. It is possible for most plants to locate
phases in the production flow where the use of process water can be drastically reduced. Some plants are reverting to recirculating cooling water systems. The water is recirculated to a cooling tower and returned in a cooled
state to the process. The initial expense is eventually recouped from lower
water and waste treatment costs and land reclamation.

An efficient waste treatment plant might include primary settling
basins, aeration chambers, and secondary clarifiers. Special instrumentation
can be provided at strategic locations to analyze the effluent flow. These
instruments can monitor the process areas that discharge into the treatment
ponds. Through their use it is possible to be promptly alerted to any change
in flow that would give early warning of a spill.

12.2 SEEPAGE CONTROL

Until a few years ago, few, if any, special precautions were taken to
control seepage from wastewater treatment ponds or lagoons. As a result
a number of Federal enforcement actions are in progress to correct ground
water contamination from the seepage of oil, petrochemicals, and hazardous
materials. Monitoring wells have been drilled to determine the extent of
vertical and lateral seepage. Damage has been so extensive that at some
locations oil well water flood techniques have been investigated as a possible
way of recovering floatable materials from contaminated aquifers.

An example of the acuteness of the problem was experienced in a non-operational refinery that developed a long record of inplant spills. However, the major problem came when the plant was in a bulk storage status rather than a refining process. The idle wastewater treatment ponds lost their water barrier, which had normally kept the waste oil in suspension above the soil forming the bottom of the pond. The water was depleted by a combination of solar evaporation and earth percolation. Once the waste oil made physical contact with the earth, it seeped down into an aquifer. Residents of a residental community were subjected to an early indication when their well water began to have a gasoline odor. Later, oily water was flushed into toilets. Eventually, during a period of high water table, a mixture of waste oil burst through the bank of a creek almost a mile away from the waste treatment ponds. Recovery wells were used to recover the waste oil from the aquifer. In one year, 200,000 gallons of oil were recovered for reprocessing. An early estimate indicated that as much as 1.5 million gallons of waste oil had accumulated in the aquifer. The cleanup action was expensive and of long duration.

From incidents as described, action taken to line holding and treatment ponds can eliminate both immediate and long-range operational problems. When the cost of lining a holding pond is related to the expense of cleaning contaminants out of an aquifer, coupled with the adverse publicity that follows such incidents, lining costs are negligible.

12.3 SEALANTS FOR WASTE TREATMENT PONDS

There are a number of spray-on sealants, which can be classified in groups as follows:

1. Aklyd Resins
2. Coal Tars and Asphaltic Products
3. Bentonites (not spray-on)
4. Exposies
5. Gunite®
6. Polyvinyl Products
7. Rubbers

8. Silicones
9. Sulphur Compounds
10. Urethanes

Testing by the Canadian Environmental Protection Services has revealed that groups 2, 6, 8, and 9 have poor resistance when exposed to petroleum products. Group 2 was found to be unsuitable for application to soil. In brief, the best suited potential products as revealed by testing were as follows.

The alternative spray-on product (Group 5) is Gunite®, a mixture of sand, cement, and water applied by pneumatic pressure through a "cement gun." The Portland Cement compound would be sprayed over reinforcing steel at a cost of $2/ft^2. Canadian testing did reveal problems with cement covering on heaving ground, such as that exposed to Arctic deep-icing conditions. Gunite® is, however, used extensively in moderate climates.

Alternative measure to control and contain leachate include clays and plasticizers, two of which are:

● Volclay-American Colloid Company, Skokie, Illinois - A sodium bentonite-based product that can be applied at 65¢/ft^2.

● Sucoat-Chevron Chemical Company - A thermoplastic molten sulphur and organic plasticizer and inorganic filter, which costs $1.75/ft^2.

The application of the spray-on material should cover the entire storage area and the sides of the retaining dikes.

There exists a wide range of flexible membranes that can be used to line settling, solar evaporation, and wastewater retention ponds. It is possible to safely store effluent containing aggressive organic materials, such as certain crude oils in liners of Neoprene fabrication.

Hydrocarbon rubber sheeting also offers good resistance to acids and bases. This material is also capable of containing effluents having a high temperature, especially when they are aggravated by chemical attack.

Synthetic rubber liners are widely used and have a long history of successful operation.

Installation can be conducted by plant personnel. It is considered advisable, however, to have the installation conducted by an experienced contractor/supplier.

The Watersaver Company, Inc., Denver, Colorado, a liner supplier, offers the following advice on membrane installation.

WATERSAVER COMPANY, INC.
Pond and Reservoir Membrane Liners
General Instructions for Jobsite Preparation

1. The earth upon which the liner will be placed must be smooth and free from sharp rocks, roots, vegetation, and other foreign material. A compacted substrate is advisable to prevent settling. Compaction around pipes and structures is especially important.

2. Check measurements and grades prior to start of liner installation. Surveyor control stakes should be left in place to assist in placing the lining panels.

3. Dig the anchor trench as shown on the shop drawing or the engineering drawings. A typical cross section is found on TDK-74, where a minimum 12" setback is shown. ALWAYS THROW EARTH FROM TRENCH AWAY FROM SIDE SLOPE.

4. The number of workers needed for liner installation will depend on the project size. A minimum crew size of 6 is required, most projects need a crew of 10 to spread panels.

5. Old tires or sand bags will be needed to keep the material in position during windy conditions. Normally 10 tires per panel up to 50 tires are required.

6. Tools and equipment not supplied by Watersaver include wiping rags, paint brushes for adhesive, rakes, and shovels. Liner panels may weight as much as 4,000 lbs. A large front end loader or forklift

will be required to assist in the spreading of the lining
material. Palleted cartons are about 84" x 36" x 36".

7. Cements and adhesives shall be kept from extreme heat and cold.

8. A Technical Services Representative is available from the Water-
    saver Company for a small fee when made part of the purchase
    agreement.

9. All PVC (Vinyl) liners must be covered with earth if an
    extended life is expected. A minimum of 12" of earth should
    be placed on the bottom and slopes. Side slopes of 3:1 or less
    are normally required to hold the earth cover. Windy conditions
    may require special rip rap considerations.

10. Driving on the liner is permitted only when the liner is first
    covered with 12" of earth. If an area is to have sustained
    traffic, 24" of cover is advised. Damage to the liner must
    be repaired as it is discovered!

11. Structures including pipes, splash pads, inlets, outlets, and
    headwalls should be finished prior to placement of the liner.
    Prefabricated corners and pipe seals are available for flashing
    the liner to structure.

This above is furnished to aid in planning liner installations.
Watersaver Company, Inc. as a supplier of materials only, does not assume
responsibility for errors in design, engineering, quantities, or dimensions.

Figure 12.1 illustrates the techniques for membrane liner installation.

12.3.1 Liner Costs

The liner cost may vary slightly by manufacturer. The following prices
are quoted only as a typical costing range.

Vinyl (PVC) Liners vary in cost from around $.135/ft^2 to $.295/ft^2,
depending on the thickness of the membrane and the quantity ordered.

Oil-Resistant Vinyl (OR-PVC) Liners range from $.35/ft^2 to $.375/ft^2
based on stipulations as above.

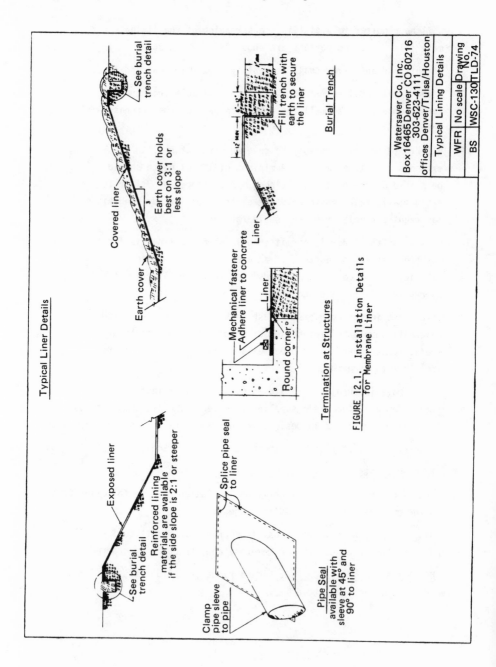

FIGURE 12.1. Installation Details for Membrane Liner

Chlorinated Polyethylene (CPE) Liners - $.35/ft^2 to $.435/ft^2.

Chlorinated Polyethylene (CPER) Reinforced Liners - $.61/ft^2 to $.58/ft^2. (Price reduction by quantity used.)

Dupont Hypalon (HYP) Reinforced Material - $.60/ft^2 to $.57/ft^2.

Neoprene (NED) 16 oz/Square Yard Reinforced 22 x 22 - 840 Nylon - $.56/ft^2.

The solvent splicing and bonding adhesive averages about $16/gallon.

Firestone Coated Fabrics Co., Magnolia, Arizona, has developed a "Fabritank," capable of containing up to 1,000,000 gallons of liquid. This tank could have application for the storage of oil-contaminated water until the oil could be recovered for reuse. The covered tank may have desirable features for use in hurricane and flood territory where the contents of treatment ponds have been "lifted" and spread over a wide land area.

The manufacturer provides the following description of his product.

This container is a lightweight rubber or plastic-coated fabric tank that is supported and protected by earthen dikes. It is used to store water for fire protection and drinking and process water for domestic and commercial uses, and as a bulk storage facility for liquid chemicals and fuels.

In a fully unfolded and unfilled condition, a Fabritank container resembles a pillowcase seamed on four sides. When it is placed in an excavation with sloped sides filled, it becomes, in effect, an integral liner with floating roof.

A filler and drain fitting is located on the bottom of the tank. As the tank is filled, the top or roof portion floats upward. It will easily and safely support the weight of a man when filled to maximum capacity.

Since the tank is never filled beyond its rated capacity, expansion and contraction of both the stored liquid and embankments are accommodated easily by the coated fabric. This eliminates the need for open venting and excludes the possibility of contamination or evaporation of the stored product.

The earthen embankments are the backbone of the Fabritank containers. They support it and hold it securely in place. To guard against erosion during rain and melted snow runoff, drains are incorporated in the design. Also, slopes are planted with grass or heavy rooted plants. This feature enhances the appearance of the installation and creates new opportunities to architecturally blend the installation with modern plant and recreational facilities.

To ensure maximum service reliability, the tank is vulcanized and fully assembled at the factory. After testing, it is folded, rolled on a mandrel, and wrapped with a protective cover for shipment. Installation is usually accomplished in one day with a supervising Firestone-designated technician and a small crew of men.

Fabritank containers are available in 12 standard sizes from 25,000 to 1,000,000 gallons. Other sizes are also made to order.

Such a tank could be used as an oil and water separator, using a settling process. An imbiber vlave drain (see Sections 4.8.1.2 and 4.10.2) would permit water drainage and oil retention for reclamation purposes. Figure 12.2 shows tank sizes and method of installation.

NOTE: Although Watersaver Company and Firestone, Inc. information was used in this report section, it should not be construed as endorsement of a partic-ular product.

12.3.2 Monitoring Leachate From Holding Ponds

It has become a standard practice to drill leachate-monitoring wells around holding ponds. The procedure practically duplicates monitoring systems for solid waste disposal sites. A typical procedure is to drill sampling wells just below the water table. A screen is provided to restrain the entry of solids into the sampling tube. The screen intersects the water table as depicted in Figure 12.3. Although this type of well is routinely used, a single well is not effective in providing information on the vertical distribution of the contaminant   The drilling of at least four wells is advisable--one upgradient to determine subterranean baseline condi-tions and two spaced down gradient, with an additional down gradient well some

EMBANKMENT FABRITANK SIZES* AND DATA

NOMINAL CAPACITY (GALS.)	TANK DIMENS. (FT.)			DIKE DIMENS.		AVERAGE EARTH DIKE (CU. YDS.)	TANK NET WEIGHT (LBS.) (APPROXIMATE)
	A	B	C	D	E		
25,000	33.2	15.2	6	22.5	1.5	549	836
50,000	41.5	17.5	8	29.0	2.0	1045	1,286
100,000	52.4	22.4	10	35.5	2.5	1855	2,045
200,000	66.4	30.4	12	42.0	3.0	3137	3,269
300,000	77.7	41.7	12	42.0	3.0	3628	4,414
400,000	87.2	51.2	12	42.0	3.0	4041	5,506
500,000	95.6	59.6	12	42.0	3.0	4402	6,563
600,000	103.1	67.1	12	42.0	3.0	4728	7,597
700,000	110.0	74.0	12	42.0	3.0	5028	8,615
800,000	116.5	80.5	12	42.0	3.0	5307	9,617
900,000	122.5	86.5	12	42.0	3.0	5568	10,609
1,000,000	128.2	92.2	12	42.0	3.0	5816	11,593

Note: A — Tank Top—Square   B — Tank Bottom—Square   C — Tank Depth   D — Earth Dike—Base   E — Earth Dike—Top

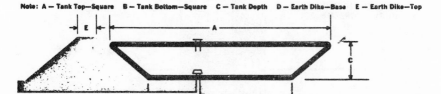

EMBANKMENT AND TANK CROSS SECTION

All tanks are oversize 10%.
*Other sizes available on request.

NOTE:
Tank dimensions A, B & C are shown to illustrate standard available sizes and capacities. Embankment dimensions D & E and average earth dike quantities are to be considered as schematic illustrations only. Since structural characteristics of soil vary, it is important that the embankment design be determined by a qualified civil or consulting engineer.

FIGURE 12.2.

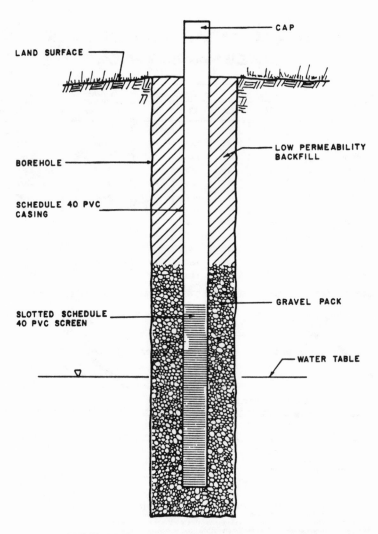

FIGURE 12.3. Typical Monitoring Well Screened
Over a Single Vertical Interval (Illustration
US/EPA)

distance away from the other two wells. If only hydrocarbon products are involved, the well need enter only the upper surface of the ground water. These wells cost between $8 to $10/ft to drill.

The US/EPA recommends the following well-drilling procedure:

● Drill a 152-to 203-mm (6 to 8 in.) diameter borehole
  with a hydraulic rotary rig to the bottom of the
  aquifer[*];
● Set 102-mm (4 in.) diameter slotted PVC well screen
  and PVC casing;
● Backfill with a gravel pack or formation stabilizer;
● Place a concrete collar around the well casing at
  ground surface to prevent downward leakage of rainwater or other fluids.

Once completed, representative samples can be either baled or pumped from the well for analytical purposes.

There is, however, a directly readout system available that should give immediate warning to a central system if hydrocarbons seep from a holding pond. The unit, a product of Pollulert Systems, Mallory Components Group, Indiana-polis, Indiana, is known as the Pollulert hydrocarbon detector. It has the capability of detecting oil in surface water, in dry sumps, and on water, including ground water. The system constantly monitors a group of detection probes at locations selected for specific applications. It flashes a warning signal whenever hydrocarbons contact the probe. At the same time, it can be equipped to activate the alarm devices, to automatically telephone supervisory personnel, or to activate pumping equipment that will collect and recover the spilled material. Early detection makes it possible for leaks and spills to be detected before large volumes of material are discharged and wasted.

---

[*] An investigation for oil seepage would not require this depth; refer to Figure 12.4.

The system is in line with desired automation practices. It is self-
cleaning. It has solid-state microprocessor-based circuits and self diag-
nosis circuitry. The pressure of hydrocarbons is detected by evaluating the
thermal conductivity of the material the probe contacts. Patents are pending
on this recently developed spill detection system. Figures 12.4 and 12.5
depict the ground sensing unit and the readout unit.

For ground-water monitoring, a four-sensor unit costs $1,295. The cost
for cable ranges from $.35/ft to $1.45/ft. To this would be added the well
drilling at $8 to $10/ft.

References:

1. Review of Spray-on or Grouting Sealants for Petroleum Product Storage
   Areas and Dykes in the North. Economic and Technical Review Report,
   EPA 3-EC-76-12. Canadian Impact Control Directorate, October 1976.
2. Watersaver Company, Inc. Sales Brochure and Letter/W. J. Slifer,
   President. March 17, 1977.
3. Procedures Manual for Ground Water Monitoring at Solid Waste Disposal
   Facilities. US/EPA 530/SW-611. August 1977.
4. Pollulert Systems Sales Brochure, Form No. P-100.
5. Firestone Coated Fabrics Company Sales Bulletin M-010 (1978).

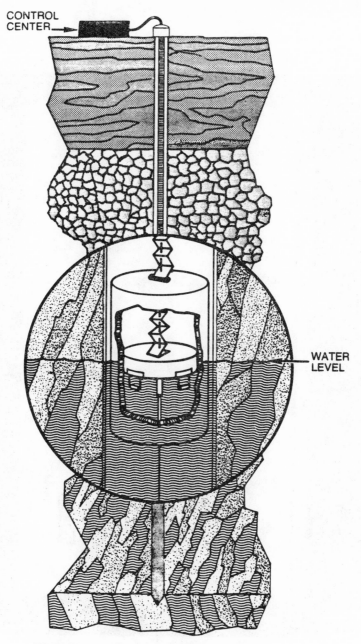

CONTROL CENTER

WATER LEVEL

FIGURE 12.4. Pollulert Ground Water Monitor
(Photograph Courtesy of Pollulert Systems)

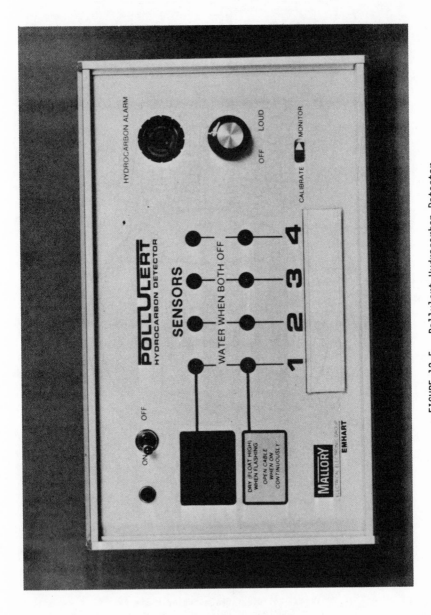

FIGURE 12.5. Pollulert Hydrocarbon Detector
(Photograph Courtesy of Pollulert Systems)

13. Personnel Qualifications and Training

13.1 GENERAL

The indoctrination and training of a new employee has considerable bearing on the worker's usefulness for his entire tenure with an organization. Training by direct work foremen is not always a satisfactory process. The foreman with many years of experience may not have the ability to transfer his knowledge in a desired manner. Frequently, the experienced foreman will try to impart 25 years of experience to a new worker in 30 minutes or less. It is almost impossible for any newly hired individual to mentally retain a long series of work instructions. The foreman frequently assumes that the new employee has mentally digested instruction when he has not. The new worker, too embarrassed to admit the situation (the foreman might think he is simple) enters a work assignment with either partial or a complete lack of knowledge. In an industrial or federal complex handling and storing flammable products, such a situation cannot be tolerated.

The industrial insurance research of H. W. Heinrich (deceased Superintendent of the Travelers Insurance Company Engineering and Loss Control Division) revealed from a study of accident causes that 88 percent of those reported resulted from human failure, 10 percent from mechanical failure, and 2 percent unpreventable acts of God, and so on. In situations where the single action of an individual can bring about a catastrophe, more intensive training techniques are warranted. This section defines, as far as practical, methods and sources of training for plant workers.

13.2 TANKERMEN AND DOCKMEN

Tankermen must pass a USCG examination before they can apply their trade. The status of tankermen is being redefined in a new Federal regulation that has not yet been released for public review. It appears that tankermen will be subjected to grading and that attendance at an industrially operated training school will be a mandatory step toward qualification.

The USCG published three instructional manuals devoted to tankerman training--two will eventually be updated to meet the content of the proposed regulation. The third publication is out of date and is no longer printed. The content should, however, be a good guide to safe practices. The publications are:

- CG-174 "A Manual for the Safe Handling of Flammable and Combustible Liquids" (To be updated)
- CG-327 "Fire Fighting Manual for Tankermen" (To be updated)
- "Oil Pollution Control for Tankermen" (No longer used or available)

Contact with Mr. Brant Houston,[*] President, Houston Marine Consultants Inc., 5616 Jefferson Highway, New Orleans, Louisiana 70123 (504/733-9013) revealed that the Maritime Training Services Division of his organization provides tankermen training. The concern provides a two-day training course every two months at a cost of $150. The course, geared toward attendees passing the USCG tankerman's examination, covers the following topics:

- Cargo handling
- Pollution prevention
- General safety
- Firefighting
- First aid

The training involves classroom instruction using lectures, slides, and overhead transparency projection as training aids.

Other known training sources include the following:

- National River Academy, Helena, Arkansas
- Western River Training Center, Greenville, Mississippi
- Harry Lundeburg School of Seamanship, Piney Point, Maryland (This is CIO Seaman's International Union School)
- National Maritime Union, New York, New York

---

[*]A retired USCG officer.

- Marine Engineers Benevolent Association,
  Baltimore, Maryland
- Masters, Mates, and Pilots Union (MITGS),
  Linthicum Heights, Maryland

The various state-sponsored maritime academies also provide oil
product transfer training, but it is not known if these courses are open
to the employees of industrial organizations. This is a factor worthy of
local investigation to curtail travel expenses.

Ideally, dockmen and tank field employees could benefit from available
tankermen courses.

Company-sponsored inhouse courses can be most valuable. The USCG
Maritime Safety School, Yorktown, Virginia has developed a unique training
aid for both tankermen and dockmen. For training of the USCG personnel the school
has developed a mobile plexiglass model of a multicompartment transportation
barge, complete with engine and pump and pipelines. A compatible loading/
unloading dock equipped with Chiksan torsion loading arms and flexible hose
lines (actually modified water hoses) is available. It is possible to load
and offload water to and from the barge for training purposes. Present
educational facilities for these categories of workers appear to be adequate.
Additional needs may develop once Federal law requires classroom training
for workers prior to their being licensed.

Many marine transportation organizations hire a tankerman on a contractual basis to offload their barges and tankers. Frequently, this contractual
source can be a lone individual who has established himself in his occupation
and has developed a series of clients who use his services on an "on-call"
basis. It is difficult to introduce these individuals into a company-operated
training program because of their independent status. Under such circumstances, it is imperative that the contractual tankerman's experience and
track record with respect for product spills be closely investigated. The
USCG states that men in this category will be automatically licensed to continue operations under a "grandfather" clause agreement.

13.3 PUMP ROOM OPERATORS

In many facilities the duties of pump room operators also include the duty of boiler operators involved in process steam generation--the pump room being immediately adjacent to the boiler room. The workers generally hold state operating engineer's licenses if a high-pressure boiler (maximum working pressure over 15 psi) is involved. The holding of a license does not necessarily indicate that the holder is fully competent with every type of pump or pumping operation. There are many engineering differences with every pumping system. All new employees should be subjected to a short probationary or apprenticeship period until they have proved complete competence and knowledge in the entire pumping system. Many states have regulations pertaining to the period of time that a boiler attendant can be absent from the boiler room. The hazards of unattendance in either the boiler or pump room approach equal proportion. An adequate staff should be available to avoid prolonged absence from any of the work assignments.

13.4 MAINTENANCE WORKERS

Maintenance workers are vital to any spill prevention program. These workers, while in a mobile status, travel the length and breadth of the plant many times during the course of a work day. They enter areas not normally seen by other plant workers. In this respect, they can become the eyes and ears of the plant in leak detection and the advance notification of potential leak sources.

The maintenance staff should have an intimate knowledge of weak links in the spill prevention chain. Alternating (to gain different viewpoints) maintenance workers, complete with a specially prepared checklist, should inspect the plant property on at least a weekly basis. The checklist should be presented to the direct work supervisor at the termination of the plant survey. The supervisor should sign and date the checklist, later indicating the date and action taken to correct a leak or potential spill source.

During plant safety meetings, the findings of each survey should be a topic of discussion by the "surveyor of the week."

13.5 TANK FIELD GAGERS

Tank field gagers should, if practical, be included in the maintenance worker cycle and be an integral part of at least the section of the safety meeting devoted to the findings of the weekly plant survey. Workers in this category should also be encouraged to report any mechanical or operating deficiencies that could contribute to a spill.

13.6 TANKER, TUG, AND BARGE CREWS

Unless they are company employees, it is difficult to train tanker, tug, and barge workers. Visiting crews should, however, be fully briefed on the plant's operational procedures. They should be introduced to the plant personnel who will be working on the loading/offloading operation. The specific duties of each individual should be established between the operating personnel. The communication system can be word of mouth, loud hailer, or two-way radio, depending on the distance involved.

As discussed previously in this manual (Section 10, Loading/Unloading Piers), a company representative or supernumerary should be assigned to monitor the entire product transfer operation, using a prepared checksheet. Through this medium the plant's loading/offloading requirements will gradually become known to both company and visiting crew members.

13.7 VEHICULAR TRANSPORTATION WORKERS

13.7.1 Company Employed

The selection of tank truck operators is an important part of the educational process. The past accident record of a prospective employee should be fully investigated through previous employers and state motor vehicle departments in past states of residency. Investigations should also be made into the excessive use of alcohol or the use of drugs.

It is preferable that an organization employ a driver instructor on a local or regional basis. This individual should interview and test the driver before and immediately after employment. Once hired, the driver instructor should ride with the new employee for possibly a week to ensure that the selected individual has the desired driving capabilities and driving habits.

Driving routes should be selected by management to circumvent road hazards, heavily traveled areas, railroad crossings, dangerous bridges, and the like. The driver should be instructed not to deviate from the selected route unless emergency situations develop.

The driver should make a daily check of his vehicle, preferably using a checklist technique. Any mechanical deficiencies should be corrected before the vehicle is permitted to leave the terminal.

Slightly more than 29 percent of reported truck accidents in which mechanical defects or deficiencies contribute to the cause are caused by tire failure; 19 percent are caused by defective brakes. Coupling (fifth wheel) defects cause 13 percent of the accidents, and wheel defects and faulty lights each contribute about 10 percent to the total accidents. Daily inspection and prompt maintenance procedures can prevent many of these accidents. Such procedures lessen the risk of cargo spills and improve safety in general.

The operator should also be instructed on procedures that can be introduced to minimize or contain a spill during highway travel. It is important that the driver has the name of a person or organization that he can call for assistance in any emergency in each town en route. This source should be called in addition to the local authorities such as fire and police. The driver should also have basic tools to contain spilled material. This should consist of a spade that can be used to dam drainage ditches or other locations needing earthen containment and a box of large, heavy-duty garbage bags that can cover storm drains. The bags can be spread over the drain and held in place with shoveled earth. This action prevents spilled hydrocarbons from entering a storm sewer that ultimately drains into a public watercourse.

Another important educational process can be introduced into driver safety meetings. Normally, the drivers sit back and wait to be "entertained" by a safety movie and a talk by some safety authority. This has become a standard procedure for safety meetings. It is suggested that, through the medium of an overhead transparency projector, each driver involved in an accident be called upon to describe the details of the incident. The remaining drivers can then be encouraged to provide peer review. In this manner many years of truck driving experience can be put to educational use. The final result provides greater benefits than safety talks-- many of which are made by individuals with little or no truck driving experience.

A common carrier driver's education process is needed to gain spill-free truck loading. Once on the highway the cargo is in the care and custody of the carrier, beyond the responsibility of the shipper. Carriers who have a good accident/spill-free record should be selected and a review of their educational and safety program is a normal procedure.

Each driver should, however, be given detailed instruction on the plant's loading procedures. Plant supervision should be closely maintained on all new drivers. This should not be relaxed until a number of loadings have been successfully completed by the visiting driver. It is also possible for plant management to monitor loading operations through strategically positioned closed circuit television cameras (CCTV). These systems are described in detail in Section 14, Plant Security.

The educational demands are dictated by the size of the plant, the number of employees, and their specific duties.

Since plants differ extensively, this section should be considered as providing only the highlights of the educational process, parts of which may be new to some organizations.

14. Plant Security

14.1 GENERAL

Nationwide surveys of plant properties indicate that most facilities need to impróve security measures. There are no known federal regulations, voluntary codes, or industrial guidelines applicable to the petroleum industry from any central source. U.S. DOE has funded extensive security research for facilities handling nuclear materials, however. Although other organizations have also set standards related to plant security, there is no consolidation of information. The Institute of Electrical and Electronics Engineers (IEEE) is a professional organization that deals with the advancement of eléctrical design, methods, standards, and codes for equipment, some of which is related to security systems. The American Society for Testing and Materials (ASTM) has a testing program for security equipment, which establishes industrial standards. The Institute of Nuclear Materials Management (INMM) is developing security standards for the American National Standard Institute. This particular organization deals with nuclear standards, which could be somewhat excessive for facilities handling only petroleum products.

The vulnerability to sabotage, vandalism, and malicious mischief is severe. During a period of civil disobedience or military attack, vast quantities of highly flammable, explosive products could be released and ignited. Most waterfront facilities are fenced on only three sides; the shoreline side of the plant and the loading/unloading piers are open to unauthorized access. Master flow control valves and drainage valves from diked areas are more often than not unlocked, presenting an easy target to the saboteur. In a similar manner, bulk tank water draw valves, normally designed to be padlocked, either lack the necessary padlock or the padlock is left on the ground near the valve.

Two sizeable spills have been investigated by the authors. In the first, an unknown person opened the valves at an asphalt plant to release between 5,000 to 8,000 gallons of heated (180°F) bitumastic product. The hot fluidized material drained into a marsh, causing considerable damage to the

242

vegetation and wildlife. The action was so "successful" that five months
later a similar act of sabotage was performed at another company-owned
location. In this incident 60,000 to 74,000 gallons of hot asphalt was
released into a wildlife habitat. Both plants were unfenced, unlighted,
unattended, and lacked a secondary means of containment around the heated
storage tanks. The natural drainage permitted the released liquid to flow
to the nearest watercourse.

In Texas City, Texas a disgruntled employee of a chemical and waste
oil disposal facility opened the unlocked valves of a waste oil storage
tank causing a 6,000-barrel spill, 200 barrels of which escaped into a
shallow water bayou.

Additional instances of vandalistic spills are reported in the Oil
Spill Intelligence Report (OSIR), Vol, II, 26 October 1979:

"Up to 42,000 gallons of kerosene spilled from a 10-inch Buckeye
Pipeline in Staten Island, New York on 18 September after vandals reportedly
dug trenches to gain access to the pipeline and attempted to tap the line
by boring into it. OSIR sources said that the vandals used the proper
equipment for tapping the line and probably knew the pressure within it. The
U.S. Environmental Protection Agency (EPA) told OSIR that the vandals left
some of their equipment behind when they fled from the site and that they
may have attempted to tap the pipe on the assumption that it contained gasoline.

"Up to 25,000 gallons of kerosene have been recovered, according to
Coastal Services Co. Inc. of Perth Amboy, New Jersey, the cleanup contractor
for Buckeye.[*] Coastal Services dug trenches and used sump pumps, 3M sorbent
pads and boom, a Seaward International Inc. Slurp skimmer, and 5 vacuum
trucks to recover oil from the Staten Island marshland where the bulk of the
oil spilled. Coastal said that the remainder of the oil either evaporated
or seeped into the ground. The cleanup cost Buckeye an estimated $50,000.

"In another recent vandalism incident, nearly 11,000 gallons of kerosene spilled from a Gilbert Distributing Co. tank in Havre de Grace, Maryland

---

[*]Buckeye Pipeline Company.

on 23 September 1979 when vandals reportedly dismantled a filter device
on a pipeline carrying oil from a Gilbert storage tank to a distribution
terminal. The kerosene overflowed the dikes surrounding the tanks, and
according to the USCG, about 6,000 gallons reached the Susquehanna River,
50 meters away. The kerosene saturated the ground between the Gilbert
facility and the river and damaged at least one private boat dock. The
Gilbert cleanup contractor, J & L Industries Inc. of Baltimore, Maryland
recovered most of the product using portable pumps, a 5,100-gallon vacuum
truck, and disposable absorbents."

The Oil Spill Intelligence Report provided additional information on
major spills resulting from vandalism:

1. Vandals reportedly opened valves at a Stockyards Service
   Co. facility in Sioux City, Iowa, on Sept. 22, 1979
   causing 74,000 gallons of oil to spill. (OSIR, 19 Oct.
   1979, p. 2);

2. Unidentified persons reportedly opened valves on four
   Pocahantas Coal Corporation tanks near Garrett, Pennsylvania on Oct. 26, 1979, allowing 32,000 gallons of diesel
   fuel to spill. (OSIR, 5 Oct. 1979, p. 2.); and

3. An estimated 30,000 gallons of oil spilled when vandals
   reportedly opened valves at a Conrail fuel storage tank
   near Pittsburgh, Pennsylvania on Sept. 12, 1979. (Oil
   Spill Intelligence Report, Vol. 1, 21 Sept. 1979, p. 3.)[*]

There are obviously many more spills that can be attributed to vandalism.
However, the few incidents listed should clearly indicate the extent of the
problem.

As an aid to introducing tighter security measures in bulk liquid storage plants, this section is devoted to procedures that should aid in reducing
"intentional" spills.

---

[*]Cahners' Publishing Company, 221 Columbus Avenue, Boston, MA   02116,
617/536-7780.

14.2 FENCING

Fencing should not be construed as maximum security; it only restricts or makes more difficult unauthorized plant entry. The fence should be considered as providing meager or minimum protection; under most circumstances, it would preclude only the entry of chilren or persons intent on malicious mischief. The storage of highly flammable, explosive materials warrants added protection, over and above a standard metal fence.

14.2.1 Gauge of Fence

There currently exists a wide variety of fences, fence posts, gates, and surmounted barbed wire/tape of different makes and manufacturers. Security fences are most commonly constructed in three gauge sizes: 11 gauge (residential fencing), 9 gauge (industrial fencing), and 6 gauge (prison security fencing). Most installations, however, use a 2-inch galvanized steel diamond mesh configuration.

14.2.2 Coatings

Organic coatings on fencing provide decoration and color and add to the installation's appearance. A variety of organic coatings are suitable for fencing applications, such as polyesters, vinyls, acrylics, and so on. Suitable formulations should be selected for optimum adherence, weatherability, and color stability. Colored fence is found mostly on residential installations, since it has only a thin coating of galvanize under the vinyl.

Zinc coatings on steel provides corrosion resistance by protecting the steel base through electrochemical sacrificial action. As long as zinc is present, the steel will remain unaffected by the corrosive action of the environment. The galvanized coated fence is available as 1.2 or 2 oz. finish coating per square foot of wire.

A commonly used fence is aluminized, having one coating 0.40 oz. per square foot of wire, which extends the life of the fence. Facilities subjected to salt air and/or extreme bad weather conditions warrant this type of construction. The fabric should be woven from 9 gauge (coated size) wire in a 2-in. mesh. The tensile strength of fabric should be a minimum of 80,000 psi. Fabric 72 or more inches high should be knuckled at one selvage and twisted at the other selvage (Figure 14.1). The minimum fence height should be 8 ft above ground (optimum security fence height is 10 ft), surmounted by some type of barbed wire or barbed tape (the latter is preferable).

14.2.3 Post Configurations

Many years of research have yielded designs that utilize a more efficient shape of steel to achieve a greater bending strength than the conventional tubular shape. This design is called the C-section post. This section is coated with two ounzes of zinc per square foot. It is well suited for drive post construction, and it does not entrap water as do conventional tubular posts if fence caps are missing.

The H-section line post configuration is best utilized as an end post when the requirement is for additional strength from an intermediate fence post. It has two ounces of zinc per square foot and provides excellent bending strength. Recommended is the C-section roll formed from steel 2.25 in. x 1.70 in. with a minimum theoretical bending strength of 314 lb under 6 ft cantilever load.

14.2.4 Terminal Posts

The terminal post design shown in Figures 14.2 and 14.3 is a good deterrent to vandalism. The sections are designed with chain link fabric weaving into the post, eliminating any nuts, bolts, or clamps. This construction has a 2-oz/ft^2 zinc coating and, unlike round posts, is self-draining and self-ventilating, thereby reducing rust deterioration.

Post settings for line and terminal posts dictate the need for approximately 36 in. of metal below grade. Concrete footings should be 10 in. x 36 in. for line posts and 12 in. x 36 in. for terminals. In some cases, C-section line posts may be mechanically driven into the earth in lieu of concrete. In drive post construction, a wedging action occurs at the bottom when the post is impacted. The bottom opens and can spread up to 1/4 in., depending on the type of soil strata. This causes a more compact and tighter

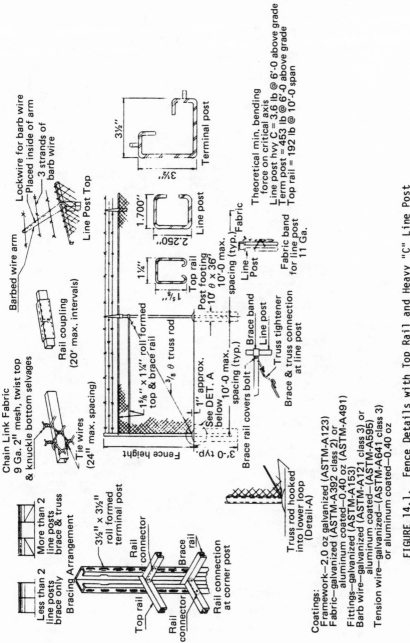

FIGURE 14.1. Fence Details with Top Rail and Heavy "C" Line Post
(Illustration Courtesy of United States Steel Corp./Cyclone Fence)

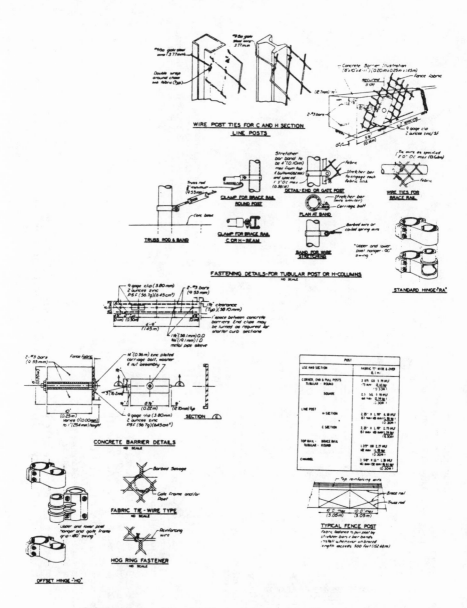

FIGURE 14.2.   Illustration of Fence Details
(Courtesy of U.S. Steel Corp./Cyclone Fence)

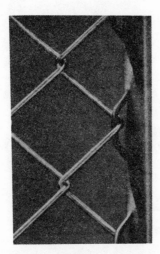

FIGURE 14.3. "No Bolt" Terminal Post
(Picture Courtesy of U.S. Steel Corp./
Cyclone Fence)

set in the surrounding soil, resulting in less movement. Furthermore,
this method eliminates pile driver-induced "mushrooming" at the top to
the post, which would hamper the use of the top caps and fittings in the
case of slide or swing gates.

Cost data for fences are shown in Table 14.1.

14.2.5 Fence Toppings

To gain added security it has been a practice to install up to three
strands of barbed wire above the fixed fence structure. The three strand
protection can, however, be easily surmounted by an individual having specific
"intent" to enter.

In 1957 the U.S. Army Engineer School in Europe conducted an engineering
evaluation of the "barbed tape concertina" (BTC) as developed by the Germans.
It was decided the the BTC design warranted modification and improvement, since
it could be crossed in a matter of seconds without the use of breaching aids.

TABLE 14.1.

Relative Cost of 2-In. Diamond Mesh Fence
(less barbed wired)

Line Fence	Fence Gauge	Fence coatings in oz/ft^2	Price per ft
9 ft	6	Zinc (2 oz)	$19.00
9 ft	6	Aluminum (0.4 oz)	19.00
9 ft	6	Zinc (1.2 oz)	18.00
9 ft	9	Zinc (2 oz)	14.00
9 ft	9	Aluminum (0.4 oz)	14.00
9 ft	9	Zinc (1.2 oz)	13.50
10 ft	6	Zinc (2 oz)	20.00
10 ft	6	Aluminum (0.4 oz)	20.00
10 ft	6	Zinc (1.2 oz)	18.75
10 ft	9	Zinc (2 oz)	15.00
10 ft	9	Aluminum (0.4 oz)	15.00
10 ft	9	Zinc (1.2 oz)	14.50

Prices are calculated estimates and may vary depending on geographical location. All prices listed include installation. Terminal posts are included in the cost unit price per foot.

Numerous Government reports by the U.S. Army[*] and the U.S. Department of Energy[**] show that conventional chain link fence topped with barbed wire or German barbed tape concertina (BTC) can be easily crossed in seconds without the use of breaching aids. In fact, these studies show that barbed wire and BTC actually make it easier to cross the fence because it provides an easy hand-hold grip.

An improvement was developed in 1970, which is known as "general purpose barbed tape obstacle" (GPBTO). The barbed material, illustrated in Figure 14.4, is manufactured by the Man Barrier Corporation, Seymour, Connecticut. The design employs a nickel-chrome stainless steel composition, which provides protection against galvanic degradation. The austenitic structure allows the materials to be hardened to spring quality, and having a brushed finish, it minimizes interference with active sensor systems (for example, microwave and infrared detectors).

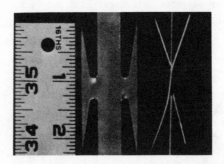

FIGURE 14.4. Barb Configuration (GPBTO Type II)
(Photograph Courtesy of Man Barrier Corporation)

---

[*]"Joint Service Perimeter Barrier Penetration Evaluation" U.S. Army Mobility Equipment Research and Development Center Report 2208.

[**] Sandia Laboratory Reports: "Barrier Technology Handbook" (Sandia 77-0777), "Entry Systems Control Handbook" (Sandia 77-1033), "Perimeter Barrier Penetration Tests Report" (Sandia 78-0241), "Intrusion Detection Systems Handbooks Vols. I, II" (Sandia 76-0554).

14.2.6 Definition of Breaching Aids and Restrictive Procedures

It is the intention of the authors not to elaborately describe the type of breaching aids used in certain tests[*] or to discuss breaching times, but rather to categorize them as follows:

- Simple Aid - one that can be concealed and carried without interfering with the ability to carry other objects.
- Compound Aid - one that can be applied by one person but cannot be concealed or interfere with the ability to carry other objects.
- Complicated Aid - one that must be transported or installed by two or more people.
- Sophisticated Aid - one that requires installation and assembly by two or more people.

Type II GPBTO as depicted in Figures 14.5 and 14.6 was U.S. Army-tested for the Department of Defense Nuclear Agency, with the following result:

FIGURE 14.5. Single Barrier   FIGURE 14.6. Double Bracket
Mounting (Photographs Courtesy   Mounting
of Man Barrier Corporation)

---

[*]Refer to Sandia Laboratories Handbooks and Reports referenced at the end of Section 14.

The DOD tests show that this material when placed on top of chainlink fence is impossible to cross without the use of breaching aids and cannot be crossed with simple breaching aids. Some compound breaching aids were applied. However, in general, the only practical crossing technique involved crews of four or more men using complicated and sophisticated breaching aids.[*]

Sandia Laboratories studies, however, contradict the DOD findings. They indicate that the material can in fact be crossed in seconds with breaching aids.[**] The bridging methods required highly visible actions by the intruders, both in carrying the bridging aids to the barrier and in crossing the barrier. Highly visible breaching aids would present a more exposed target and would be easily seen by CCTV cameras and/or guards.

Table 14.2 illustrates the configurations of the four arrays and presents the relative cost of using GPBTO and BTC in each array.

Average costs for GPBTO and BTC less installation and without array configurations as shown in Table 14.2 are as follows:

Type II GPBTO (double coil)	$4.00/ft
Type III GPBTO (single coil)	.80/ft
Barbed Tape Concertina	.60/ft

---

[*] Security Systems Analysis Report, Systems Analysis and Engineering Laboratory, Man Barrier Corporation, Seymour, Connecticut.

[**] US/DOE Barrier Technology: Perimeter Barrier Penetration Tests (Sandia 78-0241).

FIGURE 14.7. GPBTO Type II - on top 8-ft chain link fence, with middle loop and six bottom loops

FIGURE 14.8. GPBTO Type II
(Photographs Courtesy of Man Barrier Corporation)

TABLE 14.2.

Relative Cost of GPBTO and BTC

	WITH GPBTO	WITH BTC
ARRAY 1	• $115/Meter ($35/foot)	$46/Meter ($14/foot)
ARRAY 2	$247/Meter ($75/foot)	• $112/Meter ($34/foot)
ARRAY 3	• $293/Meter ($89/foot)	$224/Meter ($68/foot)
ARRAY 4	$378/Meter ($115/foot)	• $319/Meter ($97/foot)

• INSTALLED AND TESTED

(Table Prepared by Sandia Laboratories)

14.3 ENTRANCE AND EGRESS

14.3.1 The "Total System" Concept

FIGURE 14.9. The "Total Systems" Approach
(Picture Courtesy of Stanley Vemco Corporation,
Detroit, Michigan)

Figure 14.9 illustrates the total systems approach for perimeter control. Shown are two bi-parting sliding gates each with its own automatic
operator.

Entrance

As the vehicle approaches the gate entrance (Figure 14.9) the driver
places a coded card into the card reader. This transmits an impulse to
actuate the electric operator that opens the gate. A radio transmitter or
keyswitch may be used instead of a card reader or the gate can be opened via
a pushbuttom station from a guardhouse at the site. As the vehicle passes
through the opening, passing over the "inner loop," an impulse is transmitted
to hold the gate in the open position. When the loop has been cleared, the
timer, after a preset period, sends an impulse to the gate operator to close
the gate. Should the gate be closing as the vehicle reaches the loop area,
the gate reverses to the open position and the timer is automatically reset
for another full-time cycle.

Egress

From inside the fence line the vehicle approaches the gate and passes
over the inner loop, transmitting an impulse to actuate the gate operator
and open the exit gate. For controlled exit, a card reader, keyswitch, radio
control, or pushbutton (located within a guardhouse) can again be utilized.
The inner (exit) loop may not be needed if the exit is controlled by a 24-hour
guard. As the vehicle passes over the outer loop, the gate opens. When
the vehicle clears the outer loop, the gate closes.

With the "total systems" concept, security guards need not be present
24 hours a day at the gate entrance. Drivers who make regular stops at the
facility would have predesignated cards. Non-regular stop drivers would
simply stop at the main guard station and could either attain a temporary
card or the guard would open the gate using the standard three-buttom switch.

Stanley Vemco Corporation has provided information on three of their
automatic slide gate operators:

Model ASJCB

Typical Use: Medium-duty commerical installations
Features: ● 1/3 or 1/2 H.P., 115V or 230V single-phase motors
             ● Single-belt, sprocket motor speed reduction
             ● #41 output, idler sprocket and roller chain
Cost: $2,300 (includes 1/2 H.P. motor, installation, and mount-
             ing on concrete pad)

Model ASJH

Typical Use: Heavy duty, high frequency, commercial and industrial
             installations
Features: ● 1/2, 1/3, or 1 H.P. motors (available in all standard
             voltages
             ● Double-belt, chain and sprocket motor speed reduction
             ● #40 output sprocket, idler sprocket and dive roller
             chain on 1/2 and 3/4 H.P. models

- #50 output sprocket, idler sprocket and
  drive roller chain on 1 H.P. version
- MRD Timer (limits run time of operator to
  maximum of 90 seconds in any one direction.
  Prevents excessive wear on integral operator
  components in the event the gate becomes
  obstructed during the open or close cycles.
  Also delays reversal of gate 1.5 seconds
  when signaled from open button or safety
  device, reducing the shock load on both
  the gate and operator.
- Interface control for remotely located
  single and three-button stations. (Gives
  maximum remote station capability of up
  to six miles. Package is pre-wired in NEMA
  3R rain-tight enclosure.)

Cost: $2,500 (includes 1/2 H.P. motor installation, mounting
               on 8-inch thick, 3-ft^2 concrete pad. Does not include
               electrical installation.)

Model SJGFT

Typical Use: Heavy duty, oversized installations
Features:
- 2 H.P., 230V, 460V, and 550V three-phase motor
- 32:1 worm gear and "T" third reduction heavy-duty
  construction throughout
- Roller lever type limits on gates over 50 ft wide;
  standard rotary limits under 50 ft
- Torque limiter
- Hand crank for manual operation
- #50 output idler sprockets and roller

Cost: $7,500 (includes 2 H.P. motor; this unit is used with
               specially fabricated heavy duty gates.)

Should a swing gate be desired, the following operator model is recommended:

Model MSG

Typical Use: Medium-duty commercial installations

Features: ● 1/3 H.P., 115V single-phase motor

● Contactor starter

● Gear head motor, double belt final drive

● Will receive three-buttom and single-button type controls, keyswitch and radio control accessories as standard

Cost: $3,000 (includes 1/3 H.P. motor installation and mounting on concrete pad.)

Note: One operator is needed for each gate.

The price listed above is for a 10-ft gate leaf. Tables 14.3 and 14.4 list recommended capacities for slide and swing gates.

TABLE 14.3.

Recommended Capacities of Slide Gate Models ASJCB, ASJH, and SJGFT

Model	Motor	Gate Travel Speed Per Minute	Maximum Gate Weight	Maximum Overhead Gate Opening	Maximum Cantilever Gate Opening	Maximum Roller/V-track Gate Opening	Shipping Weight	Frequency of use (Complete cycles/hour)
ASJCB-3	1/3 HP	57 ft.	550 lbs.	20 ft	12 ft.	16 ft.	185 lbs.	20
ASJCB-2	1/2 HP	57 ft.	700 lbs.	20 ft.	12 ft.	16 ft.	190 lbs.	20
ASJH-2	1/2 HP	60 ft.	900 lbs.	36 ft.	20 ft.	26 ft.	205 lbs.	30
ASJH-4	3/4 HP	60 ft.	1200 lbs.	46 ft.	26 ft.	36 ft.	210 lbs.	30
ASJH-1	1 HP	60 ft.	1500 lbs.	58 ft.	32 ft.	48 ft.	230 lbs.	30
SJGFT	2 HP	45 ft.	Designed for large roller type gates.				280 lbs.	25

TABLE 14.4.

Recommended Capacities of Swing Gate Model MSG

Model	Maximum Gate Leaf	Seconds to Open Gate	Maximum Gate Weight	Shipping Weight	Frequency of use (Complete cycle/hour)
MSG	14 ft.	12 to 15 secs.	500 lbs.	180 lbs.	15

Tables Prepared by Stanley Vemco Corporation.

Figures 14.10, 14.11, and 14.12 illustrate complete fence specifications. Figures 14.10 shows the standard cantilever slide gate with malleable iron rollers, supports, mesh fabric, guide posts, and so forth. Note that the maximum design standard shown is 24 ft for single and 48 ft for double cantilever gates.

Figure 14.11 illustrates a single swing gate with a welded frame, lock keeper, lock keeper guide, posts, hinges, and so forth. This type of fence configuration would be used inside the facility for personnel passage via pathways, etc.

Figure 14.12 illustrates the double swing gates with welded frames and surmounted barb wire; lock keepers, upper and lower forks, posts, hinges, etc. Note that the three-strand method of barb wire placement has been proved ineffective in keeping out intruders without breaching aids and with minimal "malicious intent." Barbed tape configurations are recommended as mentioned earlier (refer back to Figures 14.5 and 14.6).

Table 14.5 illustrates estimated cost data for the three fence gates listed. It is suggested that the heaviest fence gauge be used where appropriate, since the gate cost is the same for either zinc (2 oz or 1.2 oz) or aluminum (0.4 oz).

Card Readers and Intercoms

Card readers used in conjunction with either a swing or slide gate fence provides one of the simplest and most efficient methods for the control of exit and entry (Figure 14.13).

Intercom systems may be used in conjunction with card readers throughout a facility; models having keyswitch capabilities are suggested for use at gate entrances (Figure 14.14).

The average cost for a four-period card reader, including installation and mounting on a concrete pad or in a footing, is $600. Again, this price is less any electrical installation. The cost for each card is around $1.50.

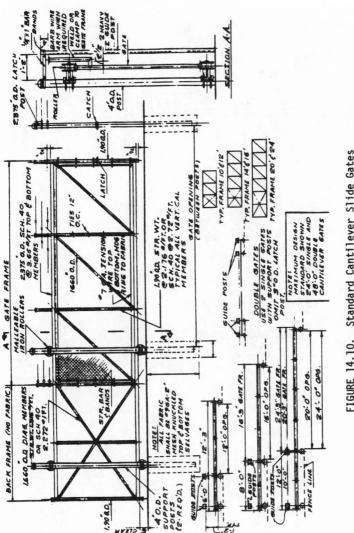

FIGURE 14.10.   Standard Cantilever Slide Gates
(Illustration Courtesy of United States Steel Corp.)

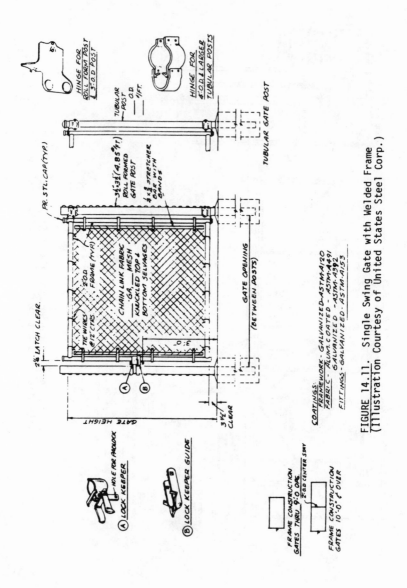

FIGURE 14.11. Single Swing Gate with Welded Frame
(Illustration Courtesy of United States Steel Corp.)

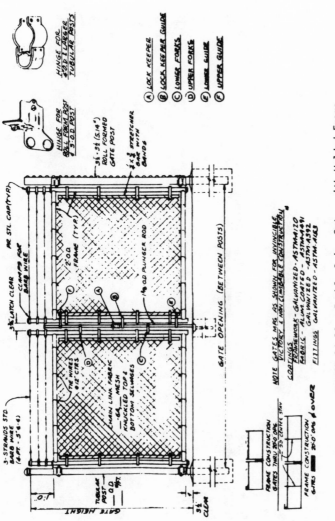

FIGURE 14.12. Double Swing Gate with Welded Frame
(Illustration Courtesy of United States Steel Corp.)

TABLE 14.5.

Slide and Swing Gate Estimated Cost Data
(Less barbed-wire)

Fence Height	Fence Gauge	Slide Gate Gate Opening	Single Swing Gate Opening	Double Swing Gate Opening	Estimated Cost	See Figure Number
9 ft	9	20 ft			$3,000	14.10
9 ft	9		4 ft		200	14.11
9 ft	9			20 ft	600	14.12
10 ft	9	20 ft			3,100	14.10
10 ft	9		4 ft		225	14.11
10 ft	9			20 ft	650	14.12

All prices include installation.

FIGURE 14.13. Card Reader - Four-
Period Card Control Unit on Gooseneck stand permits access through
the gate only to authorized card
holders. The reader has four different codes. (Illustrations Courtesy of Stanley Vemco Corp.)

FIGURE 14.14. Intercom Call Station
(on gooseneck stand) Offers point-
to-point intercommunication, a bell
pushbutton, bell signal, and indoor
call station. There is also a keyswitch in the face of the station
for local gate control.

14.4 CLOSED CIRCUIT TELEVISION SURVEILLANCE

Until a decade ago exits, entrances, storage areas, and so forth, had
to be watched by individual guards. There was not practical way to extend
man's vision to a site at which he was not physically present.

In a closed circuit television (CCTV) system, many different camera views
can be seen on a single monitor. Typically a security guard may wish to
observe several different entrances and exits, plus certain critical loading
areas, and so forth. Using a "video switch," each of the cameras' views are
seen in a designated sequence for a selected time period from one second to
a minute or more. In some instllations, it may be desirable to route the
signal from a single CCTV camera to several monitors so that people spacially
separated are able to observe the same TV picture. With CCTV systems it is
not only possible to see things beyond the reach of human eyesight, but also

to record these events for later analysis. This capacity has proved invaluable, since it is often necessary to document evidence of wrongdoing.

14.4.1 Scanning Cameras

For some applications it may be desirable to have the CCRV camera focused on a single scene--a door opening, loading platform, dock area, pumping station, etc. In other applications, however, it may be desirable to have the camera scan an area. There are a variety of motorized pan and tilt combinations that permit a view within a 350-degree horizontal and a 90-degree vertical arc of the camera mounting. In many instances the camera will continue to pan right and left in a designated arc, unless overridden by manual control.

14.4.2 Zoom Lenses

Remote-controlled zoom lenses permit the operator not only to stop the camera at one point in the arc, but also to "zoom in" on a small area to observe particular details. A zoom lens maintains the ability to keep an object or scene in focus, while changing its focal length, thus magnifying distant scenes analogous to a telephoto lens. In addition, zoom lenses are often equipped with remotely controlled or automatic irises that allow for compensation of differences in lighting conditions during the daytime, permitting opening and closing to provide the best possible picture.

14.4.3 CCTV - Near and Complete Darkness

The lens (or eye) of the CCTV camera has one other advantage over the human eye--it can see in the dark.* Lighting conditions can range from as much as 10,000 foot-candles of illumination in full sunlight to 0.001 foot-candles in partial moonlight. Specially designed cameras referred to as "LLLTV"--low light level TV cameras--are capable of seeing within the above mentioned ranges.

---

*Not absolute darkness. Visibility is fully impaired in 100 percent darkness, but in nearly all commercial or industrial applications some small amount of illumination, if only a trace of moonlight; is available. This petite quantity of light is enough for specially designed CCTV cameras.

By using infrared technology, objects invisible to the human eye can be clearly photographed. A storage area or plant exterior can be "lit" by infrared lamps. The illumination is invisible to the human eye but clearly visible to a silicon diode-type CCTV camera whose response is in the red region of the light spectrum. Incandescent or sodium vapor infrared spectrum lighting systems can be installed in warehouse areas, exterior loading/unloading areas, doorways, and so on. These lamps project infrared radiation to the area and the radiation reflection is observed by the camera and displayed on the CCTV monitor as a near-daylight picture. The nighttime vision of CCTV is of obvious importance in security applications, since the majority of all thefts, burglaries, and vandalism occur during hours of darkness.

There are specially designed exterior housings to protect CCTV cameras against variations in weather conditions through the application of blower fans, heating units, etc. Some cameras are equipped with windshield washers and wipers to keep the viewing lens clean of grime and dust. Exterior cameras can be mounted practically anywhere--on roof tops, lighting towers, and other strategic locations to gain the best viewing advantage.

Partial data for current models of CCTV units follow.

The Photoguard TV-11 CCTV System (Figure 14.15) manufactured by Mosler Systems, Wayne, New Jersey, provides live action viewing of transactions, instant replay, low cost operations using video tape recording, and time/date display. Photoguard can also be used with an instant camera to provide hard-copy photoprints of any frame selected. Additional capabilities include sequencing for up to 10 cameras, reusable magnetic tape, optional time and date recording, camera resolution of more than 600 lines, unique control center for hands-off operation and 9-, 12-, or 17-inch TV monitors. The estimated cost less installation is $5,300 (July, 1980).

The new CCTV system (Figure 14.6) introduced by the HB&W Company is designed for nighttime surveillance and can scan areas of "total" darkness up to 120 ft either indoors or out. The video monitor projects the scanned image with a detailed resolution of 700 lines per inch (twice that of home TV's). The cost of this particular system falls within the range of the Photoguard TV-11.

FIGURE 14.15. Photoguard TV-11 CCTV System
(Illustration Courtesy of Mosler Systems)

FIGURE 14.16. Infrared CCTV
(Illustration Courtesy of HB&W Company)

There are numerous CCTV systems on the market today. Prices of models
vary depending upon the number of cameras, monitors, and cathode ray tubes
(CRT's) used. To list all the models would be sesquipedal. Specific CCTV
systems are bettern discussed as part of the total security stem in combination with advanced detection systems discussed later in this report.

Figure 14.17 illustrates various cameras, monitors, and loop configurations for different kinds of commercial and industrial facilities. Naturally, each of these layouts could have been optimized with additional cameras and/ or monitors. The layouts illustrated are simply graphic illustrations of three typical CCTV installations.[*]

Poorly selected and installed CCTV systems can sometimes be worse than no system at all, since they may engender a sense of security which might prove illusory during an actual criminal intrustion. The proper CCTV system chosen should provide a 24-hour surveillance capability.

14.5 LIGHTING

14.5.1 Lamp Considerations

In designing a lighting system for any facility, lamp selection has a significant impact on the performance and cost. Depending upon the application requirements, any one of three high intensity discharge (H.I.D.) lamps may be used: high pressure sodium, metal halide, or mercury vapor. The characteristics of these sources may vary widely. The lamp performance characteristics (Table 14.6) provide a comparison of these traits. The data represent selected lamps from three manufacturers.

There are certain cases where color rendering characteristics of a light source impact heavily on the selection of a suitable lighting system. (Examples would be the preference of metal halide lamps for TV coverage or areas where color matching of certain items is important.) Although most objects appear to be steadily lighted when seen under H.I.D. sources, high speed machinery or other rapidly moving objects may appear to stand still or slow due to stroboscopic effect or "flickering" (H.I.D. lamps operating on alternating current actually go about 120 times a second, at 60 cycles). Metal halide lamps are least effected, followed by phosphor-coated mercury vapor and high-pressure sodium. The stroboscopic effect can be eliminated by operating fixtures on separate phases of a three-phase system.

---

[*] US/DOE "Intrusion Detection Handbooks," November 1976 (Sandia 76-0554) provides indepth information on CCTV systems.

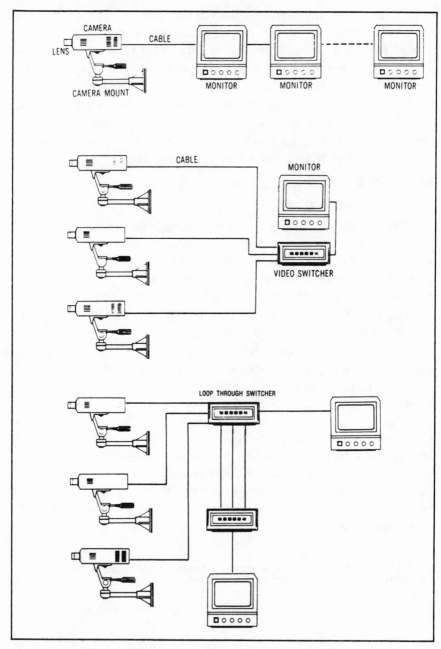

FIGURE 14.17. Camera and Monitor Configurations
(Illustration Courtesy of ADT Security Systems)

TABLE 14.6.

HI-TEK Company, Inc. Lamp Performance Characteristics

Lamp Type	Wattage	Start-Up[1] Time (min)	Restrike[2] Time (min)	Initial Lumens per Watt	Average[3] Rated Life (hr)	Initial Lumen Rating	Mean Lumen Rating
	50	3-4	1	66	24.000	3.300	2.970
	70	3-4	1	82.86	24.000	5.800	5.220
HIGH	100	3-4	1	95	24.000	9.500	8.550
PRESSURE	150	3-4	1	106.7	24.000	16.000	14.400
SODIUM	200	3-4	1	110	24.000	22.000	19.800
CLEAR	250	3-4	1	110	24.000	27.500	24.750
LAMPS	250S	3-4	1	120	24.000	30.000	27.000
	310	3-4	1	119.4	24.000	37.000	33.300
	400	3-4	1	125	24.000	50.000	45.000
	1000	3-4	1	140	24.000	140.000	126.000
METAL	175	2	10	80	7.500	14.000	10.800
HALIDE	250	2	10	82	10.000	20.500	17.000
CLEAR	400	2	10	85	20.000	34.000	25.600
LAMPS	1000	4	10-15	110	10.000	110.000	88.000
	1500	5	10-15	103.3	3.000	155.000	140.000
METAL	175	2	10	80	7.500	14.000	10.200
HALIDE	250	2	10	82	10.000	20.500	16.000
PHOSPHOR	400	2	10	85	20.000	34.000	24.600
COATED LAMPS	1000	4	10-15	105	10.000	105.000	83.000
SUPER	175HORIZ	2	10	85.7	10.000	15.000	12.000
METAL HALIDE	400HORIZ	2	10	100	20.000	40.000	32.000
CLEAR	400VERT	2	10	100	15.000	40.000	32.000
LAMPS	1000VERT	4	10-15	125	10.000	125.000	100.000
SUPER	175HORIZ	2	10	85.7	10.000	15.000	11.300
METAL HALIDE	400HORIZ	2	10	100	20.000	40.000	31.000
PHOSPHOR COATED	400VERT	2	10	100	15.000	40.000	31.000
LAMPS	1000VERT	4	10-15	125	10.000	125.000	95.800
MERCURY	100	5-7	3-6	45	24.000	4.500	3.650
VAPOR	175	5-7	3-6	49.2	24.000	8.600	7.650
DELUXE	250	5-7	3-6	52	24.000	13.000	11.000
WHITE	400	5-7	3-6	57.5	24.000	23.000	20.100
	1000	5-7	3-6	63	24.000	63.000	48.500

[1]Start up time: Time from initial energizing to 80 percent of full output.

[2]Restrike time: Time to restrike after momentary power interruptions.

[3]Average rated life: Number of burning hours at which 50 percent of a group fail, based on a 10 hour start.

When selecting a light source for a given application, certain general economic guidelines can be applied. It would be desirable to select the largest wattage lamp available and mount the fixtures as high as possible. This should result in the least expensive system with the best uniformity. High pressure sodium usually provides an economical system based upon total acquisition and operating costs and has the lowest (one minute) restrike time of all lamps recognized in Table 14.6. Metal halide and mercury vapor are next in order of system costs.

14.5.2 Outdoor Lighting Design

Lighting design dictates the need for a complete analysis to conserve energy usage and lower installation and owning costs. There are lighting companies that run computer-assisted design services for just this purpose. One company, Hi/Tek Corporation, markets their Hi/Tek ECON program, which is designed to provide a relative economic comparison and financial analysis of up to four different lighting systems, including an existing system if desired. A complete comparison of initial costs, operating costs, and total owning costs is made to determine the relative cost relationships between systems. All systems are then compared to the lowest initial cost system to determine payback periods, return on investment and benefit/cost ratios.

Most H.I.D. lamps are designed to be used in any burning position with little or no effect on light output. However, some metal halide lamps are greatly effected by off-vertical burning. Applications using metal halide lamps may be less than satisfactory if lumen correction factors are not employed. In general, it is a good practice to aim the luminaire two-thirds of the distance across the area to be lighted. Also to ensure good visual comfort the floodlight should be aimed at least 30 degrees below horizontal; if this cannot be done, the mounting height should be increased. Generally, good uniformity can be expected if luminaires are aimed so that the edge of the beam of a given fixture intersects the aiming of the adjacent fixture.

14.5.3 Floodlight Beam Spread

For security lighting, beams should overlap by one-half (Figure 14.18). This is accomplished by keeping the spacing between the fixtures less than one-half the diameter in feet of the beam on the surface.

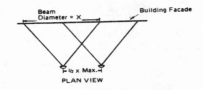

FIGURE 14.18. Beam Overlap Criteria
(Illustration Courtesy of Hi/Tek Co., Inc.)

When the luminaire is not aimed perpendicular to the surface being
lighted as shown in Figure 14.19, the horizontal beam dimension "H" may be
determined from Table 14.7 (Beam Projection Chart).

Table 14.7 is used in determining the size (in feet) of a floodlight
beam when the luminaire is aimed perpendicular to the surface being lighted,
as shown in Figure 14.20. The horizontal beam dimension "H" and vertical
beam dimension "V" for varying projection distances "P" are tabulated in
the table.

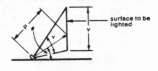

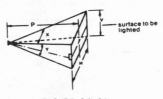

FIGURE 14.19.   FIGURE 14.20.

Techniques for Measuring Vertical and Horizontal Beam Dimensions

Note: The vertical beam dimension "V" will be larger than
that given, varying as a function
of the angle Ø.

Note: If a luminaire with a horizontal beam (X) of 130° and a
vertical beam (Y) of 100° is aimed
directly at a vertical surface 50
ft away, the horizontal dimension
of the beam would be 215 ft and
the vertical dimension would be
119 ft (see Beam Projection Chart).

TABLE 14.7.

Beam Projection Chart

Beam Spread in Degrees Hor. (H) or Vert. (V)	Beam Dimension in Feet Projection Distance P (feet)																			
	5	10	15	20	25	30	35	40	45	50	55	60	65	70	75	80	85	90	95	100
10°	0.9	1.7	2.6	3.5	4.4	5.2	6.1	7.0	7.8	8.7	9.6	10	11	12	13	14	15	15.5	16	17
20°	1.8	3.5	5.3	7.0	8.8	10	12	14	16	18	19	21	23	25	26	28	30	32	33	35
30°	2.7	5.4	8.1	11	13	16	19	21	24	27	30	32	35	37	40	43	45	48	51	54
40°	3.6	7.2	11	14	18	22	25	29	33	36	40	44	47	51	55	58	62	65	69	72
50°	4.6	9.3	14	19	23	28	33	37	42	46	51	56	60	65	70	74	79	84	88	93
60°	5.7	12	17	23	29	34	40	46	52	57	63	69	75	80	86	92	97	103	109	115
70°	7.0	14	21	28	35	42	49	56	63	70	77	84	91	98	105	112	119	126	133	140
80°	8.4	17	25	34	42	50	59	67	75	84	92	101	109	118	126	134	143	151	160	168
90°	10	20	30	40	50	60	70	80	90	100	110	120	130	140	150	160	170	180	190	200
100°	12	24	36	48	60	72	84	95	107	119	131	143	155	167	179	191	203	215	227	239
110°		29	43	57	71	86	100	114	129	143	157	172	186	200	214	229	243	257	272	286
120°	17	35	52	69	87	104	121	139	156	173	191	208	226	243	260	278	295	312	330	347
130°	21	43	64	86	107	129	150	172	193	215	236	258	280	301	322	344	366	387	409	430
140°	27	55	82	110	137	165	193	220	248	275	303	330	358	385	413	440	468	495	523	550

Table Prepared by Hi/Tek Company, Inc.

For area lighting, horizontal beam dimensions may also be determined from Table 14.7. Generally, if the width of the area is small compared to the spacing between fixtures, twin beams should be used. If the width of the area is equal to or greater than the spacing between fixtures, rectangular beam fixtures are indicated (Figure 14.21).

14.5.4 Mounting Height and Spacing Recommendations

The recommended minimum mounting heights for most lighting applications can be found in Table 14.8. This table can also be used as a guide for general lighting applications when lighting an area from the perimeter.

To use Table 14.8 locate the pole setback distance at the left and go across table to the appropriate column for width of area (at top). The minimum mounting height is where the two columns cross. (Example:

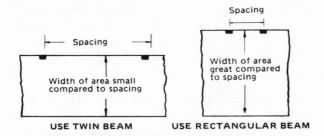

FIGURE 14.21. Light Spacing Criteria
(Figures 14.19, 14.20, and 14.21 Courtesy
of Hi/Tek Co., Inc.)

TABLE 14.8.

Recommended Floodlight Mounting Heights

POLE SETBACK (FEET)	WIDTH OF AREA IN FEET														
	20	40	60	80	100	120	140	160	180	200	220	240	260	280	300
10	20	20	20	25	25	30	35	40	45	45	50	55	60	60	65
20	20	20	25	30	35	35	40	45	50	50	55	60	65	70	70
30	25	25	30	35	40	45	45	50	55	60	60	65	70	75	80
40	30	35	40	40	45	50	55	55	60	65	70	70	75	80	85
50	35	40	45	45	50	55	60	60	65	70	75	75	80	85	90
60	40	45	50	55	55	60	65	70	70	75	80	85	85	90	95
70	45	50	55	60	60	65	70	75	80	80	85	90	95	95	100
80	50	55	60	65	70	70	75	80	85	85	90	95	100	100	105
90	60	60	65	70	75	75	80	85	90	90	95	95	100	105	110
100	65	70	70	75	80	80	85	90	95	100	105	105	110	115	120
110	70	75	80	80	85	90	95	95	100	105	110	110	115	120	125
120	75	80	85	85	90	95	100	100	105	110	115	120	120	125	130
130	80	85	90	95	95	100	105	110	110	115	120	125	130	130	135
140	85	90	95	100	105	105	110	115	120	120	125	130	135	135	140
150	90	95	100	110	110	110	115	120	125	130	130	135	140	145	145
160	100	100	105	110	115	120	120	125	130	135	140	140	145	150	150
170	105	110	110	115	120	125	130	130	135	140	145	145	150	155	160
180	110	115	120	120	125	130	135	135	140	145	150	155	155	160	165
190	115	120	125	130	130	135	140	145	145	150	155	160	165	165	170
200	120	125	130	135	140	140	145	150	155	155	160	165	170	170	175

Recommended minimum mounting height in feet.

(Table Prepared by Hi/Tek Company, Inc.)

Setback = 50 ft, width of area = 100 ft, read minimum height of 50 ft).
The values are Illuminating Engineering Society (IES) recommendations,
rounded off to the nearest 5 feet for convenience sake. Note that when
lighting an area from only one side the mounting height should be double
that shown in Table 14.8.

There are four floodlight spacing techniques that may be used, depending upon the facility's size and light placement. Although these four configurations are satisfactory by themselves, integration of one or more
illustrations is recommended for increased security.

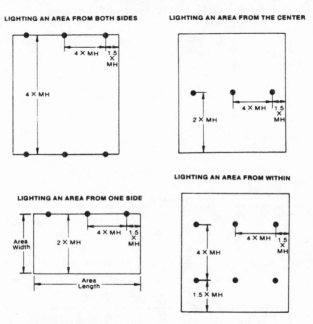

FIGURE 14.22. Floodlight Spacing Techniques
Note: For economy lighting projects, such as
minimum security lighting, spacing ratios could
be increased as follows:

      4 x MH - increase to 5 x MH
    1.5 x MH - increase to 2 x MH
      2 x MH - increase to 2.5 x MH

(Illustrations Courtesy of Hi-Tek Co., Inc.)

14.5.5 Pole Selection

In selecting the proper pole for the lighting system investigations must be made to ensure that the entire structure is suitable for the wind loads imposed by a specific location. The poles listed in this section are designed to withstand dead loads and theoretical dynamic loads developed by variable wind velocities with 1.3 gust factor. Poles that are to be located in areas of known abnormal conditions will require special consideration. In general, coastal areas require that the following wind velocities be used for applications within 30 miles of the coast.

GULF COASTAL STATES
    Louisiana, Mississippi, Alabama. . . . . .100 mph
    Texas. . . . . 90 mph
ATLANTIC COASTAL STATES
    Florida, South Carolina, North Carolina . . . . .100 mph
    Connecticut, Maine, Massachusetts, New York. . . . 90 mph
    Rhode Island, Virginia . . . . 90 mph
PACIFIC COASTAL STATES
    Oregon, Washington . . . . 90 mph
GREAT LAKES STATES . . . . 90 mph
ALASKA and HAWAII . . . . 90 mph

The wind speeds were taken from the 1975 "Standard Specifications for Structural Supports for Highway Signs, Luminaires and Traffic Signals" by American Association of State Highway Transportation Officials (AASHTO). The wind speeds are based on a 25-year mean recurrance interval at an elevation of 30 ft above the ground, as is recommended by AASHTO for light standards 50 ft or less in height. The individual designer/engineer should consider other factors, such as isolated high wind areas, in choosing the specified wind speed.

There are five basic steps in the selection of lighting poles:

1. Determine basic wind velocities to be used for given location.
2. Select bracket to be used. Determine that each tenon will carry individual fixture size and weight.

3. Determine the total projected area of equipment to
    be installed on pole. (Consists of luminaires and
    brackets.)
4. Determine the total weight, if required, by adding
    total luminaire weight to bracket weight.
5. Select the lightest duty pole with the proper tenon
    and loading capabilities to carry load determined in
    steps 2 and 3 for expected wind velocity.

As an example, if an engineer wanted to determine the correct round
steel pole and bracket to mount three Hi-Tek Corporation Model TF1000 M RA
units in a row in Savannah, Georgia, the following steps would be taken:

1. Savannah, Georgia falls in the 90-mph category.
2. Select Hi-Tek bracket number S3-B3-4P because TF1000 has a
    projected area of 4.2 ft^2. (Number S3-B3-2P cannot be
    used.)
3. Projected area per luminaire   4.2 ft^2
    Total number of luminaires   x   3
                          TOTAL   12.6 ft^2
    Projected area of selected
      bracket S3-B3-4P   + 1.9 ft^2
                     TOTAL   14.5 ft^2
4. Weight per luminaire   76 lbs
    Total number of luminaires   x   3
                    TOTAL   228 lbs
    Bracket weight   + 37 lbs
                    TOTAL   265 lbs
5. Hi-Tek pole number St-30-4H-AP would prove to be a suitable
    unit.

Ultimate selection would then be based on the three following poles[*]:

_____

[*]Models of poles are taken from Hi-Tek Corporation Catalog, "Industrial
Lighting." This should not, however, be construed as an endorsement of any
manufacturer's pole. Hi-Tek is used in this study because their sales
selection literature provided necessary report input.

● Aluminum Post - Top Poles

Cylindrical shaft poles under this category are fabricated from
6061-T6 aluminum alloy. Tapered shaft poles are fabricated from 6063-T6
aluminum alloy. Anchor base models include base covers. Embedded models
are bituminous coated inside and out for a length of 2 to 3 feet from the
base end. All units are rated 100 mph constant wind (1.3 gust factor).

● Aluminum Floodlight Poles

Shafts for these poles are fabricated from #5086-H34 marine sheet
alloy. Bases are cast from 356-T6 aluminum and include anchor bolt covers.
Shafts over 40 feet are shipped in two sections. Anchor bolt size: 8'-14'
poles, 1/2" x 12" x 3"; 10'-16' poles, 3/4" x 15" x 2"; 20'-35' poles,
1" x 36" x 4"; 35'-50' poles, 1-1/4" x 48" x 4". (Figure 14.23)

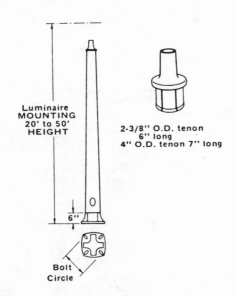

FIGURE 14.23. Aluminum Floodlight Poles
(Illustration Courtesy of Hi-Tek Co., Inc.)

● Steel Poles - Round Tapered (10 to 45 feet Anchor, Embedded, or Transformer Base Prime Painted)

All of these poles have welded, tapered shafts of hot-rolled commercial
quality carbon steel. The anchor base is of structural quality hot-rolled

steel. All poles are prime painted with an iron oxide rust inhibitive.
Nut covers provided on all poles and acorn nuts provided on pedestal poles.
Shafts over 40 feet are shipped in two sections.

Cost

The costing for lighting poles is dependent on the geographical location
and system used. Almost all contractors will include the pole cost with
the lighting system cost. Prices start at $200 per pole and can range to
$3,000, depending on mast size.

14.5.6 Outdoor Lighting Estimating Guide

The area coverage tables (Tables 14.9 and 14.10) present a way of estimating the number of fixtures needed to light an area to one initial footcandle. The tables that follow are provided only as a guide. They are not
offered in lieu of a precise lighting survey.

Area Coverage Tables

TABLE 14.9.

"TFL" Series Area Coverage (sq ft/fix)
(1 Footcandle initial)

Catalog Number	Lge.	Med.	Sm.
TFL 70S TA	2320	1740	1160
100S TA	3800	2850	1900
150S TA	6400	4800	3200
200S TA	8800	6600	4400
TFL 175M TA	5600	4200	2800
250M TA	8200	6150	4100
TFL 175H TA	3440	2580	1720
250H TA	5200	3900	2600

TABLE 14.10.

"TF" Series Area Coverage (sq ft/fix)
(1 Footcandle initial)

	Catalog Number	Lge.	Med.	Sm.
	TF 250S RA	11,000	8,250	5,500
(30,000 →	TF 250S RA	12,000	9,000	6,000
Lumen Lamp)	TF 400S RA	20,000	15,000	10,000
	TF 1000S RA	56,000	42,000	28,000
(40,000	TF 400M RA	13,600	10,200	6,800
Lumen Lamp) →	TF 400M RA	16,000	12,000	8,000
(110,000 →	TF 1000M RA	44,000	33,000	22,000
Lumen Lamp)	TF 1500M RA	62,000	46,500	31,000
	TF 400H RA	9,200	6,900	4,600
	TF 1000H RA	25,200	18,900	12,600

(Tables Courtesy of Hi-Tek/Lithonia Corp.)

How to use the area coverage tables: Each table gives square feet per fixture coverage for three size classifications: small, medium, and large as defined in the Area Size Approximator Table· at the top of Figure 14.24.

Hi-Tek/Lithonia Corporation currently markets the TF, TM, and TFL series of outdoor industrial lighting fixture. Selected models are as follows:

TF Series:

Optics--Sealed to inhibit entrance of outside contaminant. Reflector is one-piece, anodized aluminum.

Housing--National Electrical Manufacturers Association (NEMA) heavy-duty.

Light Pattern--Rectangular or twin beam.

Ballast--Constant-wattage autotransformer.

Installation--Removable power module provides easy upgrading to more efficient light sources as well as simplified maintenance and reduced cost.

Construction--Housing is strong, lightweight die-cast aluminum. Yoke is hot-dipped galvanized steel. Lens is thermal and shock-resistant tempered glass. Luminaire is enclosed and gasketed. All hardware is Series 400 stainless steel.

Listing--U.L. listed suitable for wet locations.

Photocontrol--NEMA type twist-lock (option).

Finish--Standard is American Standards Association (ASA) 70 gray baked enamel.

The TFL and TM series are very similar in appearance to the TF 1000-watt Omni-Flood light; hence, the photographs are excluded.

TFL Series - Omni-Flood, Four-Wattage

Optics--Sealed to inhibit entrance of outside contaminents. Reflector is one-piece, anodized aluminum.

Housing--NEMA heavy-duty, compact (15-3/8" wide by 18-5/16" high).

Light Pattern--Twin beam applications with two-position adjustable socket.

Ballast--Mercury Vapor: Constant-wattage autotransformer.
    Metal Halide: Peak-lead autotransformer.
    High-Pressure Sodium: High leakage reactor

AREA SIZE APPROXIMATOR	
WIDTH VS. MOUNTING HEIGHT	AREA SIZE
Width is less than 1½ x mounting height	Small
Width is 1½ x 2½ x mounting height	Medium
Width is over 2½ x mounting height	Large

Determine area size and obtain area coverage per fixture from "Area Coverage" Table for fixture type desired.
    Find quantity of fixtures required by calculating the total area to be lighted in square feet (width x length) and dividing by the area coverage per fixture.

Follow these easy steps to determine the approximate number of fixtures required:

Example: Area Size: 60' Wide x 200' Long with a 30' mounting height (pole or building mount). Light the area to 1 Footcandle, initial using #TF 400M TA fixture (400 watt metal halide).

1. AREA DETERMINED

Determine Area Type.

Width is between 1½-2½ x mounting height

$$\frac{\text{Width}}{\text{Mounting Height}} = \frac{60}{30} = 2 \text{ Medium area}$$

2. AREA COVERAGE

"Medium" area using #TF 400M fixture equals 10.190 sq. ft./fix coverage

Consult "TF" Series Area Coverage Table.

"TF" SERIES AREA COVERAGE (SQ. FT./FIX.) (1 Footcandle, initial)			
Catalog Number	Small	Medium	Large
TF 400S	10.000	15.000	20.000
TF 400M	6.790	10.190	13.600
TF 400H	4.600	6.900	9.200

3. FIXTURES REQUIRED

Determine Approximate Quantity of Fixtures by Dividing Total Area by Coverage Per Fixture.

$$\text{Fixtures} = \frac{60 \times 200}{10.190} = 1.2 \text{ Use 2 Fixtures}$$

4. SPACING CHECK

Check to Assure that Fixtures are not Spaced Greater than Five Times the Mounting Height in Either Direction.

OTHER FOOTCANDLE LEVELS

If an initial footcandle level other than 1 (one) is desired, use the following correction factors

DESIRED FOOTCANDLES:	CORRECTION FACTOR
1	1.0
2	5
3	33
4	25
5	20

Example:
For 5 Footcandles, initial:

10.190 (sq. ft./fix) x .20 = 2038 sq. ft./fix

New Fixture Quantity $\frac{60 \times 200}{2038} = 5.89$ Use 6 Fixtures

MAINTENANCE FACTORS

If maintained footcandle level is desired, multiply the area coverage per fixture by the maintenance factor.

Maintenance factor is determined by multiplying the lamp lumen depreciation factor (LLD) by the luminaire dirt depreciation factor (LDD). The lamp lumen depreciation factor is usually the mean lumen value of the lamp (see Lamp Performance Characteristics Table in the General Lighting Design Section of this catalog). The luminaire dirt depreciation factor range is described in Section 9 of the Illuminating Engineering Society Handbook. A .9 dirt factor is commonly used.

Example:
For 5 Footcandles, maintained:

2038 (sq. ft./fix) x .9 (LDD) x .75 (LLD) = 1375.65 sq. ft./fix

New Fixture Quantity $\frac{60 \times 200}{1375.65} = 8.72$ Use 9 Fixtures

FIGURE 14.24.   Criteria for Determining Quantity of Fixtures Needed (Courtesy of Hi-Tek Co., Inc.)

FIGURE 14.25.   TF Series 1000-watt Omni-Flood
with Optional Integral Slipfitter

Installation--Hinged bezel for easy access to lamp and electrical com-
            ponents.
Construction--Housing is strong, lightweight, die-cast aluminum. Yoke is
            hot-dipped galvanized steel. Lens is thermal and shock-
            resistant tempered glass. Luminaire is enclosed and gasketed.
            All hardware is Series 400 stainless steel.
Listing--U.L. listed suitable for wet locations.
Photocontrol--NEMA type twist-lock (option).
Finish--Standard is ASA 70 gray baked enamel.

TM Series -Marine Flood U.L. 595

Optics--Choice of two optical systems. Rectangular beam spread with max-
            imum candlepower at nadir and twin beam spread with two max-candle-
            powers for wider spread. Both beam spreads use a one-piece, anodized
            aluminum reflector with adjustable socket.

Housing--NEMA heavy-duty.

Ballast--Constant Wattage Peak-Lead for Metal-Halide and Auto-Regulator for
    High-Pressure Sodium. All high power factor.

Installation/Maintenance--Easily removable power module with polarized posi-
                          tive locking plug to the lamp socket, provides fast
                          access to all electrical components without opening
                          the optical chamber. Fixture is prewired with 2 ft
                          of 16-3 SO Cable.

Construction--Housing is strong, lightweight, copper-free, die-cast aluminum.
             Yoke is hot-dipped, galvanized steel. Lens is thermal, shock-
             resistant, tempered glass. Luminaire is enclosed and gasketed.
             All hardware is Series 400 stainless steel.

Cost for Hi-Tek TF, TFL, and TM Series:

The costs for the models listed are estimated at $275 to $400 less pole
and installation. Installation costs vary depending on geographical location
and union/non-union contracting. Installation costs range from $18/hour to
$25/hour.

General Electric Corporation also markets outdoor luminaires. Two
models are described as follows:

G.E. VLU Powerflood Luminaire:

Removable Ballast Tray: Permits stocking flexibility and easy conversion
from one High Intensity Discharge light source to another (Figure 14.26b).

Easy Maintenance: Access to the luminaire for relamping or cleaning
is readily accomplished by loosening two captive screws and swinging open the
front door in its hinges. This allows the maintenance man to use both hands
for his work. If total maintenance is ever required, the front door is completely
removable.

High Maintained Efficiency: Weatherproof construction protects the optical
and electrical systems from outside elements. The sealed and activated charcoal
filtered luminaire protects the ALGLAS®-finished, hydroformed reflector, providing

FIGURE 14.26a. G.E. VLU Power-   FIGURE 14.26b. Ballast Tray
flood Luminaire (Illustrations
Courtesy of General Electric Corp.)

maximum light levels for longer time periods. The filter is designed to
effectively remove particulate matter and gaseous contaminants, resulting
in fewer manhours spent for maintenance.

Versatile Mounting: There is a choice of either a heavy-gauge galvanized steel-trunnion or a cast aluminum, knuckle-type slipfitter; both
have a vertical degree scale. The trunnion version is particularly suited
for mounting on flat surfaces such as wood crossarms, concrete pads, building parapets, and so on. The knuckle-type slipfitter is designed for pole
tops; its internal wiring and clean, modern design provide excellent daytime appearance.

Easy Installation: The lightweight, rugged, heavy-duty fiberglass
reinforced polyester construction permits easy handling and fast installation.
The built-in rifle sight allows daytime aiming. The trunnion-mounted version
has a readily accessible prewired terminal board and a grommeted strain relief
bushing. The knuckle-type slipfitter has an integral wiring compartment with
pre-stripped extra length leads.

Choice of Light Sources: Economical and modern Lucalox®, mercury, or
metal halide light sources of various wattages and voltages are controlled
with quality ballasts.

Automatic Dusk-to-Dawn Control: An optional, built-in photoelectric receptacle is available for individual luminaires.

Suggested Mounting Height: 25 feet.

Cost: Estimated cost range for high-pressure sodium fixture is $400 to $450 (less pole and installation).

FIGURE 14.27. G.E. M-400A Luminaire

The G.E. M-400A Luminaire offers a shielded optical system (incorporating a faceted, hydroformed reflector and a clear flat glass lens) providing 90-degree cutoff. The unit would be used within a facility along road ways or paths where the 90-degree cutoff would drastically reduce glare in the driver's line of vision caused by the strong illumination from powerfloods. This model is available with 200-, 250-, 310-, and 400-watt Lucalox® and 400-watt mercury or metal-halide light sources. Some key features are:

Easy Relamping and Optical Assembly Servicing: One latch opens captive door for access to the entirely sealed reflector system.

Minimal Light Loss from Contaminants: Charcoal-filtered and gasketed optical assembly greatly reduces light loss that results from gaseous and particulate materials. . . and is factory-installed for high-quality sealing.

Simplified Assembly on the Ground or in the Air: Two-bolt slipfitter allows line crews to install the luminaire on its bracket on the ground, or to mount bracket and luminaire on the pole in the air with minimum effort and time.

Internal Photoelectric Control Protected from Damage: Astrodome control (optional), located inside the housing, is protected from outside dangers such as rocks, tree limbs, and/or common maintenance problems.

Optimum Lighting Efficiency: The patented ALGLAS® treated reflector is precision formed for optimum lighting efficiency. The Alglas® silicate film is chemically bonded to the aluminum interior and exterior to seal the surface. Alglas® is lightweight, nonbreakable, and glass smooth.

Suggested Mounting Height: 30 feet.

Cost: $200 (less pole and installation).

Floodlight options such as light sensing timers (automatic dusk-to-dawn controls) and QRS (Quartz Restrike Systems) are factory installed. Photoelectric cells and receptacles must be purchased per unit. Estimated cost of the cell is $18 and the receptacle is $18 per unit. Another method of dusk-to-dawn control is the time clock. It can handle five to seven luminaires, is pre-set and has been proven to be more energy and cost efficient.

14.6 GUARDS, WATCHMEN, AND WATCHDOGS

The security of plant facilities, equipment, fuels, and stored raw and finished materials can be maintained by plant employees, a plant-employed guard force, or a contractual guard service.

An essential feature of plant protection during non-operating hours is adequate guard patrols. Plant management should carefully select the proper guard personnel and equipment, as well as the layout and schedule of routes to be patrolled. If a contract guard service is employed, most of the administrative details may be developed under the contract, but this should not relieve management's responsibility to ensure that the contract service meets standards equal to a company-operated service.

The initial and continuing training of a plant guard staff should be a formal, well-documented program covering all applicable protection procedures. Personnel should be acquainted with the general nature of the facility's operations and have a specific knowledge of the inherent or special hazards of the stored products. They should be familiar with all fire protection equipment, both manual and automatic, know the location and operation of fire alarms, and the proper method of informing the fire department and/or designated company officers of the location of and directions to the fire or spill area within the plant.

Finally, the guard should maintain a shift log and prepare reports of all observations and actions made during a shift including any infraction of plant regulations. The observations can be made by roving patrols moving on a regular or staggered frequency. Clock punch systems ensure that the frequency and predesignated extent of patrols are maintained. Security personnel should be instructed to report all unsafe conditions to management. They should also have a degree of responsibility to enforce no smoking regulations in designated areas.

The guard staff should in fact be fully cognizant of the content of the plant's spill prevention control and countermeasure plan (SPCC Plan). The Environmental Protection Agency reulations on oil pollution prevention (40CFR112) require that owners and operators of certain onshore and offshore oil storage and handling facilities prepare a certified Spill Prevention Control and Countermeasure Plan (SPCC Plan). It is anticipated that implementation of properly prepared plans will significantly reduce the number of spills. The effect of implementation should be manifested in a reduction in the amount of oil reaching the navigable waters and adjoining shorelines of the United States and a lessening of the associated environmental consequences. A copy of the Regulation is included as Appendix A.

Any unusual condition that is found and cannot be corrected by the guard himself should be reported immediately to the alarm center or central station so that the situation can be remedied without undue delay.

Most security organizations are comprised of retired and former law enforcement officers and individual security specialists. Most offer security consulting, surveillance, uniformed armed and unarmed guards, radio mobile patrols (including four-wheel drive jeeps), K-9 patrols, and strike forces (in the event of an unusual occurrence, armed and unarmed personnel are ready to implement protection of facility).

Cost:

Table 14.11 lists descriptions and costs for a typical security service. These prices reflect an average cost throughout industry.

TABLE 14.11.

Security Cost Table

Quantity	Description	Unit Price
1 hour	Uniformed Security Guard (Unarmed)	$7.25
1 hour	Radio-Motorola Model MT-500	.25
1 hour	Smith & Wesson Service Revolver, Model 10	.25
1 hour	Patrol Vehicle	2.50
1 hour	K-9	1.50
	24-Hour Dispatcher Service	N/C
	On-Site Supervisory Checks	N/C
	Daily Log Reports	N/C
	Incident Reports	N/C
	Supervisor's Evaluation for Security Procedures and Job Description Done On Site Prior to Assuming Contract	N/C

Rates based on 40 hours minimum.

The service also offers the following checklist of no-cost benefits.

- Overtime: Clients pay the same hourly rate for 168 hours a week as for a 40-hour week. No matter how many guards are needed, the price is the same.
- Sick: If a guard gets sick, he is replaced at no cost.
- Insurance: One million dollars liability.
- Training: All guards are trained before hiring.
- Outside Costs: Client does not have to pay F.I.C.A., hospitalization, etc.
- Administrative Costs: None

- Seconday Costs: No sick or annual leave to pay.
- Internal Relationships: Guards can be rotated so that they don't develop too close a personal relationship with facilities employees.
- Unions: Non-union
- Disciplinary Problems: Occasionally situations arise where a guard presents a problem; he can be replaced without inquiry.
- Supervision: Round-the-clock supervision checks on the guard force, and if for some reason a guard is late, the supervisor fills in at no extra cost.
- Uniforms: Cost for each is $200, paid by the security service company.
- Options: Client has the option to cancel out within 15 days after the first month.

Note: The information provided is this section was supplied by Sting Security, Inc., Marlow Heights, Maryland.

14.6.1 Watchdogs

There is no intent to indicate that bulk oil storage plants require K-9 protection. The following information is provided to indicate the degree of security that can be attained.

There are two varieties of watchdogs: command dogs and guard dogs. Doberman Pinscher and German Shepherd (Alsatian) dogs are most commonly used.

Command dogs properly trained act only on command from their trainer. They are playful, totally tame animals that have been trained to attack and will do so upon word from the trainer. Since command dogs need their trainer present, plant management must decide whether it would be cost beneficial to pay for the most expensive command dog/trainer combination.

Guard dogs are trained to protect and/or attack with no trainer present. Trying to train a dog to be a guard dog would be a waste of time and effort, since guard dogs are bred for their specific purpose. The guard dog must be futher trained "on site" at a specific plant. Some dogs when brought to a

large facility will not be completely effective until they are familiar
with the area they are guarding.

Figure 14.28 illustrates the animal's reaction to a would-be intruder.
The individual trying to gain entrance to a facility would become suddenly
cynophobic upon seeing this vicious animal's intent. Trying to "sweet-talk"
the animal would prove worthless. All the properly trained animal knows is
that no one is to get in.

At one time, it was common practice for K-9 companies to deliver dogs
to the facility before dusk or closing and then retrieve them in the morning
or when business resumed. The K-9 companies would keep the dogs in a cen-
trally located kennel where feeding and care would take place. This practice
has been improved upon and it is now more common to have the animal live on
the premises in a fenced cage (Figure 14.29). When the plant guard has
secured the facility and upon leaving the main gate, he will pull the gate
cord, opening the cage and freeing the dog (Figure 14.30). Before business
resumes a trainer will arrive at the facility, feed, water, and secure the
dog.

Guard dogs cannot be used in areas where electronic sensing devices
are implemented and vice versa, because of the obvious number of false
alarms which would occur. Decisions on whether to use guard dogs or
electronic detection systems depend on the quality of security needs; both
have their advantages and disadvantages.

At some high-security locations, dogs are free to patrol on a regular
basis within the confines of a fence within a fence, and electronic sensors
can then be used in the plant area where the dogs are excluded.

Security dogs are available from K-9 companies for lease or direct
sale. Prices are in the range of $600 per six-month lease and $675 per one-
year lease. Direct purchase of a dog ranges from $500 to $5,000 depending
upon the breed and quality desired. (Florida and West Coast prices are 20
percent less.) Once purchased, the facility would assume full responsibility
and would need their own staff trainer. In this case, there would be no
exchange or trade-in for another animal. Leasing allows for replacement of
the dog if for some reason it does not prove effective.

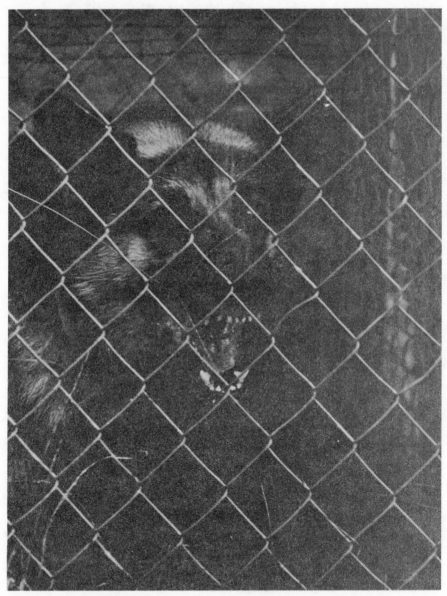

FIGURE 14.28. High Security Watchdog
(Photograph Courtesy of K-9 Security Co., Ltd.)

FIGURE 14.29.   Off-Duty Retention Cage
Note the pull cord gate opening device.
Dog has shaded area and dog house during
plant's operating hours. (Photograph
Courtesy of K-9 Security Co., Ltd.)

FIGURE 14.30. Off-Duty K-9 Retention Cage
Dog is free to roam facility. (Photograph
Courtesy of K-9 Security Co., Ltd.)

14.7 PIER, WATERFRONT, AND SHORELINE SECURITY

The weak security link in most waterfront facilities, refineries, marketing terminals, and marine loading/unloading and storage plants is the open shoreline. Illegal entry can be easily gained through this unsecured boundary by almost any person, i.e., children, saboteurs, thieves, and in some cases, disgruntled employees intent on mischief and/or trouble-making.

During a period of civil disturbance in the 1960's some plants corrected this situation by fencing the land area above the foreshore or high water mark. The metal fences in some cases extend up and onto the pier, providing access to the loading/unloading dock, with a locked gate controlling entry into the plant proper. The sides of the fence paralleling the sides of the pier should restrict side entry around the gated section of the fence. The gate is kept in the closed and locked position whenever the dock is not in service.

As far as practical, equipment on the loading dock should be fully secured. Pump controls should be locked and fuses removed to prevent unauthorized activation of the pumping system. The terminal flanges of the docklines should be gasketed and blank flanged and full bolted in position; intermittent bolt spacing should not be considered as acceptable securement.

Portable drip catch trays or pans should be either removed from the dock area (where they should not really be needed when the dock is idle) or emptied and cleared ready for future use.

Liferings and lifelines and fire extinguishers present a theft problem, since they should be maintained ready for use at any time. They can, however, be stored in marked containers where they are protected from the weather and do not present a visible inducement to theft.

When plant patrols are used, the shoreline should be frequently patrolled.

14.8 ALARM DETECTION SYSTEMS

Outdoor facilities are especially vulnerable to theft, vandalism, malicious mischief, and sabotage, since eye-witnessing (casing the site) by the intruder is accomplished rather easily. ITS Systems, Pittsburg, Pennsylvania, installed one alarm detection system that reduced a lumber-yard's losses by $10,000 annually. (Much of the theft was occuring during nighttime hours.) In this case, a buried seismic sensor was used. There are also microwave, photoelectric detection, infrared, and fence vibrator systems now in use throughout industry. Descriptions of the systems follow.

14.8.1 Seismic Sensors

Where security systems are exposed to blizzards, blowing debris, wild animals, and wind, buried seismic systems can be used. Buried seismic systems are programmed to detect footsteps and screen out other noises and vibrations.[*] Dynamic microphones or sensors placed about 12 feet apart have a range of approximately 20 feet and are linked by a cable that carries impulses generated by vibrations to a processor. When the signals reach the processor at a guard or central station, they are amplified and equalized to eliminate unrelated intruder motion frequencies. After the amplification process, signals are converted to pulses compared to a standard single thres-hold. Each pulse crossing the threshold is judged according to count and periodicity. If the pulses meet the set criteria for human footsteps, an alarm is triggered. A guard or monitor can then listen to the footsteps to determine how many intruders are present.[**]

Figure 14.31 illustrates a metal fence vibrator unit. Vibrator units are installed on every other fence post and once activated they vibrate the fence upon physical contact. The unit can be actuated without physical con-tact, however, in this instance at the Bonneville Power Administration Keeler Substation, Oregon. At certain times each morning, this metal fence intruder detection system would vibrate the fence. Investigation revealed the source to be a train passing about the same time every morning, hence setting off the alarm. Discriminator sensors might be a solution in cases such as this.

---

[*] Security Distributing & Marketing Magazine (SDM), June 1980, p. 24 "Principles of Operation for Buried Seismic Sensors."

[**] Note: The US/DOE's Office of Safeguards and Security (OSS) contends these systems are not perfected to the point where they are justifiably proven effective.

FIGURE 14.31. Metal Fence Intruder
Detection System

There exist two different philosophies in designing a system, involving
whether or not an intruder should know he has tripped an alarm. "Using
the seismic sensor, an intruder can be detected without his being aware
that he has tripped an alarm. It also restricts his being able to tamper
with the sytem. The underground sensors are not subject to varying weather
conditions and can be installed in uneven terrain, around corners, or
across landscaping."[*]

According to the article the system can also be buried under concrete
or asphalt (Figure 14.32). It can be placed before paving or narrow slits
can be cut through the concrete and sensors and cable dropped in a trench
about 18 inches deep. Some portable systems with special sensor cases are
equipped with spikes so that the unit can be driven into the ground and then
moved later.[**] Discriminator sensors are used near the source of normal
vibrations (i.e., traffic or machinery) in order to filter out noise.

----

* Security Distributing & Marketing Magazine, June 1980, Vol. 10, No. 6,
  P. 24, "Principles of Operation for Buried Seismic Sensors."
**US/DOE's Office of Safeguards & Security states that these seismic sensors
  are too fragile to withstand the shock of a hammerblow.

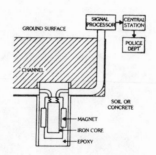

FIGURE 14.32. Buried Seismic Sensor System
(From Security Distributing & Marketing Magazine
June 1980)

14.8.2 Microwave Detection

Outdoor microwave links are line-of-sight systems that provide volu-
metric protection along a plant's perimeter. A beam of microwave energy
is radiated between a transmitter and receiver so that a person entering
the beam pattern causes a decrease in signal being detected by the receiver,
thus triggering an alarm. Unlike indoor microwave detectors that operate
on the Doppler principle, outdoor systems detect not only an intruder's
motion, but also his presence within the beam. Also, outdoor microwave
units are bistatic with transmitter and receiver in separate housings
separated by a given distance. (Indoor systems are monostatic--the transmitter and receiver in one unit.)

Beam diameters vary from 2 ft to 40 ft and can operate for lengths of
up to 1,500 ft. Microwave systems are designed to operate in snow, rain, or
fog and to allow for flying objects such as birds, paper, or leaves. The
alarm can, however, be triggered by small animals entering the coverage area
to gain maximum security. The units should be set inside the perimeter fence
at least 5 ft so that the intruder cannot vault both the fence and the beam.
The ground should be flat and free of trees, brush, weeds, or other obstruc-
tions that may restrict the coverage area. The "offset technique" --zones

that overlap each other so that triangular areas between transmitters and
the ground are covered by transmitters in another zone--is used in high
security applications (Figure 14.33).

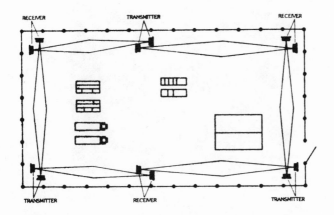

FIGURE 14.33. Recommended Microwave System Layout
(From Security Distributing and Marketing Magazine,
June 1980)

"In general outdoor microwave detection systems are not affected by
strong vibrations or radio frequency interference and can serve as a high
security system when integrated with other security measures.[*]

Figures 14.34 and 14.35 are photographs of microwave detection systems
currently being tested along with other types of systems within an electrical
substation of the U.S. DOE Bonneville Power Administration, Oregon. It is
possible for an intruder to gain undetected entrance by crawling under the
units. Configurations as in Figure 14.33 tend to control this deficiency.
The added interior perimeter protection with proper placement of multiple
transmitter and receivers should be observed.

---

[*]Security Distributing and Marketing Magazine, July 1980, Vol. 10, No. 6
P. 24, "Principles of Outdoor Microwave Detection."

FIGURE 14.34.

FIGURE 14.35.

Microwave Sending and Receiving Units

Most outdoor microwave units have a cost effective 1,500-ft coverage range. On a cost per foot coverage basis, it is about $1 a foot.

14.8.3 Photoelectric Detection

Outdoor photoelectric systems create an invisible fence made up of one or more infrared beams. A transmitter emits a pulse-modulated beam from a (LED) light source aimed at a receiver up to 500 ft away. When the conical beam is broken an alarm activates.

Figure 14.36 illustrates the perimeter configuration of transmitters (T) and receiver (R). Note how the arrows are criss-crossed to allow steady flow of signal patterns.

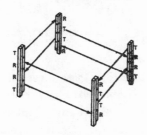

FIGURE 14.36. Photoelectric Detection System

Extreme weather conditions such as fog, smog, dust storms, and snow can cause gradual loss of signal and an unnecessary alarm. Systems can be designed to compensate for this, disqualifing any gradual loss of signal caused by adverse weather conditions. Also, flying debris, leaves, paper, or birds can be eliminated as a source of alarms by a circuit that permits the beam to be broken for a certain length of time before an alarm is triggered. The photoelectric detection system should be installed on level terrain, at least 2 ft away from a perimeter fence (Figure 14.37).

FIGURE 14.37. Infrared Intruder Detection System,
Bonneville Power Administration, Oregon

Equipment should not be positioned directly into the rising or setting
sun. In most cases, surrounding buildings, hills, or the horizon can be used
to adequately block the sun's direct rays. Infrared systems need much closer
spacing than microwave systems for efficient operation.

14.8.4 Local Alarms

The term "local alarm" includes any noise- and/or light-producing device
used to protect a selected area. It may be a siren, horn, or bell, sometimes
combined with a flashing light. These devices should be mounted high on an
outside wall or structure with full protection against weather and tampering.
The connecting cable should also be tamper resistant. Most bells have internal
battery power supplies that permit the unit to operate if line power is inter-
rupted or even if the unit itself is pulled free of the wall. One advantage
of the local alarm is its lower cost. Another is that a ringing bell lets the
intruder know that his intrusion has been detected--he might have second
thoughts about completing the job.

There may be some benefit when the intruder is aware that his presence is known or shortly will be known. Conversely, when an intruder crosses a beam detection or seismic sensor system, he does not know that an alarm has been tripped. Seconds later he is moving deeper into the protected area, possibly to be apprehended. The major benefit gained from an audible signal is the fact that a watchman or other type of security person need not stare at a panel of lights for his entire duty period. The audible system calls immediate attention to an intrusion, whereas a non-audible device is only as valuable as the time factor involved for the guard or watchman to observe the sensor's readout on the monitoring panel. The simplest unit is the local alarm that can be transmitted to a central security or police station. This demands transmission of the alarm signal over a communications link, usually leased, or through the media of telephone lines engineered for signal transmission.

14.8.5 Commercial/Central Station Connected Alarms

Central station alarms refer to security stations operated by companies that sell such services to manufacturers, banks, and other facilities in their market area. Direct connections via telephone lines to a central reporting and investigative unit are available in most communities throughout the United States. The units are monitored continually and when an alarm is received, the guard generally telephones the police and/or dispatches security personnel to the secured area. Incidents are reported as soon as possible to the customer.

Figures 14.38 and 14.39 illustrate modular security systems currently available for industrial purchase and rental. The systems provide state-of-the-art advancements in automated security. Typical modular security systems would offer the following.

- Visual Display - Cathode ray tube (CRT) screen showing keyboard command entries, alarms, and change of status with date and time.
- Command entry Panel - Keyboard controls on/off, increase/decrease, access/secure, and status summary.

FIGURE 14.38. Mosler Corporation Modular Security System BRM-2

FIGURE 14.39. American District Telegraph (ADT)
CentraScan 73 Central Security Model

- Line Printer - Hard copy printout of all commands, changes
  in status, alarm logs with times and dates.
- Control Processing Unit - Interrogates and receives response,
  micro- or mini-computers interpret alarms and operate remote
  devices.
- Closed circuit television security surveillance management
  control; video tape recording with date/time generator.
- Matrix Status Display - Constant summary, red for alarm,
  yellow for access, power failures, and line trouble.
- Programming CRT.
- Slide Projector - Allows visual display of alarm location with
  map and instructions for action.

- High Speed Printer - Full activity reporting. Records all changes of state, alarms, and logs.
- Standby Battery - Power backup keeps the central computer and the matrix display working in power failures of up to 72 hours.
- Software Programs - Energy management, load shedding, watchtour, automatic access/secure, card access, log/command sequence, self check watchlog, emergency data file, analog monitoring, and central station interaction.

These console units can be expanded to 7,000 as facility demands increase. The console units monitor not only alarms but also lights, heating, refrigeration and air-conditioning, AC power, and anything that can provide a set of normally open of closed contacts.

Costing for these modular units varies so greatly that ADT Corporation, aspecialist in security systems, could not give a fixed cost. Costing depends on location, size, and selection of monitoring stations needed for a specific facility.

14.9 SUMMATION

The price of hydrocarbon products is soaring to stratospheric heights and losses of equipment and products from theft is growing in proportion. A multimillion dollar market has developed in stolen materials and operational equipment. This is aggravated by damage to property in successful and unsuccessful attempts to gain access to short supply materials.

Thieves with ingenious methods are outwitting the oil industry and materials stolen in one state are being sold in another. The theft of oil has become a major problem challenging oil company security personnel.

Prevention measures can ensure that theft of product can be held to a basic minimum (it cannot control in-plant misappropriation of product). It is hoped that the content of this report section will enlighten plant operators to a burgeoning problem and provide a degree of preventive knowledge that will ultimately control the situation.

References:

1. Schwartz, Allan, E. (undated) Management Guide to CCTV Security Systems.
    American District Telegraph (ADT) Security Systems, 18 p.
2. Mosler Safe Company. (undated) Guide to Electronic Alarm Systems.
3. Security Distributing and Marketing. June, 1980. "Finding Profits in
    Perimeter Protection." Vol. 10, No. 6, p. ¡20-30.
4. Stanley, Arthur, T. Security Systems Analysis. Man Barrier Corporation.
    September 26, 1978.
5. Man Barrier Corporation. (undated) Prison Security.
6. Stanley, Arthur, T. 1974. Barrier Potential of Chain Link Fence.
    Report 2106. U.S. Army Mobility Equipment Research and Development
    Center. Fort Belvoir, Virginia.
7. Stanley, Arthur, T. 1973. General-Purpose Barbed Tape Obstacle.
    Report 2077. U.S. Army Mobility Equipment Research and Development
    Center. Fort Belvoir, Virginia.
8. Kodlick, Martin, R. 1978. Barrier Technology: Perimeter Barrier
    Penetration Tests. Sandia 77-0777, Sandia Laboratories, Alburquerque,
    New Mexico.
9. Industrial Risk Insurers. 1979. When the Plant Guard Takes Over.
    Hartford, Connecticut.
10. American Petroleum Institute. 1975. Precautions Against Electrostatic
    Ignition During Loading of Tank Truck Motor Vehicles. Second Edition.
    Bulletin 1003.
11. Security World. November, 1979. "Oil Industry Sleuths: Crime Preven-
    tion from Oil Rig to Pump." Vol. 16, No. 11, p. 16-19.
12. Meetings: R. J. Siclari with representatives from American District
    Telegraph (ADT), Mosler Safe Company, Man Barrier Corporation, General
    Electric Company, Hi-Tek Lighting, Sting Security Company.
13. Manufacturers' sales literature (as per each unit described).
14. U.S. Department of Energy. 1976. (Sandia 76-0554) Intrusion Detection
    Systems Handbook. Vol. I, II.

15. U.S. Department of Energy. 1978. (Sandia 77-0777) <u>Barrier Tech-</u>
<u>nology Handbook</u>.

16. U.S. Department of Energy. 1978. (Sandia 77-1033) <u>Entry-Control Systems</u>
<u>Handbook</u>.

17. U.S. Department of Energy. 1979. (Sandia 78-0241) <u>Barrier Tech-</u>
<u>nology: Perimeter Barrier Penetration Tests</u>.

18. U.S. Department of Energy. 1977. (HCP/DG540-01) <u>Nuclear Safeguards</u>
<u>Technology Handbook</u>.

19. U.S. Department of Energy. 1979. (HCP/D0789-01) <u>A Systematic Approach</u>
<u>to the Conceptual Design of Physical Protection Systems for Nuclear</u>
<u>Facilities</u>.

20. U.S. Department of Energy. 1978. (Sandia 78-0400) <u>Security Seal</u>
<u>Handbook</u>.

21. U.S. Department of Energy. 1979. (Sandia 78-1785) <u>Safeguard Control</u>
<u>and Communications Systems Handbook</u>.

Appendix A – Copy of SPCC Regulation

CHAPTER I—ENVIRONMENTAL
PROTECTION AGENCY

SUBCHAPTER D—WATER PROGRAMS

PART 112—OIL POLLUTION PREVENTION

Non-transportation Related Onshore and Offshore Facilities

Notice of proposed rule making was published on July 19, 1973, containing proposed regulations, required by an pursuant to section 311(j)(1)(C) of the Federal Water Pollution Control Act, as amended (86 Stat. 868, 33 U.S.C. 1251 et seq.), (FWPCA), to prevent discharges of oil into the navigable waters of the United States and to contain such discharges if they occur. The proposed regulations endeavor to prevent such spills by establishing procedures, methods and equipment requirements of owners or operators of facilities engaged in drilling, producing, gathering, storing, processing, refining, transferring, distributing, or consuming oil.

Written comments on the proposed regulations were solicited and received from interested parties. In addition, a number of verbal comments on the proposal were also received. The written comments are on file at the Division of Oil and Hazardous Materials, Office of Water Program Operations, U.S. Environmental Protection Agency, Washington, D.C.

All of the comments have been given careful consideration and a number of changes have been made in the regulation. These changes incorporate either suggestions made in the comments or ideas initiated by the suggestions.

Some comments reflected a misunderstanding of the fundamental principles of the regulation, specifically as they applied to older facilities and marginal operations. During the development of the regulation it was recognized that no single design or operational standard can be prescribed for all non-transportation related facilities, since the equipment and operational procedures appropriate for one facility may not be appropriate for another because of factors such as function, location, and age of each facility. Also, new facilities could achieve a higher level of spill prevention than older facilities by the use of fail-safe design concepts and innovative spill prevention methods and procedures. It was concluded that older facilities and marginal operations could develop strong spill contingency plans and commit manpower, oil containment devices and removal equipment to compensate for inherent weaknesses in the spill prevention plan. Appropriate changes were made in the regulation to simplify, clarify or correct deficiencies in the proposal.

A discussion of these changes, section by section follows:

A. *Section 112.1—General applicability.* Section 112.1(b), the "foreseeability provision", contained in 112.1(d)(4) was added to paragraph 112.1(b). As modified, the regulation applies to non-transportation-related onshore and offshore facilities which, due to their location, could reasonably be expected to discharge oil into or upon the navigable waters of the United States or adjoining shorelines.

Sections 112.1(b), 112.1(d)(4) and 112.3 are now consistent.

Section 112.1(d)(1) was expanded to further clarify the respective authorities of the Department of Transportation and the Environmental Protection Agency by referring to the Memorandum of Understanding between the Secretary of Transportation and the Administrator of the Environmental Protection Agency (Appendix).

Section 112.1(d)(2), the figure for barrels was converted to gallons, a unit of measure more familiar to the public, and now reads "42000 gallons."

Section 112.1(d)(3), exemption for facilities with nonburied tankage was extended to 1320 gallons in aggregate with no single tank larger than 660 gallons and applies to all oils, not just heating oil and motor fuel. Tanks of 660 gallons are the normal domestic code size for nonburied heating oil tanks. Buildings may have two such tanks. Facilities containing small quantities of oil other than motor fuel or heating oil would also be exempt, thus making this consistent with the definition of oil in § 112.2.

B. *Section 112.2—Definitions.* Section 112.2(1), the term "navigable waters" was expanded to the more descriptive definition used by the National Pollutant Discharge Elimination System.

Section 112.2(m), the U.S. Coast Guard definition of the term "vessel" was included. This term is used in the regulation and the definition is consistent with the Department of Transportation regulations.

C. *Section 112.3—Requirements for the preparation and implementation of spill prevention control and countermeasure plans.* A new paragraph (c) was added to § 112.3 which applies to mobile or portable facilities subject to the regulation. These facilities need not prepare a new Spill Prevention Control and Countermeasure Plan (SPCC Plan) each time the facility is moved to a new site, but may prepare a general plan, identifying good spill prevention engineering practices (as outlined in the guidelines, § 112.7), and implement these practices at each new location.

Section 112.3(a), (b) and (f) (which was § 112.3(e) in the proposed rule making) have been modified to allow extensions of time beyond the normally specified periods to apply to the preparation of plans as well as to their implementation and to remove the time limitation of one year for extensions. Extensions may be allowed for whatever period of time considered reasonable by the Regional Administrator.

Section 112.3(e) (which was § 112.3(d) in the proposed rule making) was modified to require the maintenance of the SPCC Plan for inspection at the facility only if the facility is normally manned. If the facility is unmanned, the Plan may be kept at the nearest field office.

Section 112.3(f)(1) (§ 112.3(e)(1) in the proposed regulation) was changed to include the nonavailability of qualified personnel as a reason for the Regional Administrator granting an extension of time.

D. *Section 112.4—Amendment of spill prevention control and countermeasure plans by Regional Administrator.* Section 112.4(a)(11), permits the Regional Administrator to require that the owner or operator furnish additional information to EPA after one or more spill event has occurred. The change limits the request for additional information to that pertinent to the SPCC Plan or to the pollution incident.

Section 112.4(b) now reads "Section 112.4 * * *", not "This subsection * * *"

Section 112.4(e) allowed the Regional Administrator to require amendments to SPCC Plans and specifies that the amendment must be incorporated in the Plan within 30 days unless the Regional Administrator specifies an earlier effective date. The change allows the Regional Administrator to specify any appropriate date that is reasonable.

Section 112.4(f). A new § 112.4(f) has been added which provides for an appeal by an owner or operator from a decision rendered by the Regional Administrator on an amendment to an SPCC Plan. The appeal is made to the Administrator of EPA and the paragraph outlines the procedures for making such an appeal.

E. *Section 112.5—Amendment of spill prevention control and countermeasure plans by owners or operators.* Section 112.5(b) required the owner or operator to amend the SPCC Plan every three years. The amendment required the incorporation of any new, field-proven technology and had to be certified by a Professional Engineer.

The change requires that the owner or operator review the Plan every three years to see if it needs amendment. New technology need be incorporated only if it will significantly reduce the likelihood of a spill. The change will prevent frivolous retrofitting of equipment to facilities whose prevention plans are working successfully, and will not require engineering certification unless an amendment is necessary.

Section 112.5(c), this paragraph required that the owner or operator amend his SPCC Plan when his facility became subject to §112.4 (amendment by the Regional Administrator). This paragraph has been removed. It is inconsistent to require the owner or operator to independently amend the Plan while the Regional Administrator is reviewing it for possible amendment.

F. *Section 112.6—Civil penalties.* There are no changes in this section.

G. *Section 112.7—Guidelines for the preparation and implementation of a spill prevention control and countermeasure plan.* Numerous changes have been made in the guidelines section; the changes have been primarily:

1. To correct the use of language inconsistent with guidelines. For example, the word "shall" has been changed to "should" in § 112.7(a) through (e).

309

2. To give the engineer preparing the Plan greater latitude to use alternative methods better suited to a given facility or local conditions.

3. To cover facilities subject to the regulation, but for which no guidelines were previously given. This category includes such things as mobile facilities, and drilling and workover rigs.

In addition, wording was changed to differentiate between periodic observations by operating personnel and formal inspections with attendant record keeping.

These regulations shall become effective January 10, 1974.

Dated: November 27, 1973.

JOHN QUARLES,
Acting Administrator.

A new Part 112 would be added to subchapter D, Chapter I of Title 40, Code of Federal Regulations as follows:

Sec.
112.1 General applicability.
112.2 Definitions.
112.3 Requirements for preparation and implementation of Spill Prevention Control and Countermeasure plans.
112.4 Amendment of Spill Prevention Control and Countermeasure Plans by Regional Administrator.
112.5 Amendment of Spill Prevention Control and Countermeasure Plans by owners or operators.
112.6 Civil penalties.
112.7 Guidelines for the preparation and implementation of a Spill Prevention Control and Countermeasure Plan.
Appendix Memorandum of Understanding Between the Secretary of the Department of Transportation and the Administrator of the Environmental Protection Agency. Section II—Definitions.

AUTHORITY: Secs. 311(j)(1)(C), 311(j)(2), 501(a), Federal Water Pollution Control Act (Sec. 2, Pub. L. 92-500, 86 Stat. 816 et seq. (33 U.S.C. 1251 et seq.)); Sec. 4(b), Pub. L. 92-500, 86 Stat. 897; 5 U.S.C. Reorg. Plan of 1970 No. 3 (1970), 35 FR 15623, 3 CFR 1966-1970 Comp.; E.O. 11735, 38 FR 21243, 3 CFR.

§ 112.1 General applicability.

(a) This part establishes procedures, methods and equipment and other requirements for equipment to prevent the discharge of oil from non-transportation-related onshore and offshore facilities into or upon the navigable waters of the United States or adjoining shorelines.

(b) Except as provided in paragraph (d) of this section, this part applies to owners or operators of non-transportation-related onshore and offshore facilities engaged in drilling, producing, gathering, storing, processing, refining, transferring, distributing or consuming oil and oil products, and which, due to their location, could reasonably be expected to discharge oil in harmful quantities, as defined in Part 110 of this chapter, into or upon the navigable waters of the United States or adjoining shorelines.

(c) As provided in sec. 313 (86 Stat. 875) departments, agencies, and instrumentalities of the Federal government

are subject to these regulations to the same extent as any person, except for the provisions of § 112.6.

(d) This part does not apply to:

(1) Equipment or operations of vessels or transportation-related onshore and offshore facilities which are subject to authority and control of the Department of Transportation, as defined in the Memorandum of Understanding between the Secretary of Transportation and the Administrator of the Environmental Protection Agency, dated November 24, 1971, 36 FR 24000.

(2) Facilities which have an aggregate storage of 1320 gallons or less of oil, provided no single container has a capacity in excess of 660 gallons.

(3) Facilities which have a total storage capacity of 42000 gallons or less of oil and such total storage capacity is buried underground.

(4) Non-transportation-related onshore and offshore facilities, which, due to their location, could not reasonably be expected to discharge oil into or upon the navigable waters of the United States or adjoining shorelines.

(e) This part provides for the preparation and implementation of Spill Prevention Control and Countermeasure Plans prepared in accordance with § 112.7, designed to complement existing laws, regulations, rules, standards, policies and procedures pertaining to safety standards, fire prevention and pollution prevention rules, so as to form a comprehensive balanced Federal/State spill prevention program to minimize the potential for oil discharges. Compliance with this part does not in any way relieve the owner or operator of an onshore or an offshore facility from compliance with other Federal, State or local laws.

§ 112.2 Definitions.

For the purposes of this part:

(a) "Oil" means oil of any kind or in any form, including, but not limited to petroleum, fuel oil, sludge, oil refuse and oil mixed with wastes other than dredged spoil.

(b) "Discharge" includes but is not limited to, any spilling, leaking, pumping, pouring, emitting, emptying or dumping. For purposes of this part, the term "discharge" shall not include any discharge of oil which is authorized by a permit issued pursuant to Section 13 of the River and Harbor Act of 1899 (30 Stat. 1121, 33 U.S.C. 407), or Sections 402 or 405 of the FWPCA Amendments of 1972 (86 Stat. 816 et seq., 33 U.S.C. 1251 et seq.).

(c) "Onshore facility" means any facility of any kind located in, on, or under any land within the United States, other than submerged lands, which is not a transportation-related facility.

(d) "Offshore facility" means any facility of any kind located in, on, or under any of the navigable waters of the United States, which is not a transportation-related facility.

(e) "Owner or operator" means any person owning or operating an onshore facility or an offshore facility, and in the

case of any abandoned offshore facility, the person who owned or operated such facility immediately prior to such abandonment.

(f) "Person" includes an individual, firm, corporation, association, and a partnership.

(g) "Regional Administrator", means the Regional Administrator of the Environmental Protection Agency, or his designee, in and for the Region in which the facility is located.

(h) "Transportation-related" and "non-transportation-related" as applied to an onshore or offshore facility, are defined in the Memorandum of Understanding between the Secretary of Transportation and the Administrator of the Environmental Protection Agency, dated November 24, 1971, 36 FR 24080.

(i) "Spill event" means a discharge of oil into or upon the navigable waters of the United States or adjoining shorelines in harmful quantities as defined at 40 CFR Part 110.

(j) "United States" means the States, the District of Columbia, the Commonwealth of Puerto Rico, the Canal Zone, Guam, American Samoa, the Virgin Islands, and the Trust Territory of the Pacific Islands.

(k) The term "navigable waters" of the United States means "navigable waters" as defined in section 502(7) of the FWPCA, and includes:

(1) all navigable waters of the United States, as defined in judicial decisions prior to passage of the 1972 Amendments to the FWPCA (Pub. L. 92-500), and tributaries of such waters;

(2) interstate waters;

(3) intrastate lakes, rivers, and streams which are utilized by interstate travelers for recreational or other purposes; and

(4) intrastate lakes, rivers, and streams from which fish or shellfish are taken and sold in interstate commerce.

(l) "Vessel" means every description of watercraft or other artificial contrivance used, or capable of being used as a means of transportation on water, other than a public vessel.

§ 112.3 Requirements for preparation and implementation of Spill Prevention Control and Countermeasure Plans.

(a) Owners or operators of onshore and offshore facilities in operation on or before the effective date of this part that have discharged or could reasonably be expected to discharge oil in harmful quantities, as defined in 40 CFR Part 110, into or upon the navigable waters of the United States or adjoining shorelines, shall prepare a Spill Prevention Control and Countermeasure Plan (hereinafter "SPCC Plan"), in accordance with § 112.7. Except as provided for in paragraph (f) of this section, such SPCC Plan shall be prepared within six months after the effective date of this part and shall be fully implemented as soon as possible, but not later than one year after the effective date of this part.

(b) Owners or operators of onshore and offshore facilities that become operational after the effective date of this part, and that have discharged or could reasonably be expected to discharge oil in harmful quantities, as defined in 40 CFR Part 110, into or upon the navigable waters of the United States or adjoining shorelines, shall prepare an SPCC Plan in accordance with § 112.7. Except as provided for in paragraph (f) of this section, such SPCC Plan shall be prepared within six months after the date such facility begins operations and shall be fully implemented as soon as possible, but not later than one year after such facility begins operations.

(c) Onshore and offshore mobile or portable facilities such as onshore drilling or workover rigs, barge mounted offshore drilling or workover rigs, and portable fueling facilities shall prepare and implement an SPCC Plan as required by paragraphs (a), (b) and (d) of this section. The owner or operator of such facility need not prepare and implement a new SPCC Plan each time the facility is moved to a new site. The SPCC Plan for mobile facilities should be prepared in accordance with § 112.7, using good engineering practice, and when the mobile facility is moved it should be located and installed using spill prevention practices outlined in the SPCC Plan for the facility. The SPCC Plan shall only apply while the facility is in a fixed (non transportation) operating mode.

(d) No SPCC Plan shall be effective to satisfy the requirements of this part unless it has been reviewed by a Registered Professional Engineer and certified to by such Professional Engineer. By means of this certification the engineer, having examined the facility and being familiar with the provisions of this part, shall attest that the SPCC Plan has been prepared in accordance with good engineering practices. Such certification shall in no way relieve the owner or operator of an onshore or offshore facility of his duty to prepare and fully implement such Plan in accordance with § 112.7, as required by paragraphs (a), (b) and (c) of this section.

(e) Owners or operators of a facility for which an SPCC Plan is required pursuant to paragraphs (a), (b) or (c) of this section shall maintain a complete copy of the Plan at such facility if the facility is normally attended at least 8 hours per day, or at the nearest field office if the facility is not so attended, and shall make such Plan available for the Regional Administrator for on-site review during normal working hours.

(f) Extensions of time.

(1) The Regional Administrator may authorize an extension of time for the preparation and full implementation of an SPCC Plan beyond the time permitted for the preparation and implementation of an SPCC Plan pursuant to paragraphs (a), (b) or (c) of this section where he finds that the owner or operator of a facility subject to paragraphs (a), (b) or (c) of this section cannot fully comply with the requirements of this part as a result of either nonavailability of qualified personnel, or delays in construction or equipment delivery beyond the control and without the fault of such owner or operator or their respective agents or employees.

(2) Any owner or operator seeking an extension of time pursuant to paragraph (f) (1) of this section may submit a letter of request to the Regional Administrator. Such letter shall include:

(i) A complete copy of the SPCC Plan, if completed;

(ii) A full explanation of the cause for any such delay and the specific aspects of the SPCC Plan affected by the delay;

(iii) A full discussion of actions being taken or contemplated to minimize or mitigate such delay;

(iv) A proposed time schedule for the implementation of any corrective actions being taken or contemplated, including interim dates for completion of tests or studies, installation and operation of any necessary equipment or other preventive measures.

In addition, such owner or operator may present additional oral or written statements in support of his letter of request.

(3) The submission of a letter of request for extension of time pursuant to paragraph (f) (2) of this section shall in no way relieve the owner or operator from his obligation to comply with the requirements of § 112.3 (a), (b) or (c). Where an extension of time is authorized by the Regional Administrator for particular equipment or other specific aspects of the SPCC Plan, such extension shall in no way affect the owner's or operator's obligation to comply with the requirements of § 112.3 (a), (b) or (c) with respect to other equipment or other specific aspects of the SPCC Plan for which an extension of time has not been expressly authorized.

§ 112.4   Amendment of SPCC Plans by Regional Administrator.

(a) Notwithstanding compliance with § 112.3, whenever a facility subject to § 112.3 (a), (b) or (c) has: Discharged more than 1,000 U.S. gallons of oil into or upon the navigable waters of the United States or adjoining shorelines in a single spill event, or discharged oil in harmful quantities, as defined in 40 CFR Part 110, into or upon the navigable waters of the United States or adjoining shorelines in two spill events, reportable under section 311(b) (5) of the FWPCA, occurring within any twelve month period, the owner or operator of such facility shall submit to the Regional Administrator, within 60 days from the time such facility becomes subject to this section, the following:

(1) Name of the facility;

(2) Name(s) of the owner or operator of the facility;

(3) Location of the facility;

(4) Date and year of initial facility operation;

(5) Maximum storage or handling capacity of the facility and normal daily throughput;

(6) Description of the facility, including maps, flow diagrams, and topographical maps;

(7) A complete copy of the SPCC Plan with any amendments;

(8) The cause(s) of such spill, including a failure analysis of system or subsystem in which the failure occurred;

(9) The corrective actions and/or countermeasures taken, including an adequate description of equipment repairs and/or replacements;

(10) Additional preventive measures taken or contemplated to minimize the possibility of recurrence;

(11) Such other information as the Regional Administrator may reasonably require pertinent to the Plan or spill event.

(b) Section 112.4 shall not apply until the expiration of the time permitted for the preparation and implementation of an SPCC Plan pursuant to § 112.3 (a), (b), (c) and (f).

(c) A complete copy of all information provided to the Regional Administrator pursuant to paragraph (a) of this section shall be sent at the same time to the State agency in charge of water pollution control activities in and for the State in which the facility is located. Upon receipt of such information such State agency may conduct a review and make recommendations to the Regional Administrator as to further procedures, methods, equipment and other requirements for equipment necessary to prevent and to contain discharges of oil from such facility.

(d) After review of the SPCC Plan for a facility subject to paragraph (a) of this section, together with all other information submitted by the owner or operator of such facility, and by the State agency under paragraph (c) of this section, the Regional Administrator may require the owner or operator of such facility to amend the SPCC Plan if he finds that the Plan does not meet the requirements of this part or that the amendment of the Plan is necessary to prevent and to contain discharges of oil from such facility.

(e) When the Regional Administrator proposes to require an amendment to the SPCC Plan, he shall notify the facility operator by certified mail addressed to, or by personal delivery to, the facility owner or operator, that he proposes to require an amendment to the Plan, and shall specify the terms of such amendment. If the facility owner or operator is a corporation, a copy of such notice shall also be mailed to the registered agent, if any, of such corporation in the State where such facility is located. Within 30 days from receipt of such notice, the facility owner or operator may submit written information, views, and arguments on the amendment. After considering all relevant material presented, the Regional Administrator shall notify the facility owner or operator of any amendment required or shall rescind the notice. The amendment required by the Regional Administrator shall become part of the Plan 30 days after such notice, unless the Regional Administrator, for good cause, shall specify another effective date. The owner or operator of the facility shall implement the amendment of the Plan as soon as possible, but not later than six months after the amendment becomes part of the Plan, unless the Regional Administrator specifies another date.

(f) An owner or operator may appeal a decision made by the Regional Administrator requiring an amendment to an SPCC Plan. The appeal shall be made to the Administrator of the United States Environmental Protection Agency and must be made in writing within 30 days of receipt of the notice from the Regional Administrator requiring the amendment. A complete copy of the appeal must be sent to the Regional Administrator at the time the appeal is made. The appeal shall contain a clear and concise statement of the issues and points of fact in the case. It may also contain additional information which the owner or operator wishes to present in support of his argument. The Administrator or his designee may request additional information from the owner or operator, or from any other person. The Administrator or his designee may request additional information from the owner or operator, or from any other person. The Administrator or his designee shall render a decision within 60 days of receiving the appeal and shall notify the owner or operator of his decision.

§ 112.5 Amendment of Spill Prevention Control and Countermeasure Plans by owners or operators.

(a) Owners or operators of facilities subject to § 112.3 (a), (b) or (c) shall amend the SPCC Plan for such facility in accordance with § 112.7 whenever there is a change in facility design, construction, operation or maintenance which materially affects the facility's potential for the discharge of oil into or upon the navigable waters of the United States or adjoining shorelines. Such amendments shall be fully implemented as soon as possible, but not later than six months after such change occurs.

(b) Notwithstanding compliance with paragraph (a) of this section, owners and operators of facilities subject to § 112.3 (a), (b) or (c) shall complete a review and evaluation of the SPCC Plan at least once every three years from the date such facility becomes subject to this part. As a result of this review and evaluation, the owner or operator shall amend the SPCC Plan within six months of the review to include more effective prevention and control technology if: (1) Such technology will significantly reduce the likelihood of a spill event from the facility, and (2) if such technology has been field-proven at the time of the review.

(c) No amendment to an SPCC Plan shall be effective to satisfy the requirements of this section unless it has been certified by a Professional Engineer in accordance with § 112.3(d).

§ 112.6 Civil penalties.

Owners or operators of facilities subject to § 112.3 (a), (b) or (c) who violate the requirements of this part by failing or refusing to comply with any of the provisions of § 112.3, § 112.4, or § 112.5 shall be liable for a civil penalty of not more than $5,000 for each day that such violation continues. The Regional Administrator may assess and compromise such civil penalty. No penalty shall be assessed until the owner or operator shall have been given notice and an opportunity for hearing.

§ 112.7 Guidelines for the preparation and implementation of a Spill Prevention Control and Countermeasure Plan.

The SPCC Plan shall be a carefully thought-out plan, prepared in accordance with good engineering practices, and which has the full approval of management at a level with authority to commit the necessary resources. If the plan calls for additional facilities or procedures, methods, or equipment not yet fully operational, these items should be discussed in separate paragraphs, and the details of installation and operational start-up should be explained separately. The complete SPCC Plan shall follow the sequence outlined below, and include a discussion of the facility's conformance with the appropriate guidelines listed:

(a) A facility which has experienced one or more spill events within twelve months prior to the effective date of this part should include a written description of each such spill, corrective action taken and plans for preventing recurrence.

(b) Where experience indicates a reasonable potential for equipment failure (such as tank overflow, rupture, or leakage), the plan should include a prediction of the direction, rate of flow, and total quantity of oil which could be discharged from the facility as a result of each major type of failure.

(c) Appropriate containment and/or diversionary structures or equipment to prevent discharged oil from reaching a navigable water course should be provided. One of the following preventive systems or its equivalent should be used as a minimum:

(1) Onshore facilities.
(i) Dikes, berms or retaining walls sufficiently impervious to contain spilled oil
(ii) Curbing
(iii) Culverting, gutters or other drainage systems
(iv) Weirs, booms or other barriers
(v) Spill diversion ponds
(vi) Retention ponds
(vii) Sorbent materials
(2) Offshore facilities.
(i) Curbing, drip pans
(ii) Sumps and collection systems
(d) When it is determined that the installation of structures or equipment listed in § 112.7(c) to prevent discharged oil from reaching the navigable waters

is not practicable from any onshore or offshore facility, the owner or operator should clearly demonstrate such impracticability and provide the following:

(1) A strong oil spill contingency plan following the provision of 40 CFR Part 109.

(2) A written commitment of manpower, equipment and materials required to expeditiously control and remove any harmful quantity of oil discharged.

(e) In addition to the minimal prevention standards listed under § 112.7 (c), sections of the Plan should include a complete discussion of conformance with the following applicable guidelines, other effective spill prevention and containment procedures (or, if more stringent, with State rules, regulations and guidelines):

(1) *Facility drainage (onshore); (excluding production facilities).* (i) Drainage from diked storage areas should be restrained by valves or other positive means to prevent a spill or other excessive leakage of oil into the drainage system or inplant effluent treatment system, except where plan systems are designed to handle such leakage. Diked areas may be emptied by pumps or ejectors; however, these should be manually activated and the condition of the accumulation should be examined before starting to be sure no oil will be discharged into the water.

(ii) Flapper-type drain valves should not be used to drain diked areas. Valves used for the drainage of diked areas should, as far as practical, be of manual, open-and-closed design. When plant drainage drains directly into water courses and not into wastewater treatment plants, retained storm water should be inspected as provided in paragraph (e)(2)(iii) (B, C and D) before drainage.

(iii) Plant drainage systems from undiked areas should, if possible, flow into ponds, lagoons or catchment basins, designed to retain oil or return it to the facility. Catchment basins should not be located in areas subject to periodic flooding.

(iv) If plant drainage is not engineered as above, the final discharge of all in-plant ditches should be equipped with a diversion system that could, in the event of an uncontrolled spill, return the oil to the plant.

(v) Where drainage waters are treated in more than one treatment unit, natural hydraulic flow should be used. If pump transfer is needed, two "lift" pumps should be provided, and at least one of the pumps should be permanently installed when such treatment is continuous. In any event, whatever techniques are used facility drainage systems should be adequately engineered to prevent oil from reaching navigable waters in the event of equipment failure or human error at the facility.

(2) *Bulk storage tanks (onshore); (excluding production facilities).* (i) No

tank should be used for the storage of oil unless its material and construction are compatible with the material stored and conditions of storage such as pressure and temperature, etc.

(ii) All bulk storage tank installations should be constructed so that a secondary means of containment is provided for the entire contents of the largest single tank plus sufficient freeboard to allow for precipitation. Diked areas should be sufficiently impervious to contain spilled oil. Dikes, containment curbs, and pits are commonly employed for this purpose, but they may not always be appropriate. An alternative system could consist of a complete drainage trench enclosure arranged so that a spill could terminate and be safely confined in an in-plant catchment basin or holding pond.

(iii) Drainage of rainwater from the diked area into a storm drain or an effluent discharge that empties into an open water course, lake, or pond, and bypassing the in-plant treatment system may be acceptable if:

(A) The bypass valve is normally sealed closed.

(B) Inspection of the run-off rain water ensures compliance with applicable water quality standards and will not cause a harmful discharge as defined in 40 CFR 110.

(C) The bypass valve is opened, and resealed following drainage under responsible supervision.

(D) Adequate records are kept of such events.

(iv) Buried metallic storage tanks represent a potential for undetected spills. A new buried installation should be protected from corrosion by coatings, cathodic protection or other effective methods compatible with local soil conditions. Such buried tanks should at least be subjected to regular pressure testing.

(v) Partially buried metallic tanks for the storage of oil should be avoided, unless the buried section of the shell is adequately coated, since partial burial in damp earth can cause rapid corrosion of metallic surfaces, especially at the earth/air interface.

(vi) Aboveground tanks should be subject to periodic integrity testing, taking into account tank design (floating roof, etc.) and using such techniques as hydrostatic testing, visual inspection or a system of non-destructive shell thickness testing. Comparison records should be kept where appropriate, and tank supports and foundations should be included in these inspections. In addition, the outside of the tank should frequently be observed by operating personnel for signs of deterioration, leaks which might cause a spill, or accumulation of oil inside diked areas.

(vii) To control leakage through defective internal heating coils, the following factors should be considered and applied, as appropriate.

(A) The steam return or exhaust lines from internal heating coils which discharge into an open water course should be monitored for contamination, or passed through a settling tank, skimmer, or other separation or retention system.

(B) The feasibility of installing an external heating system should also be considered.

(viii) New and old tank installations should, as far as practical, be fail-safe engineered or updated into a fail-safe engineered installation to avoid spills. Consideration should be given to providing one or more of the following devices:

(A) High liquid level alarms with an audible or visual signal at a constantly manned operation or surveillance station: in smaller plants an audible air vent may suffice.

(B) Considering size and complexity of the facility, high liquid level pump cutoff devices set to stop flow at a predetermined tank content level.

(C) Direct audible or code signal communication between the tank gauger and the pumping station.

(D) A fast response system for determining the liquid level of each bulk storage tank such as digital computers, telepulse, or direct vision gauges or their equivalent.

(E) Liquid level sensing devices should be regularly tested to insure proper operation.

(ix) Plant effluents which are discharged into navigable waters should have disposal facilities observed frequently enough to detect possible system upsets that could cause an oil spill event.

(x) Visible oil leaks which result in a loss of oil from tank seams, gaskets, rivets and bolts sufficiently large to cause the accumulation of oil in diked areas should be promptly corrected.

(xi) Mobile or portable oil storage tanks (onshore) should be positioned or located so as to prevent spilled oil from reaching navigable waters. A secondary means of containment, such as dikes or catchment basins, should be furnished for the largest single compartment or tank. These facilities should be located where they will not be subject to periodic flooding or washout.

(3) Facility transfer operations, pumping, and in-plant process (onshore); (excluding production facilities). (i) Buried piping installations should have a protective wrapping and coating and should be cathodically protected if soil conditions warrant. If a section of buried line is exposed for any reason, it should be carefully examined for deterioration. If corrosion damage is found, additional examination and corrective action should be taken as indicated by the magnitude of the damage. An alternative would be the more frequent use of exposed pipe corridors or galleries.

(ii) When a pipeline is not in service, or in standby service for an extended time the terminal connection at the transfer point should be capped or blank-flanged, and marked as to origin.

(iii) Pipe supports should be properly designed to minimize abrasion and corrosion and allow for expansion and contraction.

(iv) All aboveground valves and pipelines should be subjected to regular examinations by operating personnel at which time the general condition of items, such as flange joints, expansion joints, valve glands and bodies, catch pans, pipeline supports, locking of valves, and metal surfaces should be assessed. In addition, periodic pressure testing may be warranted for piping in areas where facility drainage is such that a failure might lead to a spill event.

(v) Vehicular traffic granted entry into the facility should be warned verbally or by appropriate signs to be sure that the vehicle, because of its size, will not endanger above ground piping.

(4) Facility tank car and tank truck loading/unloading rack (onshore). (i) Tank car and tank truck loading/unloading procedures should meet the minimum requirements and regulation established by the Department of Transportation.

(ii) Where rack area drainage does not flow into a catchment basin or treatment facility designed to handle spills, a quick drainage system should be used for tank truck loading and unloading areas. The containment system should be designed to hold at least maximum capacity of any single compartment of a tank car or tank truck loaded or unloaded in the plant.

(iii) An interlocked warning light or physical barrier system, or warning signs, should be provided in loading/unloading areas to prevent vehicular departure before complete disconnect of flexible or fixed transfer lines.

(iv) Prior to filling and departure of any tank car or tank truck, the lowermost drain and all outlets of such vehicles should be closely examined for leakage, and if necessary, tightened, adjusted, or replaced to prevent liquid leakage while in transit.

(5) Oil production facilities (onshore). (i) Definition. An onshore production facility may include all wells, flowlines, separation equipment, storage facilities, gathering lines, and auxiliary non-transportation-related equipment and facilities in a single geographical oil or gas field operated by a single operator.

(ii) Oil production facility (onshore) drainage. (A) At tank batteries and central treating stations where an accidental discharge of oil would have a reasonable possibility of reaching navigable waters, the dikes or equivalent required under § 112.7(c)(1) should have drains closed and sealed at all times except when rainwater is being drained. Prior to drainage, the diked area should be inspected as provided in paragraph (e)(2)(iii)(B),(C), and (D). Accumulated oil on the rainwater should be picked up and returned to storage or disposed of in accordance with approved methods.

(B) Field drainage ditches, road ditches, and oil traps, sumps or skimmers, if such exist, should be inspected at regularly scheduled intervals for accumulation of oil that may have escaped from small leaks. Any such accumulations should be removed.

(iii) Oil production facility (onshore) bulk storage tanks. (A) No tank should be used for the storage of oil unless its material and construction are compatible with the material stored and the conditions of storage.

(B) All tank battery and central treating plant installations should be provided with a secondary means of containment for the entire contents of the largest single tank if feasible, or alternate systems such as those outlined in § 112.7(c) (1). Drainage from undiked areas should be safely confined in a catchment basin or holding pond.

(C) All tanks containing oil should be visually examined by a competent person for condition and need for maintenance on a scheduled periodic basis. Such examination should include the foundation and supports of tanks that are above the surface of the ground.

(D) New and old tank battery installations should, as far as practical, be fail-safe engineered or updated into a fail-safe engineered installation to prevent spills. Consideration should be given to one or more of the following:

(1) Adequate tank capacity to assure that a tank will not overfill should a pumper/gauger be delayed in making his regular rounds.

(2) Overflow equalizing lines between tanks so that a full tank can overflow to an adjacent tank.

(3) Adequate vacuum protection to prevent tank collapse during a pipeline run.

(4) High level sensors to generate and transmit an alarm signal to the computer where facilities are a part of a computer production control system.

(iv) Facility transfer operations, oil production facility (onshore). (A) All above ground valves and pipelines should be examined periodically on a scheduled basis for general condition of items such as flange joints, valve glands and bodies, drip pans, pipeline supports, pumping well polish rod stuffing boxes, bleeder and gauge valves.

(B) Salt water (oil field brine) disposal facilities should be examined often, particularly following a sudden change in atmospheric temperature to detect possible system upsets that could cause an oil discharge.

(C) Production facilities should have a program of flowline maintenance to prevent spills from this source. The program should include periodic examinations, corrosion protection, flowline replacement, and adequate records, as appropriate, for the individual facility.

(6) Oil drilling and workover facilities (onshore) (i) Mobile drilling or workover equipment should be positioned or located so as to prevent spilled oil from reaching navigable waters.

(ii) Depending on the location, catchment basins or diversion structures may be necessary to intercept and contain spills of fuel, crude oil, or oily drilling fluids.

(iii) Before drilling below any casing string or during workover operations, a blowout prevention (BOP) assembly and well control system should be installed that is capable of controlling any well head pressure that is expected to be encountered while that BOP assembly is on the well. Casing and BOP installations should be in accordance with State regulatory agency requirements.

(7) Oil drilling, production, or workover facilities (offshore). (i) Definition: "An oil drilling, production or workover facility (offshore)" may include all drilling or workover equipment, wells, flowlines, gathering lines, platforms, and auxiliary nontransportation - related equipment and facilities in a single geographical oil or gas field operated by a single operator.

(ii) Oil drainage collection equipment should be used to prevent and control small oil spillage around pumps, glands, valves, flanges, expansion joints, hoses, drain lines, separators, treaters, tanks, and allied equipment. Drains on the facility should be controlled and directed toward a central collection sump or equivalent collection system sufficient to prevent discharges of oil into the navigable waters of the United States. Where drains and sumps are not practicable oil contained in collection equipment should be removed as often as necessary to prevent overflow.

(iii) For facilities employing a sump system, sump and drains should be adequately sized and a spare pump or equivalent method should be available to remove liquid from the sump and assure that oil does not escape. A regular scheduled preventive maintenance inspection and testing program should be employed to assure reliable operation of the liquid removal system and pump start-up device. Redundant automatic sump pumps and control devices may be required on some installations.

(iv) In areas where separators and treaters are equipped with dump valves whose predominant mode of failure is in the closed position and pollution risk is high, the facility should be specially equipped to prevent the escape of oil. This could be accomplished by extending the flare line to a diked area if the separator is near shore, equipping it with a high liquid level sensor that will automatically shut-in wells producing to the separator, parallel redundant dump valves, or other feasible alternatives to prevent oil discharges.

(v) Atmospheric storage or surge tanks should be equipped with high liquid level sensing devices or other acceptable alternatives to prevent oil discharges.

(vi) Pressure tanks should be equipped with high and low pressure sensing devices to activate an alarm and/or control the flow or other acceptable alternatives to prevent oil discharges.

(vii) Tanks should be equipped with suitable corrosion protection.

(viii) A written procedure for inspecting and testing pollution prevention equipment and systems should be prepared and maintained at the facility. Such procedures should be included as part of the SPCC Plan.

(ix) Testing and inspection of the pollution prevention equipment and systems at the facility should be conducted by the owner or operator on a scheduled periodic basis commensurate with the complexity, conditions and circumstances of the facility or other appropriate regulations.

(x) Surface and subsurface well shut-in valves and devices in use at the facility should be sufficiently described to determine method of activation or control, e.g., pressure differential, change in fluid or flow conditions, combination of pressure and flow, manual or remote control mechanisms. Detailed records for each well, while not necessarily part of the plan should be kept by the owner or operator.

(xi) Before drilling below any casing string, and during workover operations a blowout preventer (BOP) assembly and well control system should be installed that is capable of controlling any wellhead pressure that is expected to be encountered while that BOP assembly is on the well. Casing and BOP installations should be in accordance with State regulatory agency requirements.

(xii) Extraordinary well control measures should be provided should emergency conditions, including fire, loss of control and other abnormal conditions, occur. The degree of control system redundancy should vary with hazard exposure and probable consequences of failure. It is recommended that surface shut-in systems have redundant or "fail close" valving. Subsurface safety valves may not be needed in producing wells that will not flow but should be installed as required by applicable State regulations.

(xiii) In order that there will be no misunderstanding of joint and separate duties and obligations to perform work in a safe and pollution free manner, written instructions should be prepared by the owner or operator for contractors and subcontractors to follow whenever contract activities include servicing a well or systems appurtenant to a well or pressure vessel. Such instructions and procedures should be maintained at the offshore production facility. Under certain circumstances and conditions such contractor activities may require the presence at the facility of an authorized representative of the owner or operator who would intervene when necessary to prevent a spill event.

(xiv) All manifolds (headers) should be equipped with check valves on individual flowlines.

(xv) If the shut-in well pressure is greater than the working pressure of the flowline and manifold valves up to and including the header valves associated with that individual flowline, the flowline should be equipped with a high pressure sensing device and shutin valve at the wellhead unless provided with a pressure relief system to prevent over pressuring.

(xvi) All pipelines appurtenant to the facility should be protected from corrosion. Methods used, such as protective coatings or cathodic protection, should be discussed.

(xvii) Sub-marine pipelines appurtenant to the facility should be adequately protected against environmental stresses and other activities such as fishing operations.

(xviii) Sub-marine pipelines appurtenant to the facility should be in good operating condition at all times and inspected on a scheduled periodic basis for failures. Such inspections should be documented and maintained at the facility.

(8) *Inspections and records.* Inspections required by this part should be in accordance with written procedures developed for the facility by the owner or operator. These written procedures and a record of the inspections, signed by the appropriate supervisor or inspector, should be made part of the SPCC Plan and maintained for a period of three years.

(9) *Security (excluding oil production facilities).* (i) All plants handling, processing, and storing oil should be fully fenced, and entrance gates should be locked and/or guarded when the plant is not in production or is unattended.

(ii) The master flow and drain valves and any other valves that will permit direct outward flow of the tank's content to the surface should be securely locked in the closed position when in non-operating or non-standby status.

(iii) The starter control on all oil pumps should be locked in the "off" position or located at a site accessible only to authorized personnel when the pumps are in a non-operating or non-standby status.

(iv) The loading/unloading connections of oil pipelines should be securely capped or blank-flanged when not in service or standby service for an extended time. This security practice should also apply to pipelines that are emptied of liquid content either by draining or by inert gas pressure.

(v) Facility lighting should be commensurate with the type and location of the facility. Consideration should be given to: (A) Discovery of spills occurring during hours of darkness, both by operating personnel, if present, and by non-operating personnel (the general public, local police, etc.) and (B) prevention of spills occurring through acts of vandalism.

(10) *Personnel, training and spill prevention procedures.* (i) Owners or operators are responsible for properly instructing their personnel in the operation and maintenance of equipment to prevent the discharges of oil and applicable pollution control laws, rules and regulations.

(ii) Each applicable facility should have a designated person who is accountable for oil spill prevention and who reports to line management.

(iii) Owners or operators should schedule and conduct spill prevention briefings for their operating personnel at intervals frequent enough to assure adequate understanding of the SPCC Plan for that facility. Such briefings

should highlight and describe known spill events or failures, malfunctioning components, and recently developed precautionary measures.

APPENDIX

Memorandum of Understanding between the Secretary of Transportation and the Administrator of the Environmental Protection Agency.

SECTION II—DEFINITIONS

The Environmental Protection Agency and the Department of Transportation agree that for the purposes of Executive Order 11548, the term:

(1) "Non-transportation-related onshore and offshore facilities" means:

(A) Fixed onshore and offshore oil well drilling facilities including all equipment and appurtenances related thereto used in drilling operations for exploratory or development wells, but excluding any terminal facility, unit or process integrally associated with the handling or transferring of oil in bulk to or from a vessel.

(B) Mobile onshore and offshore oil well drilling platforms, barges, trucks, or other mobile facilities including all equipment and appurtenances related thereto when such mobile facilities are fixed in position for the purpose of drilling operations for exploratory or development wells, but excluding any terminal facility, unit or process integrally associated with the handling or transferring of oil in bulk to or from a vessel.

(C) Fixed onshore and offshore oil production structures, platforms, derricks, and rigs including all equipment and appurtenances related thereto, as well as completed wells and the wellhead separators, oil separators, and storage facilities used in the production of oil, but excluding any terminal facility, unit or process integrally associated with the handling or transferring of oil in bulk to or from a vessel.

(D) Mobile onshore and offshore oil production facilities including all equipment and appurtenances related thereto as well as completed wells and wellhead equipment, piping from wellheads to oil separators, oil separators, and storage facilities used in the production of oil when such mobile facilities are fixed in position for the purpose of oil production operations, but excluding any terminal facility, unit or process integrally associated with the handling or transferring of oil in bulk to or from a vessel.

(E) Oil refining facilities including all equipment and appurtenances related thereto as well as in-plant processing units, storage units, piping, drainage systems and waste treatment units used in the refining of oil, but excluding any terminal facility, unit or process integrally associated with the handling or transferring of oil in bulk to or from a vessel.

(F) Oil storage facilities including all equipment and appurtenances related thereto as well as fixed bulk plant storage, terminal oil storage facilities, consumer storage, pumps and drainage systems used in the storage of oil, including inline or breakout storage tanks needed for the continuous operation of a pipeline system and any terminal facility, unit or process integrally associated with the handling or transferring of oil in bulk to or from a vessel.

(G) Industrial, commercial, agricultural or public facilities which use and store oil, but excluding any terminal facility, unit or process integrally associated with the handling or transferring of oil in bulk to or from a vessel.

(H) Waste treatment facilities including in-plant pipelines, effluent discharge lines, and storage tanks, but excluding waste treatment facilities located on vessels and terminal storage tanks and appurtenances for the reception of oily ballast water or tank washings from vessels and associated systems used for off-loading vessels.

(I) Loading racks, transfer hoses, loading arms and other equipment which are appurtenant to a nontransportation-related facility or terminal facility and which are used to transfer oil in bulk to or from highway vehicles or railroad cars.

(J) Highway vehicles and railroad cars which are used for the transport of oil exclusively within the confines of a nontransportation-related facility and which are not intended to transport oil in interstate or intrastate commerce.

(K) Pipeline systems which are used for the transport of oil exclusively within the confines of a nontransportation-related facility or terminal facility and which are not intended to transport oil in interstate or intrastate commerce, but excluding pipeline systems used to transfer oil in bulk to or from a vessel.

(2) "transportation-related onshore and offshore facilities" means:

(A) Onshore and offshore terminal facilities including transfer hoses, loading arms and other equipment and appurtenances used for the purpose of handling or transferring oil in bulk to or from a vessel as well as storage tanks and appurtenances for the reception of oily ballast water or tank washings from vessels, but excluding terminal waste treatment facilities and terminal oil storage facilities.

(B) Transfer hoses, loading arms and other equipment appurtenant to a nontransportation-related facility which is used to transfer oil in bulk to or from a vessel.

(C) Interstate and intrastate onshore and offshore pipeline systems including pumps and appurtenances related thereto as well as in-line or breakout storage tanks needed for the continuous operation of a pipeline system, and pipelines from onshore and offshore oil production facilities, but excluding onshore and offshore piping from wellheads to oil separators and pipelines which are used for the transport of oil exclusively within the confines of a nontransportation-related facility or terminal facility and which are not intended to transport oil in interstate or intrastate commerce or to transfer oil in bulk to or from a vessel.

(D) Highway vehicles and railroad cars which are used for the transport of oil in interstate or intrastate commerce and the equipment and appurtenances related thereto, and equipment used for the fueling of locomotive units, as well as the rights-of-way on which they operate. Excluded are highway vehicles and railroad cars and motive power used exclusively within the confines of a nontransportation-related facility or terminal facility and which are not intended for use in interstate or intrastate commerce.

[FR Doc.73-25448 Filed 12-10-73;8:45 am]

PART 110—DISCHARGE OF OIL

AUTHORITY: The provisions of this Part 110 issued under sec. 11(b)(3), as amended, 84 Stat. 92; 33 U.S.C. 1161.

§ 110.1   Definitions.

As used in this part, the following terms shall have the meaning indicated below:

(a) "Oil" means oil of any kind or in any form, including, but not limited to, petroleum, fuel oil, sludge, oil refuse, oil mixed with ballast or bilge, and oil mixed with wastes other than dredged spoil;

(b) "Discharge" includes, but is not limited to, any spilling, leaking, pumping, pouring, emitting, emptying or dumping;

(c) "Vessel" means every description of watercraft or other artificial contrivance used, or capable of being used, as a means of transportation on water other than a public vessel;

(d) "Public vessel" means a vessel owned or bare-boat chartered and operated by the United States, or by a State or political subdivision thereof, or by a foreign nation, except when such vessel is engaged in commerce;

(e) "United States" means the States, the District of Columbia, the Commonwealth of Puerto Rico, the Canal Zone, Guam, American Samoa the Virgin Islands, and the Trust Territory of the Pacific Islands;

(f) "Person" includes an individual, firm, corporation, association, and a partnership;

(g) "Contiguous zone" means the entire zone established or to be established by the United States under article 24 of the Convention on the Territorial Sea and the Contiguous Zone;

(h) "Onshore facility" means any facility (including, but not limited to motor vehicles and rolling stock) of any kind located in, on, or under, any land within the United States other than submerged land;

(i) "Offshore facility" means any facility of any kind located in, on, or under, any of the navigable waters of the United States other than a vessel or public vessel;

(j) "Applicable water quality standards" means water quality standards adopted pursuant to section 10(c) of the Federal Act and State-adopted water quality standards for waters which are not interstate within the meaning of that Act.

(k) "Federal Act" means the Federal Water Pollution Control Act, as amended, 33 U.S.C. 1151. et seq.

(l) "Sheen" means an iridescent appearance on the surface of water.

(m) "Sludge" means an aggregate of oil or oil and other matter of any kind in any form other than dredged spoil having a combined specific gravity equivalent to or greater than water.

§ 110.2   Applicability.

The regulations of this part apply to the discharge of oil into or upon the navigable waters of the United States, adjoining shorelines or into or upon the waters of the contiguous zone, prohibited by section 11(b) of the Federal Act.

§ 110.3   Discharge into navigable waters harmful.

For purposes of section 11(b) of the Federal Act, discharges of such quantities of oil into or upon the navigable waters of the United States or adjoining shorelines determined to be harmful to the public health or welfare of the United States, at all times and locations and under all circumstances and conditions, except as provided in section 110.6 of this part, include discharges which:

(a) Violate applicable water quality standards, or

(b) Cause a film or sheen upon or discoloration of the surface of the water or adjoining shorelines or cause a sludge or emulsion to be deposited beneath the surface of the water or upon adjoining shorelines.

§ 110.4   Discharge into contiguous zone harmful.

For purposes of section 11(b) of the Federal Act, discharges of such quantities of oil into or upon the waters of the contiguous zone determined to be harmful to the public health or welfare of the United States, at all times and locations and under all circumstances and conditions, except as provided in section 110.6 of this part, include discharges which:

(a) Violate applicable water quality standards in navigable waters of the United States, or

(b) Cause a film or sheen upon or discoloration of the surface of the water or adjoining shorelines or cause a sludge or emulsion to be deposited beneath the surface of the water or upon adjoining shorelines.

§ 110.5   Discharge prohibited.

As provided in section 11(b)(2) of the Federal Act, no person shall discharge or cause or permit to be discharged into or upon the navigable waters of the United States, adjoining shorelines, or into or upon the waters of the contiguous zone any oil, in harmful quantities as determined in §§ 110.3 and 110.4 of this part, except as the same may be permitted in the contiguous zone under Article IV of the International Convention for the Prevention of Pollution of the Sea by Oil, 1954, as amended.

§ 110.6   Exception for vessel engines.

For purposes of section 11(b) of the Federal Act, discharges of oil from a properly functioning vessel engine are not deemed to be harmful; but such oil accumulated in a vessel's bilges shall not be so exempt.

§ 110.7   Dispersants.

Addition of dispersants or emulsifiers to oil to be discharged which would circumvent the provisions of this part is prohibited.

§ 110.8   Demonstration projects.

Notwithstanding any other provisions of this part, the Administrator of the Environmental Protection Agency may permit the discharge of oil into or upon the navigable waters of the United States, adjoining shorelines, or into or upon the waters of the contiguous zone, in connection with research, demonstration projects, or studies relating to the prevention, control, or abatement of oil pollution.

§ 110.9   Notice.

Any person in charge of any vessel or onshore or offshore facility shall, as soon as he has knowledge of any discharge of oil from such vessel or facility in violation of § 110.5 of this part, immediately notify the U.S. Coast Guard of such discharge in accordance with such procedures as the Secretary of Transportation may prescribe.

Appendix B–Bibliography

Bulk Handling

Menon, P. K. _Symposium on Bulk Handling of Raw Materials_. Proceeding, Calcutta. xxiv, 213 p. LC Call No: TN5 .S85 1969.

Coatings/Linings for Bulk Storage Tanks

Battelle Columbus Laboratories. _Evaluation of Methods for Measuring and Controlling Hydrocarbon Emissions for Petroleum Storage Tanks_. November, 1976. PB-262 789.

Naval Research Laboratory. _Development of Organic Coatings for Use as Linings of Bulk Fuel Storage Tanks_. 23 p. AD-A061 392/7GA.

Electrostatic Ignition During Loading/Unloading

Precautions Against Electrostatic Ignition During Loading of Tank Truck Motor Vehicles, Second Edition, 1975. American Petroleum Institute Bulletin 1003. This bulletin is designed to set forth precautions which should be taken to minimize the possibility of electrostatic ignition during loading of petroleum products into tank trucks. It is not intended to be a detailed specification or instruction for tank truck loading; nor is it intended to guarantee freedom from accidents.

Equipment Design of Refineries and Plants

72-50245. Evans, Frank. L. 1979. _Equipment Design Handbook for Refineries and Chemical Plants_. Houston, Texas. Book Division, Gulf Publishing Company, v. i, ill. LC Call No: TD690.3 .E89.

79-50251. _Energy Management Handbook for Petroleum Refineries, Gas Processing, and Petrochemical Plants_. Houston, Texas. LC Call No: TP690.3 .E89 1979.

Fire Protection In Plants

72-94065. Vervalin, Charles H. 1973. _Pollution Control in the Petroleum Industry_. Park Ridge, New Jersey, Noyes Data Corpoaration. LC Call NO: TD888.P4 J66.

317

Inplant Pipelines

Oil Pipeline Measurement and Storage Practices, Vol. III, 1955. American Petroleum Institute Publication. Contains chapters on tank construction, stripping tanks, tank gaging, tank maintenance, and quantity measurement of liquid petroleum with meters.

Marine Loading/Unloading

78-62-54. Alaska Department of Environmental Conservation. 1977. Tank Vessels and Marine Terminal Facilities for Oil and Liquefied Natural Gas: A Selected Bibliography. Juneau, Alaska Department of Environmental Conservation.

79-01454. Burklin, C. R. etal. Background Information on Hydrocarbon Emissions from Marine Terminal Operations. November, 1976. IV. PB-264 381.

Fourth New England Coastal Zone Management Conference, 4th Annual, Durham, New Hampshire. "Perspectives on Oil Refineries and Offshore Unloading Facilities." May 13-14, 1974. Mary Louis Hunter, Editor.

047435. Kilgren, Karl H., Thomas, A. Hydrocarbon Emissions During Marine Loading of Crude Oils. Proceedings Air Pollution Control Association, 71st Annual Meeting, Houston, Texas, June, 1978.

027198. International Petroleum Times. "Loading and Discharge Marine Terminals for European Crude." 1978. Pp. 15, 18, 33.

74-158901. International Oil Tanker Terminal Safety Group. International Oil Tanker and Terminal Safety Guide. London, England. Distributed by the Institute of Petroleum. LC Call No: VM455 .I46 1971.

05892. McGrath, P. New Concepts for Design of Very Large Storage Tanks. Design concepts are given which permit the extension of tank diameters and capabilities substantially beyond the maximum presently obtainable sizes, under the API standard 650 design rules. Proceedings American Petroleum Institute Refinery Department, Midyear Meeting, 41st, Los Angeles, May, 1976. Published by API (V55) P. 407-425.

72-178869. Savory, A. J. Conference on Tanker and Bulk Carrier Terminals. Twelfth Conference on Tanker and Bulk Carrier Tanker Terminals, November, 1969. Institution of Civil Engineers. 109 p. LC Call No: TC365 .C65 1969.

79-00063. National Technical Information Service (NTIS). Background information on national and regional hydrocarbon emissions from marine terminal transfer operations. PB-275 484.

75-328745. Oil Companies International Marine Forum. International Oil Tanker and Terminal Safety Guide. Second revised edition. London, Applied Science Publishers, 1974. LC Call No: VM455 .I46 1974.

79-309075. Oil Companies International Marine Forum, 1978. International Safety Guide for Oil Tankers and Terminals. London, Witherby, 1978. LC Call No: VM455 .I45 1978.

75-328745. Oil Companies International Marine Forum, 1975. International Oil Tanker and Terminal Safety Guide. Second Edition. New York, Wiley, 1975. LC Call No: VM455 .034 1975.

New England Interstate Water Pollution Control Commission. Technical Advisory Board. 1971. Uniform Guidelines for Preventing and Control of Oil Spills, and for Oil Terminal and Vessel Handling of Petroleum and Petroleum Products. Boston, Massachusetts.

*LRS78-12299. Port Safety and Tank Vessel Safety. Hearings. 95th Congress, Second Session. Washington, D.C. U.S. Government Printing Office, 1978. 571 p.

Maritime Pollution Control

International Conference on Marine Pollution. London, 1973. Preparatory meeting for the International Conference on Marine Pollution, International Maritime Consultative Organization. Contents: Minimization of accidental spillages of oil and other noxious substances from ships; draft resolution, international pollution of the sea and accidental spillages. GC 1080.173.

Wardley-Smith, J. 1976. The Control of Oil Pollution on the Sea and Inland Waters: The Effect of Oil Spills on the Marine Environment and Methods of Dealing with Them. TD 427 .P4W375.

Monobuoy Operation

77-365875. Dames & Moore. Environmental Report. Houston, Texas, Seadock, 1976. LC Call No: TD195.P4 D35.

Oil Pipeline Pumping Stations

Oil Pipeline Pumping Station Operation. Vol. IV, 1956. Covers the installation, operation, and maintenance of prime movers, pumps, and auxiliary equipment found in oil pipeline pumping stations. While internal-combustion engineers and reciprocating pumps are covered, emphasis is on the electrified station with automatic controls and centrifugal pumps which characterize the postwar installations.

*Library Research Service, Library of Congress, Washington, D.C.

Overland Pipelines

Oil Pipeline Construction and Maintenance. Vol. II, Second Edition 1973.
Presents chapter on capacity, design and specifications, and accident
prevention. Available through API.

Recommended Practices for Liquid Petroleum Pipelines Crossing Railraods
and Highways, Fifth Edition, 1980. American Petroleum Institute R.P.
1102. This recommended practice should be considered as a guide for the
design, installation, inspection, and testing required to ensure safe
crossings of liquid petroleum pipelines under railraods and highways. The
practice applies to the construction of pipelines under existing railroads
and highways and to the adjustment of existing pipelines due to the construction of new railroads and highways. Also included are nomographs
for determining the circumferential stress caused by external loads in
uncased carrier pipe with an internal pressure at railroad and highway
crossings, as well as charts showing the recommended thicknesses for
flexible casing in bored crossings.

LRS78-21415. U.S. Bureau of Land Management. Crude Oil Transportation
System. Draft environmental statement. Port Angeles, Washington to
Clearbrook, Minnesota (as proposed by Northern Tier Pipeline Company).
Washington, 1978. Assesses the environmental economic impact of the
Northern Tier Pipeline Company's proposal to transport Alaskan oil by
pipeline from a marine terminal at Port Angeles, Washington to Clear-
brook, Minnesota. Discusses the possibility of oil spills, air pollu-
tant emissions, and destruction of wildlife habitat.

Pipeline Safety

LRS76-2880. American Water Works Association Journal. "Earthquake:
Correlation between pipeline damage and geologic environment." 1976.
Pp. 165-167.

LRS79-14213. Pipeline Safety Act of 1979. Hearings, 96th Congress,
First session on H.R. 2207 and H.R. 51. May 1 and June 8, 1979.
Washington, U.S. Government Printing Office, 1979. 316 p.

LRS79-9674. Pipeline Safety Act of 1979. Hearings, 96th Congress, First
Sessions on S. 411. Washington, U.S. Government Printing Office, 1979.
246 p.

LRS78-3773. Pipeline Safety--Need for a Stronger Federal Effort. 1979
Report. Washington, D.C. 19 p.

Petroleum Marketing Terminals

4825. A Survey of Petroleum Marketing Terminals. This report was prepared under the direction of Task Force W-19, API Environmental Affairs
Department, with the assistance of its contractor, Engineering Sciences,
Inc., and the cooperation of 15 member companies of the American Petroleum
Institute. It presents the results of a survey of the physical and operating characteristics of 76 petroleum marketing terminals and of the oil
and grease content in their wastewater discharges.

Pollution Control

D'Allesandro, P. L. and Cobb, C. B. "Oil Spill Control." Hydrocarbon
Processing, p. 145-148. (March, 1976)

Department of the Navy. Oil Spill Control for Inland Waters and Harbors.
Report NAVFAX P-908, Alexandria, Virginia Naval Facilities Engineering
Command (January, 1977)

67-19834. Beychok, Milton R. 1967. Aqueous Wastes from Petroleum and
Petrochemical Plants. London, Wiley, 1967. LC Call No: TP690.8 .B4.

Department of Trade. 1974. Manual on the Avoidance of Pollution of the
Sea by Oil. V. 1, 22 pages. London HMSO.

Garrett, M. J. and Smith, J. Wardley. Oil Spills from Tankers. Report
of the symposium on prevention of marine pollution from ships. Acapulco,
Mexico. U.S. Office of Environment and Systems, 1976. Washington, D.C.
(CGWEP1-77).

Iammartino, N. R. Chemical Engineering. "Oil Spill Control Nears for Two
Pesky Problems." Pages 76-80, May 10, 1976.

74-172600. Interstate Oil Compact Commission. 1966. Research Committee
Subcommittee on Water Problems Associated with Oil Production in the
United States. Oklahoma City, Interstate Oil Compact Commission. LC
Call No: TD427.P4 I52.

Joint Conference on Prevention and Control of Oil Spills. Proceedings at
Washington, D.C. 1971. GC1080.C6.

Joint Conference on Prevention and Control of Oil Spills. Proceedings at
New York. 1969. GC1085.J6.

GC1085.L585. Little, Arthur D., Inc. 1971. Regulations, Practices, and
Plans for the Prevention of Spills of Oil and Hazardous Polluting Substances.
Volume I. Washington, D.C. U.S. Government Printing Office.

Milgram, Jerome H. 1974. Evaluation of the Strength and Safekeeping Ability of Pollution Control Barriers. J. F. O'Dea, Massachusetts Institute of Technology Washington USCG Office of Research and Development. Available through NTIS, V393.A3055 No. CG-D-55-75.

Oil Conference on Prevention and Control of Oil Pollution. San Farncisco. National conventioneer issued at the Conference on Prevention and Control of Oil Pollution, 1975--instrumentation, equipment, and supplies exposition product ideas book. Allison Park, Pennsylvania. Publishers for Conventions, Inc.

Wardley-Smith, J. The Control of Oil Pollution. 1976. London, Graham and Trotman, Ltd.

Little, Arthur D., Inc. 1974. Learning Systems, Guide to Water Cleanup: Materials and Methods. Cambridge, Massachusetts.

Port and Coastal Structures

Johnson, C.T. Coastal Structure. Special conference on the design, construction, maintenance, and performance of port and coastal structures. Alexandria, Virginia, March 14-16, 1979. Pages 537-546.

Spill Prevention: General

4283. Environmental Research Annual Status Report. Since a continuing concern of the petroleum industry is the protection and improvement of the environment, API sponsors a substantial number of scientific research projects in the environmental area. This report summarizes such projects which have been recently completed or are on-going. An API Publication.

4041. Industrial Oil Waste Control (API-ASLE Handbook). A collection of articles including "Industry's Oil Waste Problem," "Standards for Effective Oil Waste Control," "Control of Oil Waste at the Source," "Oily Wastewater Treatment," and "Disposal of Oil Wastes."

International Conference on Tanker Safety and Pollution Prevention. London, England. Tanker safety and pollution prevention: Final act of the conference with attachments, including the protocol of 1978 relating to the International Convention for the safety of life at sea, 1974 and the protocol of 1978 relating to the International Convention for the Prevention of Pollution from Ships, 1973. London: Intergovernmental Maritime Consultative Organization. VK 200.1775 1978.

Joint Conference on Prevention and Control of Oil Spills. New York, 1969. Proceedings. American Petroleum Institute. GC 1080 .J64 1969.

Joint Conference on Prevention and Control of Oil Spills. Washington, D.C. Proceedings. American Petroleum Institute. GC 1080 .J64 1971.

Joint Conference on Prevention and Control of Oil Spills. Washington, D.C. Proceedings. American Petroleum Institute. GC 1080 .J64 1973.

Environmental Protection Agency. Oil Pollution of Rivers and Harbors. Book Catalog V. II, 1974.

72-75239. Jones, H. R. 1973. Pollution Control in the Petroleum Industry. Park Ridge, New Jersey, Noyes Data Corporation. LC Call No: TD88.P4 J66.

Little, Arthur D. Inc., 1971. The Prevention of Spills of Oils and Chemicals into Baltimore Harbor and Environment. Report to Maryland Environmental Service. Cambridge, Massachusetts. GC 1211 .L58.

Milgram, J. (Sea Grant Program-MIT.) Being Prepared for Future Argo Merchants. Rockville, Maryland, National Oceanoic and Atmospheric Administration. April, 1977.

Oil Spill Prevention Institute. 1978. Test and Research Facilities of the Oil Spill Prevention Institute. The Shipbuilding Research Center of Japan.

4225. Oil Spill Prevention Primer. This booklet outlines in a general way procedures designed to reduce the risks of an oil spill. It deals with controls whcih may be found in most industry operations. API Publication.

4271. Oil Spill Control Course (December 1975). The purpose of this course is to provide the trainee with information and training necessary for handling an oil spill within the capabilities of manpower and equipment at a company facility. API Course.

Oil Spill Conference. New Orleans, 1977. Sponsored by American Petroleum Institute. "Prevention, Behavior, Control, and Cleanup."

Oil Spill Liability and Compensation. Hearings from 95th Congress, Committee on Commerce, Science, and Transportation. June, 1977. Washington, U.S. Government Printing Office. 346 p. LRS 77-11270.

National Petroleum News. January, 1980. "Brooklyn Spill: 17 Million Gallons Under the Sidewalks of New York," Pages 34, 35, 58, and 59.

72-171291. Presentation for the Inquiry into Pollution Control of Petroleum Refineries in British Columbia, Vancouver, BC. 1972. LC Call No: TD899 . P4 P73.

Potter, Jeffrey. 1973. "Disaster by Oil. Oil Spills: Why They Happen, What They Do, How We Can End Them." GC1085 .P67.

Sittig, Marshall. Oil Spill Prevention and Removal Handbook. 1974. Noyes Data Corporation, Park Ridge, New Jersey.

Sittig, Marshall. Petroleum Transportation and Production: Oil Spill and Pollution Control. Noyes Data Corporation. Park Ridge, New Jersey.

Smith, Wardley, J. 1979. The Prevention of Oil Pollution. John Wiley and Sons, New York. TD .P4P83.

Voluntary Environmental Activities of Large Chemical Companies to Address and Control Industrial Chemicals. US/EPA. PB-271 907/8BE.

74-179973. Institute of Petroleum, London. 1971. Gas Evaluation-- Tanker and Terminal Safety. LC Call No: VM455 G.37.

Trentacoste, Nicholas P. 1980. Spill Prevention, Control and Counter- measure Practices at Small Petroleum Facilities.

Storage Tank Seepage

European Model Code of Safe Practice for the Prevention of Ground and Sur- face Water Pollution by Oil from Storage Tanks and During the Transport of Oil. London: Applied Science Publishers. 1974. TP 692.5 .E95. Prepared by a working group formed following a meeting of representatives of European technical organizations held at the Institute of Petroleum, London.

Tank Bottom Failures

063507. Edwards, H. R., Diesterte, R. J. Petroleum Engineer. May, 1979. "How Reinforced Plastic Lining Repairs Petroleum Storage Tanks." This article discusses some of the reasons for tank bottom failure, gives case history examples, and provides detailed procedures for coating steel bottoms with fiberglass reinforced plastics. Volume 51, n. 6., pages 112-126.

Card, J. C. 1975. Marine Technology. "Effectiveness of Double Bottoms in Preventing Oil Outflow from Tanker Bottom Damage Incidents." Pages 60-64.

RP 1004. Bottom Loading and Vapor Recovery for MC-306 Tank Motor Vehicles, Fourth Edition, 1977 (Supersedes Third Edition, 1975.) The objective of this recommended practice is to provide an industry standard for bottom loading and vapor recovery of proprietary and hired carrier DOT MC-306 tank vehicles at terminals operated by more than one supplier. It is intended to guide the manufacturer and operator of a tank vehicle as to the uniform features that should be provided to permit loading of a tank vehicle with a standard 4-in. adapter.

Tank Design

78-903839. Indian Standard Institution. 1977. Indian Standard Code of
Practice for Design, Fabrication, and Erection of Vertical Mild Steel
Cylindrical Welded Oil Storage Tanks. First Revision. New Delhi, India.
LC Call No: TP692.5 I.52.

Tank Foundations

058959. Penman, A. 1977. Soil Structure Interaction and Deformation
Problems with Large Oil Tanks. International Symposium on Soil Structure Interaction. University of Ruorkee, India. Pages 521-536.

Transportation of Hazardous Materials

European Model Code of Safe Practice for the Prevention of Ground and Surface Water Pollution by Oil from Storage Tanks and During the Transport
of Oil. London: Applied Science Publishers. 1974. TP 692.5 .E95.
Prepared by a working group formed following a meeting of representative
of European technical organizations held at the Institute of Petroleum,
London.

Fawcett, Howard H., Wood, Wm. S. 1979. Toxic Chemicals and Explosives
Facilities (Handling and Transport of Hazardous Materials). ACS Symposium
Series 96, p. 263-272. 79-03738.

Great Lakes Basin Committee. October 18-19, 1978. Transportation of
Hazardous Materials in the Great Lakes Region: Recommendations for the
Future. Great Lakes Basin Committee; Standing Committee on Transporta-
tion, etal. Seminar proceedings. 79-04063.

LRS79-3517. Hazardous Materials Transportation. A review and analysis of
the Department of Transportation's regulatory program. Prepared for the
Committee on Commerce, Science and Transportation. U.S. Senate. Washington, U.S. Government Printing Office, 1979. 251 pages.

LRS79-4856. Hazardous Materials Transportation Act Amendments. Report to
accompany H.R. 3502. Including cost estimate of the Congressional
Budget Office. Washington, U.S. Government Printing Office, 1979. 12 pages.

LRS76-14380. Hazardous Materials Transportation Act. 1976. Hearing 94th
Congress, Second Session on S. 2991. March 4, 1976. Washington, U.S. Government Printing Office. 139 pages.

LRS76-8176. Hazardous Materials Transportation Act Extension. 1976. Joint
hearing before the Subcommittee on Surface Transportation and the Subcommittee on Aviation of the Committee on Public Works and Trasnporta-
tion, House of Representatives, 94th Congress, Second session on H.R.
13124. May 10, 1976. Washington, U.S. Government Printing Office.
33 pages

089892. Hirota, Yoshiro. 1978. Japan Railway Engineer. "Safety Measures for Transportation of Hazardous Materials." Volume 18, N. 2, pages 20-21.

LRS77-21721. National Technical Information Services (NTIS) 1974. A Model Economic and Safety Analysis of the Transportation of Hazardous Substances in Bulk. Final Report. Washington, U.S. Maritime Administration, Office of Domestic Shipping. 266 pages.

LRS79-4562. Rail Transport of Hazardous Material. Hearing, 95th Congress, Second Session, March 20, 1978, Part 1. Washington, U.S. Government Printing Office, 1979. 158 pages.

RP 500C. Recommended Practice for Classification of Areas for Electrical Installation at Petroleum and Gas Pipeline Transportation Facilities, 1966. (Reaffirmed, 1974.) This recommended practice classifies areas within petroleum and gas pipeline transportation facilities for the installation of electrical equipment. Classified areas include pump stations, compressor stations, storage facilities, loading racks, and manifold and pipeline right-of-way areas where flammable liquids and gases are handled. The classifications contained herein were developed with consideration being given to a uniform system for pipelines which, by their very nature, are subject to various degrees of public exposure. Consideration also was given to the increasing application of centralized control with the consequent increase in unmanned or semiattended facilities. It includes statements and recommendations for classification of areas based on experience of the pipeline industry. It does not constitute, and should not be construed as a code of rules or regulations.

RP 1111. Recommended Practice for Design, Construction, Operation and Maintenance of Offshore Hydrocarbon Pipelines, First Edition, 1976. This recommended practice sets forth criteria for the design, construction, testing, operation and maintenance of offshore hydrocarbon pipelines. These criteria cover engineering considerations for the movement by pipeline of substances commonly encountered in hydrocarbon production and transportation operations. It is intended for application in all climatic regions. The design, construction, inspection and testing provisions of this recommended practice are not intended to be applied to offshore hydrocarbon pipeline systems designed or installed before its issuance. The operation and maintenance provisions of this recommended practice are generally suitable for application to existing facilities, as soon as such application is practicable. This recommended practice does not apply to production facilities forms. Neither does it cover the transportation of cryogenic substances or large storage facilities, tanker loading arms, monobuoy piping, swivel fittings, flexible hoses and related equipment API Publication.

LRS78-15042. Transportation Law Journal, Volume 19, No. 1, 1978. "Recent Developments in the Transportation of Hazardous Materials." Pages 97-120.

PB-276 734/1G1. United States Environmental Protection Agency. Manual for the Control of Hazardous Material Spills. Volume 1: Spill Assessment and Water Treatment Techniques.

Tanker Safety

Large Tankers--Our Energy Lifelines, 1979. This 24-page brochure describes Very Large Crude Carriers (VLCC), their structure, use, and general operation. The booklet also contains information on pollution prevention, crew training, tanker safety, tanker regulations and deepwater ports. API Transportation Department.

Vapor Control

75-304166. American Petroleum Institute. Symposium on Evaporation Loss of Petroleum from Storage Tanks, Chicago, 1952. LC Call No: TP692.5 .S9 1952. Subjects include oil storage tanks and evaporation control.

Bulletin 2522. Comparative Test Methods for Evaluating Conservation Mechanisms for Evaporation Loss, 1967. This bulletin presents relatively simple test procedures that have practical application in the determination of evaporation loss. These procedures may be used when the precise measurements necessary for a complete and accurate evaluation of evaporation loss provided by the comprehensive test methods of API Bulletin 2512 are not available or applicable. API Publication.

Bulletin 2518. Evaporation Loss from Fixed-Roof Tanks, 1962. This bulletin contains the correlation and evaluation of test data from companies concerned with evaporation loss from fixed-roof tanks. The information has been used to develop methods of estimating breathing and working losses from gasoline and crude oil tanks. Also included is a loss calculation summary and sample calculations. API Publication.

Bulletin 2517. Evaporation Loss from Floating-Roof Tanks, 1962. A method for estimating standing-storage and withdrawal evaporation losses from floating-roof tanks is presented along with a description of metallic and nonmetallic floating-roof seals. API Publication.

Bulletin 2516. Evaporation Loss from Low-Pressure Tanks, 1962. Breathing, working and leakage losses encountered in low-pressure tanks (atmospheric to 15 psig) are discussed in this bulletin, which also provides equations for calculating these values. API Publication.

Bulletin 2514. Evaporation Loss from Tanks Cars, Tanks Trucks, and Marine Vessels. Testing methods and correlations for loading, unloading and transit losses from tank cars, tank trucks, and marine vessels are explained. API Publication.

Bulletin 2513. Evaporation Loss in the Petroleum Industry--Causes and Control, 1959. (Reaffirmed 1973.) Sources of evaporation loss, the factors affecting evaporation and the equipment and procedures for controlling evaporation loss are discussed. API Publication.

Bulletin 2512. <u>Measuring Evaporation Loss from Petroleum Tanks and Transportation of Equipment, 1957</u>. In this first of a series of bulletins that the Committee on Evaporation Loss Measurement has issued on the subject, technical details are given for several methods of measuring actual evaporation losses. API Publication.

<u>Evaluation of Hydrocarbon Emissions from Petroleum Liquid Storage</u>. US/EPA PB-286. 190/4BE.

Bulletin 2514A. <u>Hydrocarbon Emissions from Marine Vessel Loading of Gasolines, 1960</u>. This bulletin presents information on hydrocarbon emissions that occur during the loading of gasoline into marine vessels. The data were obtained during the loading of 71 ship compartments and 12 barge compartments. API Publication.

Bulletin 2523. <u>Petrochemical Evaporation Loss from Storage Tanks, 1969</u>. This bulletin presents procedures utilizing the test results and correlations for gasoline evaporation loss given in API Bulletin 2518 and provides a method for determining the evaporation loss of petrochemicals. Also included are a basis for the new correlation and a sample calculation to demonstrate the method presented. API Publication.

Bulletin 2519. <u>Use of Internal Floating Covers and Covered Floating Roof to Reduce Evaporation Loss, 1976</u>. This bulletin describes methods to reduce evaporation from a free liquid surface in a fixed-roof tank by separating the liquid and vapor regions and summarizes the characteristics of floating covers and steel pans. API Publication.

Bulletin 2515. <u>Use of Plastic Foam to Reduce Evaporation Loss, 1961</u>. Evaluation of effectiveness, testing methods for measuring loss, loss reduction in static and working service, disappearance of foam particles and reduction in corrosion rates are covered. API Publication.

Bulletin 2521. <u>Use of Pressure-Vacuum Vent Valves for Atmospheric Pressure Tanks to Reduce Evaporation Loss, 1966</u>. This bulletin describes the use of pressure-vacuum vent valves to reduce evaporation loss of petroleum and petroleum products stored at essentially atmospheric pressure in above ground fixed-roof tanks and variable-vapor-space systems. It also presents factors to be considered when selecting vent valves and serves to increse the awareness of operationals and maintenance requirements. API Publication.

Bulletin 2520. <u>Use of Variable-Vapor-Space Systems to Reduce Evaporation Loss, 1964</u>. This bulletin describes the various types of variable-vapor-space systems used for storing gasolines and presents methods for determining the expansion volume required when economic studies indicate such a system is justified.

Underground Storage

LRS77-18494. <u>Mining Magazine.</u> "Storing Oil in Rock Caverns." Volume 137, November, 1977, p. 541.

Hydrocarbon Emissions from Tanks

058954. Harrier, Robert D. <u>Hydrocarbon Emissions from Fixed-Roof Tanks.</u> API, Pacific Coast Chapter Meeting, Bakersfield, October 25-27, 1977. 14 pages.

Petroleum Marketing Terminals

4285. <u>A Survey of Petroleum Marketing Terminals.</u> This report was prepared under the direction of Task Force W-19, API Environmental Affairs Department, with the assistance of its contractor, Engineering Sciencies, Inc., and the cooperation of 15 member companies of the American Petroleum Institute. It presents the results of a survey of the physical and operating characteristics of 76 petroleum marketing terminals and of the oil and grease content in their wastewater discharges.

Other Noyes Publications

HANDBOOK FOR
FLUE GAS DESULFURIZATION SCRUBBING
WITH LIMESTONE

by

D.S. Henzel
B.A. Laseke
PEDCo Environmental, Inc.

E.O. Smith
D.O. Swenson
Black & Veatch Consulting Engineers

Pollution Technology Review No. 94

This up-to-date and thorough handbook provides guidance for the selection, installation, and operation of limestone flue gas desulfurization (FGD) scrubber systems. The book covers all of the stages of the project from inception, through design, procurement, operation, and maintenance of the system.

Of the many available processes for FGD, the limestone wet scrubbing process is widely used and is continually being improved by numerous technological advances. The limestone wet scrubbing process generates wet sludge as a "throwaway" product or gypsum as a recoverable by-product. The book deals extensively with optional process features and recent innovative modifications that enhance the efficiency of a system.

The emphasis throughout is on practical applications. For example, the discussion of system design and performance provides the kind of information requested by regulatory agencies. This information can also be applied in evaluating preliminary studies and recommendations of a consulting architectural/engineering firm. Further, it can be used in developing detailed equipment specifications and assessing the performance predictions of various scrubber suppliers.

The condensed table of contents given below lists **chapter titles and selected subtitles.**

ISBN 0-8155-0912-X (1982)

424 pages

Other Noyes Publications

RESOURCE RECOVERY PROCESSING EQUIPMENT

by David Bendersky, Daniel R. Keyes, Marvin Luttrell,
B.W. Simister, Mary Simister, Denis Viseck

Midwest Research Institute

and George M. Savage, Geoffrey R. Shiflett

Cal Recovery Systems, Inc.

Pollution Technology Review No. 93

This book provides information useful for the selection and operation of resource recovery processing equipment for existing and future waste-to-energy systems. The need for conservation, materials recovery, and better waste disposal techniques has stimulated considerable interest and activity in the recovery of energy and other resources from municipal solid wastes. The book reviews the state of the art of resource recovery processing equipment and identifies research needs required to advance the technology of materials recovery, and waste-to-energy systems.

The methodology used to establish the state of the art included a literature search, a survey of present and planned recovery systems, the compilation of available information on equipment used in these systems, visits to plant sites, and discussions with system designers, system operators, equipment manufacturers, and other knowledgeable people.

Part I covers the present state of the art, and additional research needs. Parts II and III are devoted to in-plant tests and equipment evaluation aimed at satisfying some of the designated needs.

A condensed table of contents containing **part titles, chapter titles and selected subtitles** is given below.

ISBN 0-8155-0911-1 (1982)

360 pages.

Other Noyes Publications

HAZARDOUS WASTE LEACHATE MANAGEMENT MANUAL

Alan J. Shuckrow   **Andrew P. Pajak**   **C.J. Touhill**

Touhill, Shuckrow and Associates, Inc.

Pollution Technology Review No. 92

This manual describes various management options available for controlling, treating and disposing of hazardous waste leachate. Leachate generated by water percolating through hazardous waste disposal sites could contain significant concentrations of toxic substances. Proper leachate management practices are essential in order to avoid contamination of surrounding soil, groundwater, and surface water.

There is little past experience in the area of hazardous waste leachate, therefore the manual draws heavily upon experience in related areas. Sufficient information is provided so that the user can readily identify potential treatment alternatives for a particular situation. Logical thought processes are developed for arriving at reasonable process trains for given leachates.

The manual starts with a discussion of factors influencing leachate generation, details leachate characteristics, and then identifies principal options for processing the leachates. Treatability data, by-product and cost information supplement processing descriptions. Other sections cover monitoring, safety, contingency plans and emergency provisions.

The condensed table of contents below lists **chapter titles and selected subtitles.**

ISBN 0-8155-0910-3 (1982)

379 pages

Other Noyes Publications

ACID RAIN
INFORMATION BOOK

by

Frank A. Record   **David V. Bubenick**   **Robert J. Kindya**

GCA Corporation

This book, based on a study by *GCA Corporation*, discusses the major aspects of the acid rain problem which exists today; it points out the areas of uncertainty and summarizes current and projected research by various government agencies and other concerned organizations.

Acid rain, caused by the emission of sulfur and nitrogen oxides to the atmosphere and their subsequent transformation to sulfates and nitrates, is one of the most widely publicized and emotional environmental issues of the day. The potential consequences of increasingly widespread acid rain demand that this not altogether surprising phenomenon be carefully evaluated. This review of over 350 literature sources reveals a rapidly growing body of knowledge, but also delineates major gaps in understanding which should be narrowed and unanswered questions which should be resolved.

The book is organized by the logical progression from sources of acid rain precursors such as sulfur oxides and nitrogen oxides from fossil fuel combustion in industrial processing, electric power production, and the transportation sector; to atmospheric transport, possible chemistry of transformation, and deposition processes; to effects, both adverse and beneficial, on aquatic and terrestrial ecosystems, construction materials, and humans. Possible mitigative measures and further research needs are considered. Finally, a chapter is included on current and proposed research by the U.S. Department of Energy, the U.S. Environmental Protection Agency, and the Electric Power Research Institute.

Listed below is a condensed table of contents including **chapter titles and selected subtitles.**

ISBN 0-8155-0892-1 (1982)   **228 pages**

SMALL-SCALE RESOURCE RECOVERY SYSTEMS

Edited by A.E. Martin

Pollution Technology Review No. 89

This book, based on studies by *SCS Engineers* and *Systems Technology Corp.,* details the performance and economics of small-scale resource recovery systems and the types of operations which might employ them. "Small-scale," as covered here, implies systems which handle up to 250 tons per day of solid waste. Resource recovery provides an attractive alternative to landfill disposal because of its potential for resolving the solid waste problem while at the same time supplementing the nation's energy and raw material supplies.

There has been a severe lack of information on small-scale resource recovery systems, such as those which could be employed by industrial plants, smaller communities, institutions, office buildings, and multiunit residences. When smaller operations consider resource recovery systems for their waste disposal, they may not be able to economically apply most of the existing waste disposal technology and they lack funds to develop new technologies.

This book contains several studies prepared for the U.S. EPA in an effort to bring together the available technology for small-scale recovery systems. Various types of systems are investigated including thermal processing, mechanical separation, and biological processing. The end products of these systems range from combustible gases and solids, which can be sold as fuels, to ferrous metals, glass, aluminum, and marketable organic materials. A **condensed table of contents** is given below.

ISBN 0-8155-0885-9 (1982)

HAZARDOUS WASTE INCINERATION ENGINEERING

by T. Bonner, B. Desai, J. Fullenkamp, T. Hughes, E. Kennedy, R. McCormick, J. Peters, and D. Zanders

Monsanto Research Corporation

Pollution Technology Review No. 88

The engineering guidelines contained in this book are a compendium of the available literature on current state-of-the-art technology for hazardous waste incineration. They are intended to be used as a source of information for operational decisions and as a reference in the preparation of permit applications for hazardous waste incineration facilities.

A sizable fraction of the millions of tons of industrial waste material generated in the United States each year is considered hazardous (approximately 57 million metric tons in 1980). Incineration has recently emerged as an attractive alternative to other hazardous waste disposal methods such as landfilling, ocean dumping, and deep-well injection.

The advantages of incinerating hazardous wastes are several: toxic components can be converted to harmless or less harmful compounds; volume can be greatly reduced; heat recovery is possible as a means of saving energy; and incineration provides ultimate disposal, thereby eliminating the possibility of problems resurfacing at a later date. Because of these advantages, incineration may become a principal technology for hazardous waste disposal in the near future.

The various chapters of the book detail waste characterization, current commercial technology as well as emerging technology, incinerator design, and overall facility design. Listed below is a condensed table of contents giving **chapter titles and selected subtitles.**

ISBN 0-8155-0877-8 (1981)

385 pages

HOW TO DISPOSE OF OIL AND HAZARDOUS CHEMICAL SPILL DEBRIS

Edited by A. Breuel

Pollution Technology Review No. 87

This book describes various techniques which can be used to dispose of the collected debris from oil and hazardous chemical spills. It is based on research prepared by *SCS Engineers* and *CONCAWE* (the European oil companies' international study group for conservation of clean air and water).

Engineering constraints and equipment requirements for handling and disposal are evaluated. Debris management aspects considered are storage, transport, treatment, reprocessing, and disposal. Hardware and processing systems are identified and conceptual transport/disposal plans are developed. A literature review and several case studies are included. U.S. and European technology are covered.

Disposal is a complex problem. Spill incidents which have occurred in recent years have highlighted the difficulties in dealing with the large quantities of emulsions and debris which are collected during control and cleanup operations. Close cooperation between authorities and industry is a necessity at every stage, to ensure disposal in an environmentally acceptable, cost-effective, and energy-conserving manner.

Each case is different and no specific rules can be formulated, but guidance is given on the choice of options and their state of development. The condensed table of contents listed below gives **chapter titles and selected subtitles.**

ISBN 0-8155-0876-X (1981)

420 pages

OIL SPILL CLEANUP AND PROTECTION TECHNIQUES FOR SHORELINES AND MARSHLANDS

Edited by A. Breuel

Pollution Technology Review No. 78

When a major oil spill occurs, it usually involves contamination of coastal or inland shorelines and marshlands, which can result in serious environmental and economic damage. Such damage can be significantly reduced if proper protection and cleanup actions are taken promptly.

This book provides a systematic easy-to-apply methodology for assessing the threat or extent of contamination and choosing the most appropriate protection/cleanup procedures for each shoreline or marshland contamination event.

Oil spill on-scene coordinators and local officials, as well as petroleum, shipping and chemical industry personnel should find this book directly applicable during prior planning and oil spill operations.

A condensed table of contents of the three parts of the book, including **important subtitles,** is given below.

ISBN 0-8155-0848-4 (1981)

404 pages

HANDBOOK OF
TOXIC AND HAZARDOUS CHEMICALS

by Marshall Sittig

"Protecting against potential public health hazards requires widespread knowledge about commercial chemicals. We need to know more about the effects they will have, and, most importantly, how we can minimize the risks posed by them."—*from the Foreword by Bill Bradley, U.S. Senator from New Jersey.*

This handbook presents concise chemical, health, and safety information on about 600 toxic and hazardous chemicals, so that responsible decisions can be made by chemical manufacturers, safety equipment producers, toxicologists, industrial safety engineers, waste disposal operators, health care professionals, and the many others who may have contact with or interest in these chemicals due to their own or third party exposure. This book will thus be a valuable addition to industrial and medical libraries.

Essentially the book attempts to answer six questions about each compound (to the extent information is available):

(1) **What is it?**
(2) **Where is it encountered?**
(3) **How much can one tolerate?**
(4) **How does one measure it?**
(5) **What are its harmful effects?**
(6) **How does one protect against it?**

Included in the book are **all** of the substances whose allowable concentrations in workplace air are adopted or proposed by the American Conference of Governmental Industrial Hygienists (ACGIH), **all** of the substances considered to date in the Standards Completion Program of the National Institute of Occupational Safety and Health (NIOSH), **all** of the priority toxic water pollutants defined by the U.S. Environmental Protection Agency (EPA), and **most** of the chemicals in the following classifications: EPA "hazardous wastes," EPA "hazardous substances," chemicals reviewed by EPA in Chemical Hazard Information Profiles (CHIPS) documents, and chemicals reviewed in NIOSH Information Profile documents.

The necessity for informed handling and controlled disposal of hazardous and toxic materials has been spotlighted over and over in recent days as news of fires and explosions at factories and waste sites and groundwater contamination near dump sites has been widely publicized. In late 1980 the EPA imposed regulations governing the handling of hazardous wastes —from creation to disposal. Prerequisite to control of hazardous substances, however, is knowledge of the extent of possible danger and toxic effects posed by any particular chemical. This book provides these prerequisites.

The chemicals are presented alphabetically and each is classified as a "carcinogen," "hazardous substance," "hazardous waste," and/or a "priority toxic pollutant"—as defined by the various federal agencies, and explained in the comprehensive introduction to the book.

Data is furnished, to the extent currently available, on any or all of these important categories:

Chemical Description	**Routes of Entry**
Code Numbers	**Harmful Effects and Symptoms**
DOT Designation	**Points of Attack**
Synonyms	**Medical Surveillance**
Potential Exposure	**First Aid**
Incompatibilities	**Personal Protective Methods**
Permissible Exposure Limits in Air	**Respirator Selection**
Determination in Air	**Disposal Method Suggested**
Permissible Concentration in Water	**References**
Determination in Water	

An outstanding and noteworthy feature of this book is the Index of Carcinogens.

ISBN 0-8155-0841-7 (1981)

729 pages

Other Noyes Publications

HAZARDOUS CHEMICALS DATA BOOK 1980

Edited by G. Weiss

Environmental Health Review No. 4

Instant information for decision-making in emergency situations by personnel involved with chemical accidents is the prime purpose of this compilation of about 1,350 hazardous chemicals. The book, prepared in clear, concise, easy-to-locate format, should become an invaluable source on any library or laboratory shelf; is intended for use by scientists, engineers, managers, transportation personnel, or anyone who may have contact or require data on a particular chemical.

There is a large amount of pertinent data provided for each chemical, and examples of a few of the headings are:

Toxicity and Health Hazards	Flammability and Explosion Hazards
Hazards Relating to Water Pollution	Chemical Reactivity
Spill and Leak Procedures	Extinguishing Media
Special Protective Equipment	Special Precautions

The volume is set up in two sections, alphabetically arranged. The first section is based on the Department of Transportation's Chemical Hazard Response Information System. The second section is based on Material Safety Data Sheets obtained from the Oak Ridge National Laboratory. The use of a convenient all-inclusive index furnishes the responsible individual with the desired information in a minimum of time.

A partial list of 54 hazardous chemicals taken from the index of about 1,350 chemicals, is shown below.

Acetal	**Allyl Alcohol**
Acetaldehyde	**Allylamine**
Acetamide	**Allyl Bromide**
Acetic Acid	**Allyl Chloride**
Acetic Anhydride	**Allyl Chloroformate**
Acetone	**Allyltrichlorosilane**
Acetone Cyanohydrin	**Aluminum Chloride**
Acetonitrile	**Aluminum Fluoride**
Acetonylacetone	**Aluminum Nitrate**
Acetophenone	**Aluminum Sulfate**
Acetylacetone	**p-Aminoazobenzene**
Acetyl Bromide	**p-Aminoazobenzene Hydrochloride**
Acetyl Chloride	**2-Amino-1,4-Dimethylbenzene**
Acetylene	**4-Amino-1,3-Dimethylbenzene**
Acetyl Methylamine	**5-Amino-1,3-Dimethylbenzene**
Acetyl Methylurea	**o-Aminodiphenyl**
Acetyl Peroxide Solution	**Aminoethylethanolamine**
Acetyl Thiourea	**2-Aminopyridine**
Acetyl Urea	**Ammonia, Anhydrous**
Acridine	**Ammonium Acetate**
Acrolein	**Ammonium Benzoate**
Acrylamide	**Ammonium Bicarbonate**
Acrylic Acid	**Ammonium Bifluoride**
Acrylonitrile	**Ammonium Carbonate**
Adipic Acid	**Ammonium Chloride**
Adiponitrile	**Ammonium Citrate**
Aldrin	*plus about 1,300 other chemicals*
Alkylbenzenesulfonic Acids	

ISBN 0-8155-0831-X

1188 pages

HAZARDOUS CHEMICAL SPILL CLEANUP 1979

Edited by J. S. Robinson

Pollution Technology Review No. 59

There is a delicate balance in nature involving a perfect blending of chemical, physical and biological elements. Once this balance is upset by a spillage of chemicals, it may take years to restore the original environment. Meanwhile, untold damage results. The deleterious effects, however, can be lessened if there is a rapid response reinforced by expertise in proper procedure. It is the intent of this book to provide the specific know-how needed to cope with chemical spillage quickly by correct containment, countermeasures, and cleanup.

After an introductory overview of legislation, Robinson offers practical advice and response techniques for different kinds of substances, including preferred methods of disposal and restoration. Managers in all fields with problems of potential spills should avail themselves of this information now.

The partial table of contents below lists chapter headings and **examples of some** subtitles.

ISBN 0-8155-0767-4

406 pages

Paper Money

Yasha Beresiner

A Collector's Guide to
PAPER MONEY

5D STEIN AND DAY/*Publishers*/New York

First published in the United States of
America, 1977
Copyright © 1977 by Yasha Beresiner
First printing

Stein and Day/Publishers/Scarborough House,
Briarcliff Manor, N.Y. 10510

Library of Congress Catalog Card Number:
77-80372

Filmset and printed in Great Britain by
BAS Printers Limited, Over Wallop,
Hampshire

To Laz
with love and in friendship
Tempus Fugit!

Contents

Abbreviations for Publications

*BCD	*Bulletin du Centre de Documentation pour L'Etude du Papier Monnaie*
BNR	*Bank Note Reporter*
C&M	*Coins and Medals*
*CC	*World Paper Currency Collector*
CCRT	*Cheque Collectors Round Table*
*CCW	*Coin Collecting Weekly*
CM	*Coin Monthly*
*CMC	*Coins, Medals and Currency*
*CS&C	*Coins, Stamps and Collecting*
CPMJ	*Canadian Paper Money Journal*
CW	*Coin World*
IBNS	*International Bank Note Society*
LANSA	*Latin American Notaphilic Society*
NSB	*Numismatic Scrap Book*
*MCBN	*Modern Coins and Bank Notes*
PM	*Paper Money*
WC	*World Coins*
*WNJ	*Whitman Numismatic Journal*
*WPMJ	*World Paper Money Journal*

*no longer published

Acknowledgements

In the many years that I have been a notaphilist, I have read and enjoyed articles and books written by many dozens of collectors. Most of them are mentioned in the bibliographic listings in this book. To them, who made my hobby so pleasurable and, indeed, made this publication possible, go my sincerest thanks.

Innumerable friends and fellow collectors have helped unreservedly with information and comments and without their assistance this work would never have been undertaken. I should like to mention specifically: Colin Narbeth who has been to me a guiding light throughout my endeavour; Albert Pick whose advice has been invaluable; I have received particular help from Geoffrey Grant regarding England, John Wall regarding Persia, Mario Perricone regarding Italy and David Keable regarding Scotland; Peter Males prepared the index.

Maurice Barnett, John Rea and my brother Yos Beresiner have checked through the manuscript for me while Sandra Pott has kept, incredibly, a constant smile in spite of the 'ordeal' of typing and retyping this material.

NB Where two pictures appear on the same opening they are reproduced approximately in proportion, unless otherwise indicated.

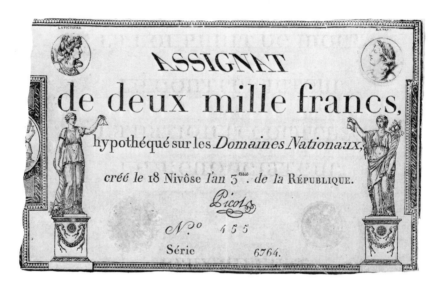

Preface

Less than a decade ago there were no more than a hundred enthusiasts who could be considered serious paper money collectors. Today, in 1976, the ten or so international societies who serve the notaphilist boast a total membership in excess of thirty-five thousand collectors, and this does not include many thousands more whose interests may still be only latent.

The major factor influencing the tremendous growth in the popularity of this hobby has been the increasing number of notaphilic publications making an appearance on the market in recent years. In addition to most numismatic periodicals devoting space to paper money, catalogues for specialised collectors are now available for the majority of the countries of the world. Relatively few books have appeared covering the hobby as a whole and most of these have been introductions directed at the newly-interested collector.

The purpose of this book is to bridge the gap between books for the beginner and those for the seasoned collector who has already selected a theme on which to base his collection. My aim is to cover every aspect of the hobby in a general way. The emphasis that has been placed on the select bibliographic listings should compensate to a certain extent for this fact. The bibliography lists, in addition to books, the relevant articles readily available on special themes. The libraries of the notaphilic societies mentioned in Appendix III stock not only sets of their own journals, but often each others', and many of the professional periodicals that are made available on the market. Further research on any special subject will have to be undertaken by the individual, guided, hopefully, by the contents of this book.

Chapter III, more than any other, will no doubt undergo some criticism. The notaphilic market is a fluid one. The discovery of a hoard of notes can plunge the value of a rare one to ten per cent of its former worth. Such are the perils of a 'new' hobby. Every attempt has been made to give a reliable indication of price, but the collector is asked to bear in mind that the pricing was undertaken in January 1976. There are bound to be price changes due to the increasing popularity of the hobby as well as those due to inflation, during the time between writing and publication. Similarly, the dollar equivalent constantly changes. In giving both the pound and dollar values of a note, a convenient rate of exchange of £1 = $2 has been used – the approximate rate at the time of writing.

YASHA BERESINER
London, January 1976

Normal stamp production processes were used by the Russians early this century to produce 'currency stamps'. The reverses indicate that they are to be used as legal tender. (270 × 342 mm)

Previous page: French assignats have come in many shapes and sizes, with a wide range of signatures, which makes the formation of a complete collection a challenging task. The higher denominations are attractive and can be expensive. Assignats of the third year of the Republic circulated in 1795. (128 × 215 mm)

Specimen notes of the Bank of England are extremely rare because they were only released to the other national Central Banks. (130 × 214 mm)

A Collector's Approach to Paper Money

For over two hundred years paper money has been gradually and increasingly replacing metallic currencies. Lower weight and lesser volume established paper as a convenient medium in times of emergency. Simpler and cheaper manufacturing processes led governments to adopt paper currency as an essential and convenient monetary measure.

Such adoption, however, followed a period of experimentation which frequently brought about a sad state of affairs in the economies of innumerable countries. The almost immediate abuse of paper issues by individuals, private bankers and even authorised institutions led to the early realisation that if the public interest was to be safeguarded, issues would have to be controlled. The paper note that replaced the coin was equivalent to credit replacing money. The French Revolution and the British monetary policies in her American colonies at the end of the eighteenth century are blatant examples of the dangers of unrestrained issues.

The introduction of paper money marked a new type of financial experiment in a world dominated by coins. Its evolution has been natural and logical. It may have brought with it many maladies but as many problems have been alleviated. Coinage has constantly been losing its intrinsic value. Gold coins are no longer legal tender and they have been replaced by silver ones in most countries. Silver, too, is being replaced by lesser metals. Nickel and its alloys, copper and even aluminium are now used for the smaller coins. Furthermore, many a collector prizes the emergency stamp monies, the cardboard tokens and the porcelain and ceramic 'coins', that have been used as substitutes in the recent past. A few steps down the gold ladder indeed! Paper money is here to stay. Convertibility, whereby a note issue was redeemable *in specie* (metallic coin), is a thing of the past in most countries.

The term 'notaphily' used throughout this book is beginning to be accepted gradually, if somewhat reluctantly, by all. It was devised by Kenneth R. Lake and in his own words, notaphily is 'the generic term of paper money collecting, from the Latin *nota* (a note) and Greek *philos* (love); the adjective is notaphilic, and a collector is a notaphilist'. There has been a great deal of controversy on the use of the word mainly because of its mixed derivations, but if we consider that it took some twenty years before the word philately was fully accepted by the collector, we may be wise to accept this new term, with its limitations, and patiently await its adoption by the notaphilic brotherhood.

Notaphily is now irrevocably accepted as an integral part of numismatics. Recent years have seen an immense upsurge in the popularity of what was considered a new

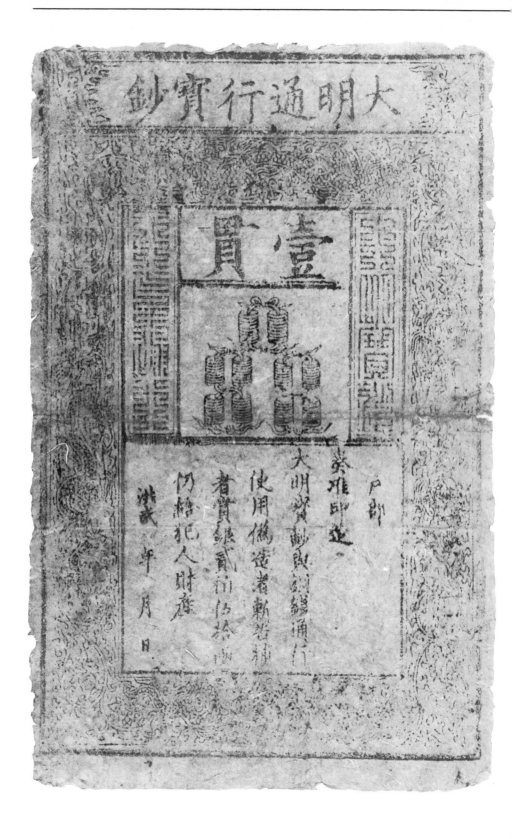

hobby only a decade ago, and many coin collectors and a large number of novices find latent interests coming to the surface.

The words 'bank note' have become almost synonymous with 'paper money'; the latter term is, however, very deliberately used here. A 'bank note' is a paper money issue placed into circulation by a bank. There have been, as we shall see, paper monies issued by treasuries, local governments, military forces, individual enterprises and even private persons.

Chinese notes from the fourteenth-century Ming Dynasty (the earliest dated paper money a collector can hope to acquire) were printed on the bark of the mulberry tree. These are identified as true 'paper' issues, but no such unequivocal statement can be made about leather or silk or mere cloth and yet all three, and a range of even more unusual materials have been used as paper substitutes. The term 'paper money' must, therefore, be considered in its widest possible sense.

A coin is, of course, made of greater substance than paper; it has the intrinsic value of the metals from which it is composed; but the artistic quality, the infinite amount of information and the 'collectability' of a paper note must often supersede that of a coin. Not only the date but the place of issue also appears on paper money. Signatories, issuing authorities, names of the printer and engraver, redemption and counterfeiting clauses may also be found. Add the colour and art work, and clearly such a quantity of information could never appear on the limited surface of a coin.

The educational value of collecting paper money cannot be overstressed. World economic history has left a record of its upheavals in inflationary paper money issues; a physical evidence to support the millions of words written on the subject. Although its origins do not go as far back as those of the coin, paper money does record important national historical events, wars, revolutions, and emergency situations. Geographically, too, many nations now no longer existing have their paper money issues as the only remaining relic of their presence in our civilisation.

Starting a Collection

No collector can expect to approach absolute completeness of any collection and he must, therefore, in the early stages decide on a framework, a specialised subject within the wide spectrum of possibilities. Individual items of paper money can be extremely attractive. As one becomes more knowledgeable about notes, the temptation increases to obtain every piece from every source until a jackdaw type of collection is accumulated.

Start by purchasing one (and make a point of this being your only one) indiscriminate purchase of a job-lot. Endeavour to obtain as wide a range of different issues as your budget will allow. Many dealers advertise such job-lots and for about £5 ($10) you should expect some fifty different – albeit common and inexpensive – notes in good condition. Use this as your starting point and inspect the issues closely. See whether you can identify the country, the bank of issue, the date, and then find a

Opposite
The Chinese Ming note has become a classic notaphilic item and its value, as a collectors' piece, has tripled in the past few years. (220 × 335 mm)

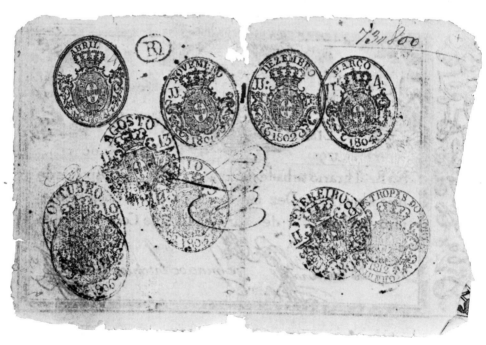

The Portuguese 'War of the Two Brothers' early in the nineteenth century gave rise to a series of issues that circulated for nearly thirty years. These notes, found in good condition, should be treated with some suspicion. Some were overstamped on the reverse as many as twelve times, to be revalidated. (96 × 148 mm)

source of information in a notaphilic book or in ordinary historical and economic books you may encounter in your local library. From there your imagination and instinct will lead you to your choice. With the aid of associated literature you will be able to decide on which aspect of the hobby you wish to concentrate. It will guide you in being more selective and allow you time to gain the knowledge that will lead you throughout your collecting life. You need not fear that your initial purchase will turn out to be a waste; you will find that every note you ever purchase will have, at the least, some exchange value and certainly an academic one.

Collecting Topics

The natural inclination of a new collector is to specialise in the paper money issues of the country in which he resides. A great number of countries in the world, when they issue new notes, declare the previous issue to be invalid. The people are given the opportunity to 'redeem' notes of the old issue in their possession by exchanging them for the new notes, but, after a specified date the old notes are demonetised and therefore valueless, except to the collector. This is not the case in England or the USA. Every issue of the Bank of England, since her establishment in 1694, and all paper money issues of the United States since 1861, are redeemable at face value by the Bank and the US Treasury respectively. The well established American and English collectors have set a trend which has increased the cost of all but the very modern issues of these two countries; thus, to begin a collection with the paper money issues of your own country can be an expensive affair if you are an Englishman or an American! (This equally applies to a number of additional countries, notably Canada and Scotland.)

There are many collecting topics, other than the 'one-country-collection'. The widest scope is afforded by a collection based on 'one-note-per-country'. The wide range of possibilities of such a collection becomes apparent when we consider that it would include notes of many nations now no longer in existence, such as Biafra and Bohemia, others issued by governments in exile (eg Kossuth of Hungary), and even some by dissidents within the boundaries of a sovereign state (eg the Spanish Civil War issues). Another popular topic is the 'historical collection'; a specified period in world history is chosen and represented through its paper money issues. A sizeable collection of the Russian Revolution notes or of the issues of the American colonial period can be formed under this heading. The Second World War is also a historical period which is very popular as a result of a number of highly authoritative publications on the subject. Early notes, large-sized notes and emergency issues are just a few more examples of collecting subjects.

'Thematics' is also gaining ground with the paper money collector. There are two distinct ways of starting a thematic collection. The first is obvious: a defined topic or object is chosen, be it sport, vessels, birds, children or whatever, and as many distinct issues as possible with the same subject are sought. The second is the abstract way, where a common factor, rather than a specified topic, connects the subject matter. Such a collection may be based on misprints and errors, or one-denomination notes, or a specified pattern in the serial numbers and letters.

Where to Search

The most obvious place to look for paper money is at the professional dealer. The society you join will put you in touch with other collectors and organised mail auctions will afford many opportunities. There are additional paper money auctions which take place at regular intervals, organised by professional business enterprises. If you live in or near London you can attend the paper money auctions that are held at regular intervals by Stanley Gibbons (Currency) Limited. Equally well known are those of Almanzar's in Texas and Henry Christenson's in New Jersey, USA. A number of established rules govern public auctions and a strict code of ethics is adhered to by the auctioneer. Although many an opportunity may present itself at an auction, the reverse is also true. A note which you could possibly have purchased at a reasonable price in a shop may be priced tenfold at an auction. A day or so before the auction the material is placed on display, or otherwise made available for inspection by the potential buyers. Illustrated catalogues with estimated prices are also printed and distributed. Both deserve careful consideration.

It has often been stated that the true value of a collector's item is the price that the collector himself is prepared to pay for it. This applies whether you are purchasing in an auction or privately. A bank note for which you would gladly pay several pounds in excess of its catalogued value, may be worth no more than a few pence to another collector and is probably worth nothing at all to the man in the street. Dealers, societies, advertisements and auctions are all important sources of supply.

All National Swiss Bank notes are still redeemable and many are very colourful. The high denomination notes are expensive as they fetch prices above their face value. (130 × 215 mm)

Housing and Display

Housing a collection must be based on a simple and practical system. It must allow for additions without inconvenience and be easy to handle; it must also be well displayed and easily accessible. There are many ways in which a collection can be housed. The number of notes will determine the best method to be used. Initially, envelopes containing the notes listed by countries can form a simple filing system when placed in alphabetical order in a suitable box. When a more elaborate method is later adopted this system can remain in use for housing duplicates. Some collectors use the normal loose-leaf stamp or photograph albums by mounting their notes with photo corners. Sometimes stamp stock albums with strips are used for smaller notes, such as notgeld, where individual items can be mounted overlapping each other. These three methods have one main disadvantage: the reverse of the notes cannot be seen. Although this may be unimportant in relation to many of the early notes which were printed on one side only, the majority of paper money issues have elaborate and beautiful designs on the back which should be displayed.

When a collection consists of a relatively small number of notes of high values or particular interest, they may be framed and hung on the wall.

The most popular system used by the majority of collectors is the transparent vinyl plastic sheets, in special binders. There has recently been a great deal of controversy regarding this method because of the apparent damage that may be caused to some notes if the oily substance in the plastic sheets gets on to the notes. This has proved to be true only when a collection in plastic sheets is left untouched for several years. If, on the other hand, the sheets are 'aired' at regular intervals, it would seem that no such damage is caused. Furthermore, separating the plastic edges with plain white paper will ensure that the substance is absorbed outwards causing no damage to the contents. There is great danger if the sheets are exposed to excessive heat or dampness; they then become brittle and can harm the notes inside. Unless your intention is to store your collection without looking at it for several years, you can be safe in using ordinary transparent plastic sheets.

Condition and Grading

Condition in the notaphilic field is not as important a matter as it is in the philatelic and numismatic fields. Where a dirty stamp, irrespective of its rarity, would be almost worthless to the stamp collector, an early paper money issue of relative scarcity can be included in your collection in any reasonable condition. It is, of course, desirable to collect as perfect specimens of paper money as possible.

Modern bank notes can usually be obtained in crisp condition; the problem with legal tender or redeemable notes is that they are worth at least their face value and the high denominational notes will be expensive to obtain. In principle, all issues after the Second World War, whether demonetised or not, should not be included in your collection unless they are in very good condition. It is, however, the exception that makes the rule. The last three decades have seen some extremely rare notes appear on the market. The first Israeli fractional notes of 1948, known as the 'carpets' (because of the mosaic design on the vertical notes) are rare specimens; because of the small quantity issued and the brief period during which they circulated. The French issued a series of military notes for use by the troops during the 1956 Suez crisis. The 50, 100 and 1,000 franc notes, overprinted for use in the Middle East, are valued at well over £150 ($300) each. Even rarer are the notes introduced with the Currency Reform of 1948 by the German Federal Republic. A new currency unit was established and the 50 Deutsche mark note, portraying the head of a female in the centre, was withdrawn from circulation only a few days after being introduced. Should you come across any of these or similarly rare modern notes your considerations as to condition should be secondary.

Unsigned and unissued notes frequently make an appearance on dealers' and auction lists. These must necessarily be in perfect condition as they will never have seen circulation. But also bear in mind that on a number of exceptional occasions, unsigned and undated notes were circulated because of the emergency situation at hand. Runaway inflation during the Russian Revolution early this century led to ever-increasing paper issues. When the notes were delivered by the printing company, the normal practice of dating and signing them was dispensed with. The note issues went into circulation straight from the printing presses!

Philatelists and numismatists place far more emphasis on the price difference between a perfect specimen and a lesser one than a notaphilist does. Paper money is 'mishandled' by force of circumstance. It is folded into wallets, stuffed into pockets and it easily collects dirt. The average circulation period of a Bank of England note is three months. This bears no comparison to stamps, which, of course, do not circulate from hand to hand nor to coins which last for many decades before they are withdrawn and replaced. In notaphily a fifty-year-old note is ancient! The vulnerability of paper prevents such issues from reaching us in any better than 'good' condition. Old notes in crisp condition are only encountered when discovered in quantity, never having circulated. Some individual crisp notes over a hundred years old appear when passed on from collector to collector through generations. As a general rule, if you encounter a note you need and the price is reasonable, buy it, irrespective of its condition. The opportunity to replace it with a better specimen will arise in the future.

Until a few years ago the grading system used in notaphily was identical to that used in determining the quality of a coin. The system is based on a descriptive scale of seven conditions: Uncirculated (UNC), Extremely Fine (EF), Very Fine (VF), Fine (F), Very Good (VG), Good (G) and Poor (P). The words are self-explanatory. In numismatics, these terms are quite precise, but that is not the case for paper money issues. Dealers differ in their interpretation of individual terms. The word 'good', for example, is quite deceptive in that a 'good' note is truly a 'worn' note (a term which is still occasionally used). Furthermore, new qualifications began to appear on dealers' lists. 'Nearly Extremely Fine' and 'Almost Uncirculated' are two examples which can be very confusing.

In 1969 a London dealer, Douglas Bramwell, devised a numerical system for bank note grading. It is now more or less established, although some continue to persist in the use of the earlier terminology. Bramwell's system is based on a table of damage numbers where the condition of a note is divided into five main headings. (See Appendix I.)

In each section, a damage number is allocated, to be finally added together and deducted from one hundred. Thus a precise percentage of the condition of a note is reached. When you correspond with fellow collectors, some of them possibly several thousand miles away, your integrity is at stake. You will enjoy a high reputation and a valuable relationship if the descriptions and values you give to your notes are accurate. The Bramwell system for valuation, when strictly followed, is an excellent guide to the condition of any paper money issue.

Restoring Notes

As a rule: don't. In notaphily reparing and restoring a note is not considered unethical practice as it would be in the case of philately, but you must refrain from the temptation to clean and iron out every creased note that comes your way. Bank notes printed by reputable companies are made of durable and high quality paper, but they are still prone to damage. Should you decide to treat a note, there are a few important rules to remember. The most common practice among collectors is to iron a note to straighten the edges and smooth the surface. This can be done; but it is essential that the iron does not come into direct contact with the surface of the note. Use blotting paper, put the note between two sheets and iron both sides. The iron must not be too hot. Great care must be taken otherwise the note will lose colour and become unnaturally shiny. This may occur even if the note does not suffer burns.

A dirty note may be cleaned by simply washing it with a neutral soft soap in cold water. This should never be done to notes bearing handwritten signatures. Ink runs!

Never expect perfect results, nor persist in trying to obtain them. Where notes have transparent plastic tapes stuck to them, the tape can be easily removed by allowing wet blotting paper to rest over the area for about ten minutes. Do not place any pressure on it and the tape will easily peel off after that period of time.

In spite of all the above, constantly remember, *the less you tamper the better*.

Research

One initial problem encountered in collecting is the language barrier. But by learning the basic characters of the important languages, and with the aid of authoritative catalogues, deciphering relevant texts will not be too difficult. The only difficulty may arise where two or more languages appear on the same note. Bi-lingual notes are common; even tri-lingual notes, from many of the Middle Eastern countries in the 1940s, are encountered: Cypriot notes showing English, Greek and Turkish; Lebanese and Syrian issues with Arabic, French and English. Most fascinating, however, are the truly multilingual notes. The 1919 Russian accounting notes were nicknamed 'Babylonians' because of the multi-lingual text on them. Similarly the Austro-Hungarian Empire formally adopted eight languages, and all eight appear on the issues of the Central Bank of the Empire; they include Czech, Polish, Ukranian, Italian, Ruthenian, Croatian, Serbian and Rumanian. Even more exotic are the thirteen dialects in which the words 'one rupee' are printed on the Government of India issues, still in circulation today.

Be observant and you will encounter a great deal more information just by looking at the note in hand; pay attention to the quality of the paper, the watermark, the printing and the anti-forgery devices in the design. The vertical metal filament on current Israeli bank notes spells out the name of the Central Bank – in morse code! Some Scottish notes have visible magnetic strips on their issues, used for computerised sorting of the notes as well as forgery detection. If you are research-minded, you will find the information on the face (obverse) and back (reverse) of a note a great deal to go on. The obverse usually bears the main 'legend' (name of bank, city of issue and date), design and the vignette of a portrait.

There are five basic aspects to researching notes:

The Economic Aspect involves discovering the authority by which the note was issued; the standing and history of the issuing body; the change in the currency denominations, and so forth.

The Historical Aspect has even wider scope: a large number of notes were issued as a result of war and revolution. Details of the circumstances of such issues, and other emergency situations which have led to inflationary issues, are rich pastures for the researcher. The Russian and French Revolutions, Germany between the Wars, equatorial Africa under the French; Brazil under Portugal, are just a few examples of periods in history amply represented by paper money issues.

The Political Aspect that can be researched is closely identified with the historical but it is also a field on its own. Political aspirants have used paper money as a first step in establishing their authority over a country or a limited geographical area. The 'Arias' notes of Panama, which circulated for only one week, were intended to emphasise the people's objection to American sovereignty over the Canal Zone. Francisco 'Pancho' Villa in Mexico and Guillermo Cervantes in Peru issued

Revalidation of these Mexican issues was authorised by a decree in December 1914. The provisional Government's seal was overstamped on the notes to indicate the new authority under which they were to circulate. (82 × 187 mm)

notes showing their names and military titles. Research is bound to lead to new discoveries.

The Geographical Aspect is not as easy as it may seem. It is not always simple to allocate a specific bank note to a sovereign state. The Ottoman and Austro-Hungarian Empires issued notes, Biafra and Bohemia and Moravia have also done so, but these countries cannot be pinpointed on a modern map. Here the historical and geographical aspects meet. 'Theresienstadt', for example, is the only indication of the location of the German concentration camp in Czechoslovakia, where internal money circulated among the imprisoned Jewish community; 'Puerto Plata' gives no clue to being a city in the Dominican Republic, yet it is the only geographic indication to be found on the notes of 'El Banco de la Compañia de Credito de Puerto Plata'. Research on the geographic aspect of a note cannot be solely limited to the modern atlas.

The Artistic Aspect, finally, can be the most difficult to research but the most gratifying. Many European issues indicate on the margin of the note the name of the engraver and printer, and at times, that of the artist. Many countries reproduce on their currencies world famous paintings by their national artists with Spain, Portugal, Peru, the Netherlands among them. A visit to the Prado Museum in Madrid will allow you to purchase, at a few pence, reproductions on postcards of almost every portrait and landscape depicted on Spanish notes during the course of this century.

The current paper money issues of Peru, in circulation since 1968, were printed by Thomas de la Rue, but the artistic designs were prepared by Dr G. Suarez Vertiz of Lima. An eccentric, young-hearted old man, he spends months on end in the Andes, studying and painting features of the native Indians of Peru.

Publications

One must think of building a library from the very early stages of a collection. A notaphilic library must necessarily include a great deal of incidental material relevant to the subject matter, as well as notaphilic publications and books (see Bibliographical listings). An eye should be kept open for early books, many of them on banking rather than paper money; these are fast becoming collectors' items in their own right; dealers' price lists and auction catalogues – with the 'prices realised' – should be carefully kept as reference material. Many national and local papers are now writing regularly about the hobby. It is advisable to form a special file for paper clippings. Here you should also include your own notes and comments. Jottings taken down at a numismatic lecture, or the address of a specialised collector, could turn out to be essential information.

Reference material is the soul of a collection. Many catalogues can be cheaply purchased. Some, because they are pioneer works, are not comprehensive; but will undoubtedly be improved upon in the future. Irrespective of your own specialisation, consider including all notaphilic publications in your library. By the time you decide to diversify you may find that a book you need has gone out of print.

RAMON CASTILLA

The hobby is new and publishers go to print with limited quantities. Even today some five-year-old catalogues can no longer be obtained.

Security

There are several physical aspects relating to the safety of notes. Housing has been considered. It is important that a collection be kept in a comfortable temperature, irrespective of the quality of the container in which the notes are kept. Dampness has a habit of penetrating through most barriers and should be avoided. Even more harmful, however, is excessive heat. In your own house care should be taken not to place your albums over storage heaters or in cupboards through which hot water pipes run.

Some collectors like to ensure the safety of their collections by placing them in bank vaults; it must be remembered that not all bank vaults are atmospherically suitable for storing documents. It would be best to wrap the collection as a whole in plastic bags before placing them in the vaults. The financial security of your collection can, of course, be covered by insurance. You will need to be fairly specific as to the values. The premium for a comprehensive coverage should be in the region of one per cent of the total value insured.

One last word of advice regarding safety. Avoid, if possible, using your home address in your role as collector. Although the numismatist is under a greater danger of being robbed, enterprising criminals exist the world over! The publicity being given to our hobby may just tempt someone to try his hand at bank notes. Many collectors use their office addresses or Post Office Box numbers. Some join societies requesting to remain anonymous to the membership, while others are so security-minded that they are reluctant even to disclose the fact that they collect. They vehemently object to joining any societies and they even attend auctions by proxy!

Opposite
General Ramon Castilla is among the popular figures in Peruvian history. Modern Peruvian notes show him as President in the 1850s. The portrait was drawn by Suarez Vertiz. (67 × 156 mm)

The History of Paper Money

When the word 'currency' was first used early in the eighteenth century, it encompassed every possible commodity generally acceptable as a medium of exchange in trade, be it paper or metallic money, rum in Australia or fish hooks in Alaska! In today's economic terms, bills of exchange, cheques and other forms of credit cannot be considered equivalent to circulating legal tender coins and bank notes, although they are 'money' in one sense. As late as the beginning of this century, bills and credit documents were used as means of payment in everyday transactions. The notaphilist may not be immediately concerned with material other than paper money, but he must bear in mind that all money items, from sea shells to the million pound interbank transfer chits, are relevant to the study of the hobby.

The essential precursor to the establishment of a currency other than metallic was the invention of paper. This is attributed to the Minister of Agriculture of the Han Dynasty of China, in AD 105, a man named Tsai Lun. It is probably no more than coincidence that the country in which paper was invented was also the first on record to use paper for currency. The word itself has its origins in the Latin word *papyrus* which was invented about 2500 BC, shaped in the form of sheets made out of water reed. Its use as writing material became almost immediately apparent. The Greeks and Romans used papyrus extensively, well before the invention of parchment around 190 BC. By then skins of young animals were being cut into irregular shapes and used as a medium of exchange in commercial transactions. The Carthaginians are said to have used leather as currency. Unidentified objects were wrapped in pieces of leather and circulated to represent coinage. These monetary items were issued under the authority of the state.

The best recorded use of leather as currency, however, takes us back to China. More than a hundred years before the advent of Christianity the Chinese Emperor Outi, or Wu Ti, (140–86 BC) under the Han Dynasty, ran into a financial crisis. At the suggestion of his Prime Minister he appropriated the white deer throughout his Empire. A new decree was then issued whereby the custom of covering one's face when entering the court was improvised upon: visiting noblemen, courtiers and princes had to do so with the skin of the white deer, now obtainable only from the Royal park, and at no small expense. The skins were often passed from one nobleman to another and although they probably never entered general circulation the implications are quite clear.

The action taken by Emperor Outi was an early form of banking monopoly – just as the Bank of England, today, has a monopoly on the paper it uses for its bank notes.

Banking, unrelated to currency, dates a lot further back than the Han Dynasty. The term itself is relatively modern. The word 'bank' originates in Italy from the late eighteenth-century practice whereby financiers set up their benches – *banchi* in Italian – in the market place. Here money exchanged hands and small financial transactions took place. A money changer whose activities failed had his bench broken in half and the Italian word for broken, *rotto*, led to the adoption of the word 'bankrupt' in the English language. The concept of banking, however, whatever it may have been called at the time, is over five thousand years old.

Egyptian historians claim that 'cattle banks' were the first to be established about 3900 BC; cows and oxen were accepted on deposit, because they were the agreed medium of exchange in the community. Firmer evidence of the existence of banks, however, are the receipts, contracts and other documents discovered in Assyria and Babylonia. The engraved and baked clay tablets on which transactions were recorded date back to 2500 BC at least. The examples in the British Museum include a number of tablets discovered about 700 BC, in an earthenware jar in Hillah a few miles from Babylon. These belong to the Egibi and Son Company and it would appear that the family acted somewhat like the National Bank of Babylonia. An early transaction of one of the tablets dates it to the '4th day of Sivan, 8th year of Darius' and indicates a 'loan of 2/3 of a mann of silver . . . at an interest of one shekel monthly upon the manna'. The 'manna' contained sixty shekels; an almost exorbitant rate of interest (2.5 per cent per month), but the simplicity and business-like appearance of the transaction are admirably surprising.

During this same period there existed in Babylon the Rab-Dinab – the official Court Treasurer responsible directly to the King. He sealed the gold and silver collected from the community, and issued the precious metals. The Rab-Dinab also had the authority to prohibit the placing into circulation of gold and silver. This clearly indicates a controlled 'banking mentality'.

From the fourth century BC the Greeks are on record with diverse financial dealings undertaken by public bodies and private entities. The activities of the *trapezitae*, or bankers, were initially limited to changing money for foreigners, but gradually evolved to include accepting deposits on interest. Evidence of strong Greek influence in Italian banking is found in the works of early Latin writers about finance. Roman law, as early as the second century AD, appointed public notaries whose function was to record payments in discharge of debts. These duties gradually evolved to include clear bankers' functions. The necessity for these financial institutions can be appreciated when one considers the geographical areas covered by the Empire and the impracticability of transporting heavy metallic currency from one region to another.

Early literary works make repeated reference to the use of letters of credit. Stratocles, for example, purchased a bill in Pontus which was drawn in Athens by Iceratus on his own father. The bill was further guaranteed by Pasion. Similarly, Cicero wrote a letter to Atticus in Athens, enquiring whether his son Marcus, due in Athens to complete his education, need carry money with him, or whether alternatively, it was possible to 'procure a letter of credit' on Athens.

Trading and commerce were important activities in Rome, and banking declined when the Roman Empire began to fail in the middle of the first millennium. From this period, trade throughout the civilised world began to develop with increasing sophistication.

A banking watershed was reached in Europe in the fourteenth and fifteenth centuries. What has been claimed to be the very first national bank, that of Venice, had already been founded in 1157; but there is a great deal of scepticism as to whether this was a bank in the true sense of the word. Historical records indicate that in the second half of the sixteenth century the Bank of Venice began to take deposits – a practice that appears to be a pre-requisite to allowing an institution to be named a 'bank'.

The Bank of Barcelona, on the other hand, was the first 'real' bank. It opened its doors in 1401 and accepted deposits, discounted bills, exchanged money and had the backing of the city funds to guarantee all amounts of monies entrusted to it. The bank of St George, in Genoa, followed suit in 1407.

These early banking institutions ran parallel with many of the financial activities that were being undertaken by individual entrepreneurs. Commercial fairs of the period had a decisive influence on the establishment of an accepted system of finance, handled by individuals who acted, one may say, as 'clearing bankers on the move'. Their main duties involved the collection and delivery of currencies. Simultaneously, many European cities were becoming business centres for commercial activities. Each city developed its own banking system, while individual financiers established themselves as specialists in certain commodities.

The fourteenth century, for example, saw a great increase in the international wool trade. A number of banking concerns were involved in the financial aspect of wool-trading alone. Italy at the time was foremost among the trading nations of the world and the bankers with the revenues due were involved in making loans to the Vatican and other heads of state throughout Europe. The fairs became important business occasions, and this led to an interesting form of 'barter-on-paper'. Those present at the fair bought items from each other and the amounts due in payment were recorded in the seller's book, countersigned by the purchaser. At the end of the fair, the individual traders off-set debits against one another and balances were settled in cash. Gradually settlements of balances became payable by bills. These were non-negotiable documents; the debtor, in his own writing, committed himself to payment of a certain sum on a certain date to his creditor. This was the forerunner of the written bill of exchange as we know it today.

The origins of the deposit bank began in Italy. Merchants, who began losing money, through the depreciation in the value of coinage, deposited coin which could be accurately valued and had the sums credited to their names. These amounts were known as 'bank money'. Receipts signed by the banker were given to the depositor and these began to circulate as currency. They were equivalent to the seventeenth-century British goldsmiths, the fathers of English banking and bank notes.

By the sixteenth century banking activities were well covered by laws, trading rules and commercial practice. Through the sixteenth and seventeenth centuries further development of the written commitment took place. The bill of exchange was made transferable and the beneficiary could now pass it on to another creditor. Thus the doctrine of negotiability became established. The seventeenth century also saw the establishment of the modern banks in Europe. The Bank of Amsterdam opened in May 1609; the now famous Swedish 'Stockholm's Banco' (first to issue bank notes in Europe) followed suit in 1661 and the Bank of England came into being as a result of an Act of Parliament, in June 1694.

China

Europe was about to follow, nearly ten centuries later, in the footsteps of the economic wizards of the Chinese Empire. Chinese historians put claim on the first paper money issues *per se*. They were circulated throughout the dynasty in the seventh century AD.

The City of Glasgow Bank issued notes, cheques and bills of exchange, which complement each other very attractively in a collection. (Bill: 120 × 235 mm; cheque: 78 × 198 mm)

明 太 祖

Emperor T'aitsu Hung Wu Ming portrayed on a postcard reproduced from an early print. During his dynasty, *circa* 1368–99, the well-known Ming notes were first issued. (149 × 103 mm)

The money in question was known as *fei'-ch'ien* which translated means 'flying money', reference no doubt, to the easy transportability of paper issues. The 'flying money' was really no more than a certificate of deposit. There is, unfortunately, no physical evidence of the existence of these notes. The earliest issues of which specimens exist are the Treasury issues under Emperor Ching Tsung, during the T'ang Dynasty. The first of these were circulated in AD 825 and thereafter new and different issues were placed into circulation by successive Emperors.

Several examples of notes issued by Emperor T-tsung, dating back to AD 860, were discovered in 1833 in a private Chinese collection. They are the earliest notes in notaphilic history still in existence. By the time the Ming notes came into existence in the fourteenth century, China had already experienced complex financial evolution. 'Bank notes' both of government and private origin had already been circulated before the catastrophic Chinese inflation of the twelfth century.

The economic crisis of this period was not equalled until the German hyper-inflation of the 1920s, eight centuries later. China had learnt her lesson. Paper issues were abolished for the next two hundred and fifty years! In these early times paper money already gave an indication of the cancerous effects that it could have on a nation's economy.

The abolition of paper money was to be repeated at the end of the Ming Dynasty, which saw the termination of an era of considerable strife in the Chinese Empire. When the Ta Ch'ing Dynasty was set up by the Manchus in 1644, paper issues were abolished for a second time. They did not make a reappearance until the Taiping Rebellion in 1850.

A word of caution must be given here. In the eighteenth and nineteenth centuries, Ming notes were falsified. These are, however, easily identifiable because they were considerably smaller in size and the material on which they were printed was totally different in texture to the mulberry bark on which the originals were reproduced. Furthermore, because Emperor Hung Wu's title appears on all of the notes issued throughout the dynasty, it is impossible to determine exactly the date of issue of individual pieces. They all date between 1368 and the early seventeenth century.

The United States of America

The nearest equivalent to the Bank of England in the USA was the Bank of the United States. It was established as the national bank under a federal constitution in 1791. By this time the United States had experienced a chaotic financial period, to which paper money had contributed only further confusion. The break of the colonies from British rule in 1775 brought an almost absolute freeze on foreign trade and to finance its economy the new-born government had to resort to paper money. The Continental Congress authorised the issues in 1775 and they entered circulation shortly after the Declaration of Independence was signed, the following year. Today's colloquial term 'not worth a Continental' is ample witness to the fate that awaited the notes in question. By 1781 the 'Continental currency' was inflated beyond commercial use; the issues were suspended.

Then state chartered banks appeared on the scene. The earliest of these was the Bank of North America. It opened its doors in Philadelphia, less than three months after the British were defeated by Washington in Yorktown on 19 October 1781. The Bank issued its own notes redeemable *in specie* and for the first time a workable currency system was established. Issues by individual money lenders and the government and colonial 'bills of credit' lost public faith. The successful financial activities of the Bank of North America encouraged a number of new banks to be formed.

Unlike the Bank of England, the Bank of the United States encountered one unsurmountable difficulty: policital opposition. At the end of the twenty-year charter period authorised by Congress, the Bank was unable to continue operating. Thus in 1811, the United States was without a government banking authority. The economy was being primarily run with treasury issues, which were effectively interest-bearing promissory notes. With one exception they never circulated as currency money. The exception was the 1815 small denomination notes bearing no interest; but they too remained in circulation for only a brief period of time. The only other currency in circulation consisted of 'loan notes', issued by a large number of banks to their borrowers. The ever-increasing issues of these notes, without central banking control, began to cause an inflationary trend which was again endangering the whole of the economy. Finally, in 1816, Congress chartered a second Bank of the United States.

The note issues of the Bank of the United States were well trusted by the populace. They were referred to as 'good as gold' and freely circulated throughout the country. Unlike the notes of the many smaller banking enterprises, they circulated at face value. The second Bank of the United States successfully carried out the responsibilities of restraint which led to the smooth functioning of the monetary system. It was, however, a short-lived success. In the 1830s under the leadership of Andrew Jackson, criticism began to mount; small banking institutes bitterly complained of discrimination. Jackson's supporters claimed the Bank to be undemocratic and monopolistic. The opposition became insurmountable and in 1836 the bank ceased its activities. The huge number of local and private banks that had set up business by then only added to the resulting economic chaos in the country.

In 1838 payments were resumed through bankers established in New York. For the next twenty-seven years, until the enactment of the National Bank Act of 1863, there was no federal responsibility for the circulating medium. Initially, this was left to the banks chartered under state laws and private banking enterprises. Although in 1787 only three state chartered banks were in existence, by the end of the next decade, almost sixty such banks operated throughout the Atlantic coast of the United States. Soon the note issues of each individual bank began to circulate at a discount, whenever they were being negotiated in another state; the farther the distance, the higher the discount. As a result the period between 1790 and 1860 saw the increasing availability of periodicals known as 'bank note reporters' and 'counterfeit detectors'. The aim of these regular publications was to inform the public of the rates of discount applicable to the bank notes which originated outside the state in which they circulated. The 'counterfeit detector' consisted of instructions for ready identification of forged notes. Some of the later counterfeit detector booklets were published by permission of the Treasury and included reproductions of United States currency taken from original plates. The best known of these, *Heath's Detector*, has become an expensive collectors' item.

Through the period in question, a wide divergence in local legislation existed. Some states had no legislative regulations relating to bank notes, whereas others enacted almost prohibitive laws in an endeavour to control internal economic conditions. From 1836, a free banking system was gradually established allowing businessmen to set up a bank as they would set up any other commercial venture. 'Free banking' quickly spread throughout the United States. It brought about some well-founded bank note issues, but others were extremely frail. The less scrupulous among these became known as 'wildcat banks'. Their notes are known to collectors as 'broken bank notes'.

Simultaneously many state institutions, under official supervision, began to establish their own banks.

On 12 February 1863, the National Bank Act was passed (revised in June 1864), which provided for a uniform national currency to replace the wide range of varied state bank notes in circulation. Of greater consequence, however, was the subsequent action taken by Congress. A tax of ten per cent was levied, from August 1866, on the notes issued by any bank. State banks were thus forced to redeem their issues. Many were, thereafter, formed by federal charter and known as national banks. Under this title, they were authorised to issue their own notes, guaranteed by government bonds. United States currency began to circulate at an equal rate everywhere.

The National Bank Act of 1863 coincided with the establishment of the Bureau of Engraving and Printing of the United States. A year later the Bureau of the Treasury Department was established, but it was only in 1869 that Congressional legislation brought into being the Bureau as a distinct entity within the Treasury Department. By October 1877 all United States currency was being printed by this entity.

The circumstances surrounding the 'Continental' issues of 1775 repeated themselves soon after the outbreak of the American Civil War in April 1861. In July of the same year, an Act of Congress authorised the Treasury Department to issue legal tender notes. In spite of the fact that payment *in specie* was suspended, the public accepted the issues in good faith. These notes comprised the first bank notes of the United States: the 'demand notes'. The term 'greenbacks' originated from these early issues. In the following year, 1862, the 'legal tender notes' were placed into circulation and hence banking practice and legislation in the United States was established as a co-ordinated national activity.

The next most important development in the United States banking history was the Federal Reserve Act of 1913. Almost all of the large banks in the United States co-ordinated their activities in a membership system. This Act authorised the creation of twelve Federal Reserve Districts each with a Federal Reserve Bank. The banks were compelled by law to subscribe to the Federal Reserve system. Other banks had the option of joining the system if they so wished.

The monetary aspect of the Federal Reserve Bank is such that capital stock is owned by the member banks of the district. The latter are obligated to subscribe to the capital stock of the Federal Reserve bank in their district. The current Federal Reserve notes – the modern 'greenbacks' (which constitute the bulk of the money in circulation throughout the United States) bear witness to the success of the Federal Reserve Act.

The Bank of England

The Bank of Stockholm, established in July 1661, was the first European bank to issue bank notes proper. These issues, however, were not government backed and the enterprise was forced to close its doors within a few years. The title of 'the first western bank to issue national bank notes on a continuous basis' falls to the Bank of England.

Medieval England's economic standing had not kept up with the pace of the large European commercial centres. European financiers were mainly concerned with making loans to Royalty and financing international transactions. The money they used for the loans consisted of the deposits made with them by merchants as well as their own capital. The bank note was unknown and the catalyst to the transactions, the instrument used, was the non-negotiable bill of exchange.

Trade began to expand in England during the Tudor period. The London scriveners were the first to act as money lenders; they were followed by the goldsmiths, who were finally instrumental in establishing paper currency in England. These affluent businessmen acted as bailees, accepting deposits in the form of coins, jewellery and other personal valuables for the purpose of safe-keeping. Against these valuables they issued receipts, undertaking to return the item in

The Federal Reserve notes were authorised by an Act of December 1913 and denominations from 5 to 1,000 dollars were issued. The notes were redeemable in gold until 1934 at the Treasury in Washington DC or in 'lawful money' at any Federal Reserve Bank. (190 × 77 mm)

question on demand. The receipts were often transferred from the owner to other individuals, in normal trading transactions. It was logical for the later goldsmiths to begin issuing promissory notes which were *not* backed by specific deposits but more like loans, and these seventeenth-century promissory notes can be claimed to be the first paper money issues intended for circulation.

It was an Act of Parliament in 1694 which led to the foundation of 'the greatest commercial institution that the world has ever seen'. But the Bank was brought into being by a bill the aim of which, clearly, was not the formation of an everlasting financial body. The purpose of the bill, brain-child of Scotsman, William Paterson, was 'for granting to Their Majesties . . . the sum of fifteen hundred thousand pounds towards carrying on the War against France'. The sum needed for the war was to be raised through a public loan. The bill was unanimously passed through Parliament on 21 June 1694 – but did not specify that the Bank had powers to issue paper money! By 25 June, over £100,000 had been subscribed by wealthy merchants supporting the government. The Bank of England was set up; and this greatest of institutions had at the head of its court of directors two grocers – the Governor of the Bank, Sir John Houblon and the Deputy Governor, Mr Michael Godfrey.

Within the first week of its existence, the Bank issued three forms of paper money: the 'sealed bill', an interest-bearing promissory note given to depositors in return for cash; the 'Running Cash note', also in the form of a receipt for deposits, but made out to the bearer, thus allowing it to be transferred; and the 'Accomptable note', a certificate of deposit, but not intended to be passed from hand to hand. The accomptable note allowed the depositor to 'draw' notes. All of these note issues were written out by hand.

The total number of bank notes in circulation at the time of the most serious crisis faced by the Bank in the mid-eighteenth century was £3 million. These notes were in printed form and appeared in fixed denominations. By the end of the Napoleonic Wars, in 1795, the total circulation had jumped to £13½ million and by 1821 to £26 million. The Bank Charter Act of 1844, in order to keep control of these notes, fixed a fiduciary (without gold backing) issue limited to £14 million. The law, however, allowed for increase. By 1919, after the Government issued its own 'Treasury notes' (for £1 and 10s.), the number of notes placed in circulation by the Bank of England was nearly £70 million. (The 'currency notes' of the Treasury totalled over £300 million.) The Currency and Bank Notes Act of 1928 authorised the bank of England to issue, for the first time, £1 and 10s. notes with full legal tender status.

At the outbreak of the Second World War, gold held by the Bank's issuing department was transferred to the Exchange Equalization Account and as a result the issues of bank notes increased to £580 million, reaching £1,350 million by the end of the war. In 1973 the amount in circulation was in excess of £3,000 million!

In 1724 the privilege of manufacturing paper for the Bank was entrusted to Henry Portal and the family link continues to this day. A year after entering into agreement with Portals, the Bank of England decided to issue partially printed notes. Copper plates were engraved for the printing of the denomination, while the remainder of the note was written in by hand. It was not until over a hundred years later, in 1853, that hand-written figures and words were to be finally eliminated from the bank note.

The printing and engraving of notes until 1791 had been undertaken by private individuals. James Cole took over the engraving for the Bank of England in 1721 and the printing ten years later. He continued with the Bank until his death in 1748 when his son, George Cole, succeeded him. In 1791 George Cole's business was transferred to within the Bank's premises. Only some two thousand notes were printed daily before the printers moved into the Bank. With the introduction of the low denomination notes in 1797, there was a great increase in demand and by 1805 over thirty thousand notes were being printed each day.

The practice of the Bank had been to enter into private contracts with individual printers. This was changed in 1808, when Garnet Terry was sworn into the Bank's service at a substantial salary, and he gave up his private business. Since that date, the printing of Bank of England notes is undertaken by the Bank's staff in its own department.

Private Banking in the United Kingdom

Country banking in the UK had been growing slowly in the early eighteenth century; when Edmund Burke, the Irish-born statesman, said in 1750, that 'there are not twelve country banks in the whole of England', he was undoubtedly overlooking the part-time activities of many merchants. From the mid-eighteenth century, private banking began to gain momentum. When the Bank of England was prohibited from redeeming its note issues in gold in 1797, country banking began to

The Tweed Bank was a typical English provincial bank founded in 1800, which went bankrupt in 1842. The stamps on the right-hand side indicate the dividends paid to the creditors. (109 × 185 mm)

flourish. The additional government sanction allowing bank notes of less than £5 gave rise to private note issues which circulated in abundance.

It was not until 1826 that a banking concern could be established on the basis of company law. Until that date, the existing banks were based on partnerships, often family owned, and without government control over note issues. The existing weakness of the economy, brought about by the several crises which periodically swept through the country, caused the collapse of many of the country banks. Between 1791 and 1818 alone, over one thousand banks suspended payment. In 1826 the Bank Act was passed through Parliament and joint-stock banks were allowed to establish themselves. The Bank of England, however, kept its monopoly within a sixty-five mile radius of London for the next seven years.

In 1833 joint-stock banks were allowed to operate in London but could not issue notes and suffered from considerable additional legal difficulties until the Bank Charter Act of 1844. By now the Bank of England had a network of branches established throughout the country. Its bank note issues circulated far and wide and began to replace private note issues. The Bank Charter Act of 1844 is a landmark in British financial history. It checked the circulation of the notes of banks already in existence by limiting the quantities issued; the Act also forbade note-issuing rights to any new banks which were to be established; and finally it provided for the eventual closure of many private banks which lost their right of issue when they were absorbed into other banks.

Towards the end of the century the Bank of England began co-operating rather than competing with the commercial banks. The latter continually increased their balances with the Bank of England. Many British banks in the meantime were established specifically to operate abroad.

In the same way many foreign banks began to open branches in London. This led to the amalgamation of several small banks. Between 1890 and 1918 the British banking system consisted of a small number of very large banks. With the outbreak of the First World War, only thirteen joint-stock banks, operating through a system of branches, were in existence in England and almost no local banks had survived. In 1917 and 1918 a climax was reached when a number of large banks merged. But there have been few amalgamations leading to the formation of 'giants' since the Great War.

Central Banks

There are two modern systems which operate side by side in most countries: the commerical and central banking systems.

The main purpose of a commercial bank is to earn profit for itself and its shareholders. Commercial banks are involved in all aspects of normal commercial transactions. Today it is very rare for a commercial bank to issue notes. This function is normally the monopoly of a Central Bank.

A Central Bank can be defined as the commercial bank's banker. This is the governmental institution charged with the function of regulating all of the money matters within a country's economy. Most Central Banks are state owned, but they may be private institutions (the Bank of England was only nationalised in 1946). In any case they are always subject to state control. They, in turn, must have the means of controlling the commercial banks within the country. International loans and the government's fiscal negotiations are undertaken by the Central Bank's officials. Thus the Central Bank has a general influence on the use of money and the economy.

England has a strong centralised banking system, and the 'influence' of the Bank of England on commercial banks has been mainly in the form of supervision. The many thousands of banks in the USA, however, are independent, and their supervision is granted to the Central Bank by law; the Federal Reserve Bank, for instance, issues bank currency in conjunction with Treasury issues and other government notes.

A Central Bank's main duty, today, is that of note issuing. But this need not be the general rule. The central or state bank of the USSR, for instance, acts as a deposit bank whereas the legal tender notes and coins are issued by the State Treasury.

Currency Names

It is of interest to note the derivations of some of the names we give to our currency denominations, most of which originally referred to coinage and were adopted for paper money when it came into use. Since coins were valued by weight, some names are also those of the standard weight used at the time. For example, the Spanish 'peso' comes from the word 'weight', and has been adopted by many Latin American countries. The English 'pound' was an Anglo–Saxon weight, and the Latin equivalent was the 'libra' from which many modern denominations have derived their names, including the French 'livre'. The French 'franc' can be traced back to John II of France who, in 1360, had gold coins struck on which the Latin legend included the words 'of France' translated as *francorum*; the people came to refer to the coins as francs simply from an abbreviation of this legend.

The American dollar derives from a coin called the 'thaller' which was struck in Germany in the sixteenth century, following the discovery of a silver mine in Bohemia. The name was introduced to America by immigrants and was officially adopted as a name by Congress in 1784. By 1792 the content of gold and silver in a dollar had been formally defined and it became the official currency of the land thereafter.

An A to Z of Countries

This chapter is intended as a general guide to the issues and values of the paper currency of the world, existing and defunct. It is not intended to be a complete catalogue of *all* the issues of any one country. The issues of some dependencies have been collectively covered under the name of the present or one-time mother country. Thus, of the seventy or so Commonwealth countries, British colonies and protectorates, nearly half have been considered under the heading 'British Commonwealth'. Similarly, Yugoslavia, for instance, includes general coverage and values of the note issues of Serbia, Montenegro and Croatia.

The select bibliographies show the sources of the information given in the chapter. Where the specific issues of a country include emergency notes, these have been only briefly mentioned in this chapter and covered in greater detail in the next. This applies equally to the technical aspects of some notes, which have been separately considered in Chapter V. Reference to the index will guide the reader to the different chapters in which issues of a specific country have been discussed.

Values
The reader is asked to bear in mind that values are applied in a general way and that they refer to the lower denomination of an issue. The notes with a denomination of over 100 are almost always more expensive.

The price lists, auction catalogues and other material consulted for the purpose of pricing the notes were dated before January 1976. Allowance has been made for an increase in price in the immediate future. Where dollar equivalents are given in brackets, the approximate rate of £1 = $2 has been used.

Condition
The prices given, unless otherwise stated, relate to a bank note in good collectable condition. The word 'good' throughout this chapter is used in its normal sense and not as a grading term.

Catalogues
The bibliographical listings after each country sometimes consist entirely of articles that have appeared in numismatic and notaphilic journals, if this is the only available material. Four basic works have been omitted to avoid repetition. These are:

Pick, A. *Catalogue of European Paper Money since 1900*, New York, 1971
   Catalogue of Paper Money of the Americas, New York, 1973
Sten, G. K. *Bank Notes of the World*, Vol. I, A–C and Vol. II, C–K, California, 1967

A

ADEN (Now People's Democratic Republic of Yemen)
See British Commonwealth

AFGHANISTAN (Aryana)
Currency: 100 puls = 1 afghani
(Also rupees and caboulies)

The early notes of Afghanistan, dating back to the beginning of the century, were printed on very low quality paper and many were destroyed before they were withdrawn from circulation. Although the notes are comparatively rare, they can still be purchased for £10 ($20). Many later issues, up to 1936, can be purchased for even less. The low denomination modern issues, portraying King Mohamed Zahir, are still obtainable at a few pence or cents.

NB Afghanistan used the Muslim calendar.

Bibliography:
> Anon. *Doulate e Padshahi ye Afghanistan da Afghanistan Bank*, CC, Summer 1968
> Panish, E. C. K. *Catalogue of Afghan Bank Notes*, WC, January and February 1968

ALBANIA (Shqipni or Shqiperia)
Currency: 20 qindtar = 1 lek(e) since 1946
5 leke = 1 franga (ari = gold)

The first Albanian issues were put into circulation when the country was under siege by Montenegrin troops in 1913. They are notes that rarely come onto the market at prices below £20 ($40).

The issues under Italian rule, with designs almost identical to Italian ones, are fairly common at about £4 ($8). Later issues of the Republic itself, excluding the 1945 overprinted notes which are highly valued, do not exceed the same price range.

Bibliography:
> Bobba, C, *Cartamoneta Italiana*, Asti, Italy, 1971 (4th edition)

ALGERIA (Algérie)
Currency: 100 centimes = 1 dinar since 1962
(Also francs before 1962)

With the exception of the 1873 and 1903 type notes, there are not many rare Algerian issues. The most highly priced should not exceed £12 ($24). The Banque de l'Algérie notes were overprinted 'Tunisie' after 1904. Until that date the notes circulated both in Algeria and Tunisia.

The notes of the 1942–44 period are very easily obtainable in all denominations.

Pre-Second World War notes almost invariably are found in uncollectable condition. Those that are near crisp are more expensive but still quite cheap at £5 ($10).

Emergency issues, circulated by municipalities during the First World War, are scarce items but due to a lack of demand they can be purchased for as little as £2 ($4) each.

Bibliography:

> Philipson, F. *Paper Money of Algeria and Tunisia,* I and II, CM, January and February 1973
> Toy, S. *World War II Allied Military Currency,* Arizona, 1969

ANGOLA (Luanda)

Currency: 100 centavos = 1 escudo since 1956
100 centavos = 1 angolar, or escudo Angolano until 1956
1,000 reis = 1 milreis until 1918

The Ministry of Finance notes of Angola were authorised by a law dated 1867 and the Banco Nacional Ultramarino issued notes in Angola as early as 1866. As in other overseas provinces of Portugal, many of the issues of Angola are Portuguese notes overprinted with the name of the province. The notes of the Banco Nacional Ultramarino before 1918 are in the milreis denomination and overprinted 'Loanda'.

The higher denominations in extremely fine condition are highly priced. The notes overprinted 'Angola', which appeared after 1918 and circulated until 1937, are generally cheaper, priced at below £10 ($20).

The Banco de Angola issues since 1937 are also fairly common, but the higher denominations become relatively more expensive. The 1,000 angolares of 1951 is priced at about £65 ($130).

Emergency issues are invariably primitive in appearance – the 1920 Albanian skutari notes are a typical example. (58 × 84 mm, actual size)

Although the Banco de Angola continues to be the government issuing body today, between 1921 and 1948 another issue emphasised Angola as a Portuguese province. These were the notes of the 'Republica Portughesa – Provincia de Angola'; they are extremely attractive and are still in the £8 ($16) range or less.

Bibliography:
> Rebelo de Sousa, L. M. *O Papel Moeda em Angola*, Luanda, 1969

ANTIGUA

Currency: 20 shillings = 1 pound sterling (government and UK bank issues)
100 cents = 1 dollar

Antigua is among the elite British colonies that has some early paper money issues fetching the highest prices on the market. The Government of Antigua notes of 1914 for 4 shillings and 10 shillings are particularly prized items. The same Government issue (never placed into circulation) of 1938 is equally rare. Other pre-1950 issues are those of the Royal Bank of Canada and Barclays Bank DCO, originally intended for circulation in Jamaica. The price range is above £100 ($200) for all these notes.

The notes in Antigua after 1951 were the issues of the British Caribbean Territories, and from 1965 those of the East Caribbean Currency Authority.

ARGENTINA (also the Falkland Islands)

Currency: 100 centavos = 1 peso

The note issues of Argentina after 1880 are those of the Republic and many can be obtained at between £3 ($6) and £8 ($16). The earliest issues of the Confederation headed 'El Gobierno Nacional' and dated either 1857 or 1859 are also surprisingly common and most do not exceed £25 ($50).

The more expensive Argentinian notes are those of the individual provinces (excluding the issues of the Province of Buenos Aires). The notes dated before the formation of the Federation, between 1826 and 1850, fetch about £30 ($60).

The Bank of London and the River Plate notes are, understandably, highly popular. Although a hoard of the issues of the Montevideo branch in Uruguay were recently found, the same bank's notes, issued by the Cordoba and Rosario branches in the Argentine, are far more expensive; the higher denominations reach the £40 ($80) level. The lower denominations are characteristically unattractive in comparison to the colourful larger sized issues of Latin America in general.

Many private banks had branches in other capitals and notes were thus redeemed outside the country. The legend on some Argentine notes, reading 'Pesos Bolivianos' or 'Plata Boliviana', signifies the quality of silver in which the notes were redeemable. It does not necessarily indicate that the notes circulated in Bolivia. Most early Argentinian notes were printed by the San Martin Printing Company in Buenos Aires and the printer's name appears in small print on the side of the notes.

Falkland Islands (South Atlantic – British since 1833)

An extremely interesting issue, reported but not yet verified, is that of the Military and Political Argentine Commandant of the Falkland Islands. These issues for 1 and 5 pesos, dating back to 1829, are reported to be headed 'Islas Malvianas'. Otherwise the British Notes of the Government of the Falkland Islands have been the only circulating notes. These are popular Commonwealth issues and some of the earlier notes with George V's portrait fetch over £60 ($120).

Bibliography:

Beresiner, Y. L. *Custom House Acts as Treasury*, CC, 27 February 1971

Matz, A. C. *Overprints on South American Notes*, CC, Summer 1966

Pick, A. *National Bank and the Guarantee System in the Argentine*, LANSA, June 1973

AUSTRALIA (also Tasmania)

Currency: 20 shillings = 1 pound sterling

100 cents = 1 dollar since 14 February 1966

Early notes of Australia, between 1817 and 1880, do not enter the market and therefore cannot be realistically priced. The only known specimens of such notes are to be found in a number of Australian public institutions. There are no common Australian notes of the last century. The cheapest would fetch £100 ($200) or more.

Buenos Aires in the 1840s was a separate Argentinian state and as it was in favour of Confederation, it used bank notes for propaganda purposes. The heading on this note states 'Long live the Argentine Confederation! Death to the savage unitarians!' (84 × 87 mm, actual size)

Notes of this century are not easily obtainable either. The decimalised dollar issues since 1966 fetch a price nominally over their circulating value, but previous issues of the Commonwealth of Australia, portraying George VI, can only be purchased at several pounds over their face value. The earlier 'sovereign' notes and those redeemable in gold are priced between £12 ($24) and £40 ($80), depending on the denomination, and the 1912 issues will fetch up to £30 ($60).

A large number of private Australian banks issued notes. Those between 1910 and 1914 were overprinted by the government and are expensive. All the issues by the provinces are in the £40 ($80) bracket. Collectors of Australian notes must consider additional issues.

Tasmania

As with Australia, Tasmanian note-issuing banks were all founded in the last century, the majority between 1823 and 1840. There is no exception to the scarcity of these; all are priced above £100 ($200). Many are unobtainable and those that can be purchased are almost invariably found in a dilapidated state. This, however, does not seem to detract from the high value given to them by collectors.

The Bank of New Zealand had some notes which were overprinted for use in Australia and circulated there. Again these are highly sought after and must be among the rarest issues, particularly those dating to the last century.

Bibliography:

Freehill, M. *Australian Star Replacement Notes*, IBNS, June 1972

Hyman, C. P. *An Account of the Coins, Coinages and Currency of Australia*, Sydney, 1893 (reprint London, 1973)

McNiece, R. V. *Coins and Tokens of Tasmania, 1803–1910*, Hobart, 1969

Skinner, D. H. *Rennik's Australian Coin and Bank Note Guide*, Adelaide, 1969 (4th edition)

Tomlinson, G. *Australian Bank Notes 1817–1963*, Melbourne, 1963

AUSTRIA (Österreich)

Currency: 100 kreuzer = 1 gulden (florin) since 1717
4 pfennig = 1 kreuzer
100 heller = 1 krone since 1892
10,000 old kronen = 100 groschen = 1 schilling since 1924

Rather surprisingly, the earliest paper issues of the Austro-Hungarian Empire are not too difficult to come by. They date back to the Seven Years War between 1756 and 1763 and were first issued by the City of Vienna in very large quantities. They are known as the Banco Zettel notes and were an emergency replacement measure. The higher denomination will fetch about £35 ($70) in good condition. Specimen notes of these issues on hard cardboard have made an appearance. The prices of specimen notes are always subject to the individual collector's interests. Some specimen notes, particularly early ones, have fetched prices in excess of £300 ($600).

The most difficult Austrian notes to come by are the odd-denomination notes (30,

90,600 gulden etc.) issued by the 'K.K. Zentral-Staatskasse' in 1848. The note issues of the Austro-Hungarian Bank of this century, which have the denomination on the notes in ten languages, are priced under £3 ($6).

Modern Austria's notes date only as far back as 1919; some are early Austro-Hungarian Bank notes overprinted 'Deutschösterreich' in 1919/20. These issues are witness to the first German involvement in Austrian affairs. The Austrian National Bank was formed following the currency reform of 1924, and issued notes until 1938. German currency circulated in the country after this date.

The Second World War also saw some Russian notes issued by the Allied Military Administration. With few exceptions, most of these notes can be obtained at below £12 ($24).

The Austrian National Bank was reactivated in 1945, and continues to issue legal tender notes.

The Austrian Empire also issued notes during the occupation of Venice, in the First World War. These are the 'Casa Veneta dei Prestiti' issues. The lower denominations, up to 50 lira, are common and priced at about £3 ($6). The higher, that is the 100 and 1,000 lira notes, are much rarer and valued at about £25 ($50) and £150 ($300) respectively.

Bibliography:

Austrian National Bank *Österreichische Nationalbank 1816–1966*, (in Austrian), Vienna, 1966

Jones, D. J. *Austrian Currency*, CM, May 1970

Kupa, M. *Paper Money of Austria-Hungary*, Budapest, 1965

Pick, A. *Casa Veneta dei Prestiti*, IBNS, Autumn 1969

Pick, A. & Richter, *Österreich Banknoten und Staatspapiergeld an 1759*, Berlin, 1973

Spajic, D. B. *Austro-Hungarian and Bulgarian Notes*, IBNS, June 1965

AZERBAIDZHAM (Azerbaijan) – see USSR

AZORES – see PORTUGAL

B

BAHAMAS

Currency: 20 shillings = 1 pound sterling
100 cents = 1 dollar since 1965

The only issues made by a private bank, that of the Bank of Nassau, are the rarest of the Bahamas issues. £659.6 shillings of this Bank's notes remain unredeemed as of 1970 and the notes are highly sought after. Prices in excess of £200 ($400) have been paid for them.

The notes issued this century are those by the authority of the Currency Acts of

1919, 1936 and 1965; the 4 shilling notes of 1953 tend to be overpriced, as fractional curiosities. These and previous issues of the Government are priced in direct relation to the denomination and date. The £5 1919 note is the highest and earliest and therefore the most expensive of the issues, priced at about £350 ($700).

With the exception of the 1965 issue, the notes of the Bahamas after decimalisation in 1965 are the common modern dollar notes. When in crisp condition they are not priced at more than a small premium over their face value.

BAHREIN (Bahrain)

Currency: 1,000 fils = 1 dinar

Until 1965 the notes circulating in Bahrain were those issued by the Reserve bank of India. There is very little information regarding these issues. Mr Richard Leader, a recognised expert in the field, says that British influence in the area involved led to the distribution in the Gulf of the Indian issues, printed in England and duly overstamped. They are all rare items. The only notes of Bahrain itself are the issues since 1964, which include the $\frac{1}{4}$ and $\frac{1}{2}$ dinar notes of the Bahrain Currency Board.

BANGLADESH (Formerly East Pakistan)

Currency: 100 paisa = 1 taka

Immediately after independence, the Bangladesh Government overstamped Pakistani 1, 5 and 10 rupee notes while waiting for the printing and delivery of its own taka issues. The overprinted notes were formally withdrawn from circulation on 8 June 1972. It would seem that some of these notes were illegally rubber stamped and are Bangladesh counterfeits but remain of interest as contemporary forgeries. The originals fetch up to £10 ($20).

The modern notes of the Bangladesh Bank are obtainable at just over face value.

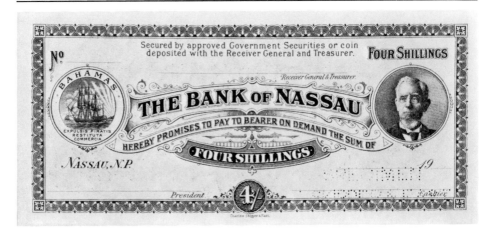

There are few surviving specimens from the private banks of the Bahamas. The 4 shilling Bank of Nassau specimen is extremely desirable for this reason, and also because it is an odd denomination. (87 × 188 mm)

BARBADOS

Currency: 20 shillings = 1 pound sterling
100 cents = 1 dollar
100 cents = 1 British and East Caribbean dollar (1950 to 1973)

Barclays Bank, the Royal Bank of Canada and the Canadian Bank of Commerce issued notes for circulation in Barbados. The 100 dollar denominations are the rarest, well above the £200 ($400) price range.

The only recorded pound sterling issue is dated 1917 but it is not quite as rare as the Canadian 100 dollar note.

Issues from 1909 to 1949 decrease in value in direct relation to date.

In 1951 the British Caribbean Currency Board began its issues, and all notes previously in circulation were withdrawn. The Central Bank was established in 1973.

The new issues are comparatively common; they are priced according to their denomination.

BELGIUM (also Belgian Congo and Katanga)

Currency: 100 centimes = 1 Belgian franc
1 belga = 5 francs

Belgian issues are all bilingual: French and Flemish. The National Bank of Belgium (Banque Nationale Belgique – National Bankvan Belgie) has been the note issuing authority throughout Belgium's monetary history. The earliest notes date back to the 1850s, fetching prices over £180 ($360) each.

All notes were dated and issued by type, until 1922. There is, therefore, a wide range available to the collector. Issues dated after the Great War do not cost more than a few pounds for the higher denominations. Nor are the several Treasury notes, and the war issues, scarce. The general rule is that high denomination notes (up to 1,000 francs – with one 10,000 issue dated 1920) are the expensive ones.

Congo

In 1885 the Congo Free State was established and annexed to Belgium in 1908. Ruanda-Urundi was added to the State in 1916 and only seven years later the capital was changed from Boma to Leopoldville. The notes of interest to the collector are those issued in the franc denominations, prior to the establishment of the Republic of Congo in 1960.

The Banque du Congo Belge notes are bilingual and extremely attractive. The most expensive of these notes is the 10,000 franc note, issued in the 1940s but not dated. Only a small number of these were printed and most of them were redeemed before they were declared invalid. They are now valued by collectors at about £150 ($300). The Banque du Congo Belge (et du Ruanda-Urundi, after 1952) is the only bank that issued notes between 1913 and 1960. Notes dated after 1952 are easy to come by. Some of the earliest notes, on the other hand, such as the issues between 1924 and 1940, will fetch over £15 ($30).

Katanga

The short-lived Republic of Katanga was one of the Congo's six provinces under President Moise Tshombe. The first notes issued in July 1960 by the National Bank of Katanga portray Tshombe, though in 1962 this portrait was replaced by a local scene. Denominations from 10 to 1,000 francs were issued. Because Katanga was in existence for only three years the notes are very popular, but are still relatively cheap. They fetch approximately £65 ($130) per set.

Bibliography:

Hoche, J. *Notes of the Belgian National Bank*, IBNS, Summer 1962
  Paper Money of the Congo, CM, October 1972
Loeb, W. M. *New Issues*, WNJ, October and December 1964
Maes, A. *Portraits of Kings and Queens on Belgian Money*, IBNS, Autumn 1969
Morin, F. *Catalogue des Billets de Banque Belges de 1900 à 1966*, (in French) Brussels, 1967
Philipson, F. *Two Faces of the Congo*, IBNS, Summer 1969

BERMUDA

Currency: 20 shillings = 1 pound
100 cents = 1 dollar since 6 February 1970

Most issues of the former Belgian Congo have native scenes as their design and the text is printed in both French and Flemish. (100 × 158 mm)

In 1970 the Government of Bermuda decided to decimalise; notes as of that date are easily obtainable at face value. A number of sets of specimen notes of the dollar issues have entered the market priced at £40 ($80) per set. They are extremely picturesque.

The rarest notes of Bermuda belong to the Merchants Bank of Halifax. These last-century notes were issued in the dollar denomination but overprinted in pounds and shillings. Issues of this bank are valued well above £200 ($400).

The Bermuda Government issued notes in 1915 and 1927, which are rare and valued at over £100 ($200); the subsequent notes issued by the Government, however, till 1964, become gradually less expensive.

The portraits of the monarchs appearing on the notes are a reliable guide to date, and thus help the collector to value the notes.

BIAFRA

Currency: 20 shillings = 1 pound

In its short history Biafra issued two series of notes in denominations from 5 shillings to £10. The first series of 1967 consisted of two notes only, the 5 shilling and the £1; which can be obtained as a pair for about £10 ($20).

The second series, printed in Switzerland two years later, is more easily obtainable. The 5 shillings and 10 shillings are priced at under £2 ($4) each. The £5 and £10 range between £5 ($10) and £8 ($16) each, and the £1 is very common.

Towards the end of the War, Biafran agents in Geneva openly sold bank notes to the public in different denominations at premium prices. Biafran notes are therefore not very rare.

Although Biafran notes when issued by the Bank of Biafra were at par with the

The Republic of Biafra had notes printed in Switzerland. Many were sold to collectors in complete sets. These attractive notes are not particularly rare. (80 × 155 mm)

pound sterling, within a few months inflation had taken a grip on the economy and brought about a situation where one single egg was priced at £2!

BOHEMIA (also Bohemia-Moravia) – see CZECHOSLOVAKIA

BOLIVIA

Currency: 100 centavos = 1 boliviano
   1,000 bolivianos = 1 peso boliviano since 1962

The first government issues of the Banco Boliviano in the last century were taken over by the Banco Nacional de Bolivia which then changed its name to El Banco de la Nación Boliviana and in 1928 became the Banco Central de Bolivia.

The Italian-printed 1 boliviano 1911 issue of the Banco de la Nación Boliviana is an exception in the series; where the Central vignette of Hermes appears on the ordinary note here it is replaced by a watermark. In the 1973 'Farouk Auction' the note fetched $200. This is the only scarce note of the series.

A beginner should be able, easily and economically, to form a good representative collection of the issues of the Central Bank spending no more than £20 ($40) on the most expensive. The 1945 series are the commonest.

The earlier notes of the Banco Nacional dating back to 1870 are more expensive. It is difficult to obtain any of these notes in EF or better condition and they are priced in the £15 to £30 ($30 to $60) range.

Regarding the confusion over the use of the words 'Plata Boliviana' or 'Moneda Boliviana' on some notes, refer to 'Argentina' in this chapter.

Bibliography:
   Matz, A. C. *Latin American Notes*, IBNS, Autumn 1962
   Matz, A. C. *Overprints on South American Notes*, CC, Summer 1966
   Philipson, F. *Bank Notes of Bolivia and Chile*, CM, August, 1971
   Seppa, D. A. & Almanzar, A. S. *The Paper Money of Bolivia*, Texas, 1972

BRAZIL

Currency: 100 centavos = 1 cruzeiro
   1000 reis = 1 milreis (up to 1942) = 1 cruzeiro

Most collectors will have come across the cruzeiro issues of the 'Republica dos Estados Unidos de Brazil', ranging from 1 to 10,000 cruzeiros. Some bear the legend 'Valor Legal' while other issues of the same denomination read 'Valor Recebido'. Different printers have produced identical notes; overprints of new values are abundant and even errors have been determined and repeated in different issues. A complete collection of these issues would consist of some sixty notes. The lower denominations are priced at under £1 ($2).

The hand-signed issues are the rarer, and the 200 cruzeiro hand-signed note is considered to be among the scarce Brazil issues, valued at £40 ($80) or more. The higher denomination cruzeiro notes do not fetch more than £15 ($30) each in mint condition. These are all undated but they circulated from 1943 to 1970.

The earlier milreis issues, and particularly those of the Empire, are far more attractive and expensive, some early ones rising to the £50 ($100) range.

Private bank issues, such as those of the Banco do Brazil and Banco do Bahia, are also difficult to come by and fetch above £25 ($50).

A general guide is that all issues of the Empire dating up to 1889 are the more expensive notes at £10 ($20) to £50 ($100). The period of the Republic, dating from 1890, saw a large number of private banks up to the first decade of this century; their notes are in the £35 ($70) to £100 ($200) range. Later notes are easier to come by. The new Brazil issues of May 1970 are the first to be officially issued by the Banco do Brazil and are so headed.

Bibliography:
Banco Central do Brazil *Iconografia do meio Circulante do Brasil*, (in Portuguese) Brasilia, 1972

Goncalves, A. A. *Catalogo de Cedulas e Moedas Brasileiras do Padrao Meontario 'Cruzeiro' 1942–1967*, (in Portuguese) Saõ Paulo, 1969

Monteiro, F. *Cedulas Brasileiras da Republica*, (in Portuguese) Rio de Janeiro, 1965

Seppa, D. A. *The Paper Money of Brazil*, Oak Park, 1971

Stephenson, T. *Brazilian Money of Necessity*, LANSA, Vol. 1, February 1973

Trigueiros, F. *Dinheiro no Brasil*, (in Portuguese) Rio de Janeiro, 1966

BRITISH COMMONWEALTH, ASSOCIATED STATES, COLONIES, PROTECTORATES, ETC.

In view of the great number of collectors of British notes of every kind, it is felt that a reference section may be devoted here to the subject matter as a whole. Many of the countries listed are separately covered in this chapter and this is indicated by an asterisk.

Hand-signed notes are uncommon today. Brazil's 200 cruzeiros, issued in 1953, has six variations. That printed by De La Rue's is the rarest in the series. (66 × 157 mm)

Definitions (Whitaker):

The Commonwealth: A free association of sovereign states (thirty-two in 1973), together with their dependencies and including the Associated States of the East Caribbean.

Colony: A territory belonging by settlement, conquest or annexation, to the British Crown.

Protectorate: A territory not formally annexed, but in respect of which, by treaty, grant, usage, sufferance and/or other lawful means Her Majesty has power and jurisdiction.

Protected State: A territory under a ruler which enjoys Her Majesty's protection; over whose foreign affairs she exercises control, but in respect of whose internal affairs she does not exercise jurisdiction.

Commonwealth Countries:

(An asterisk indicates that the country has been separately mentioned.)

British Isles	*Channel Islands (Guernsey and Jersey), *England, *Isle of Man, Northern Ireland, *Scotland, *Wales
*Australia *Canada *New Zealand	In 1931 the Statute of Westminster clarified their legal position; they had long been self-governing independent states.
Antigua	Member State 1967
Bahamas	Member State 1973
*Bangladesh	Member State 1972 (Republic 1972; independent, originally as East Pakistan 1948, through partition from India in 1947)
*Barbados	Member State 1965
Basutoland	now Lesotho
Bechuanaland	now Botswana
Botswana	Member State 1966 (Republic; formerly Bechuanaland Protectorate)
British Guiana	now Guyana
Ceylon	now Sri Lanka
*Cyprus	Member State 1957 (Republic 1960)
*Fiji	Member State 1970
Gambia	Member State 1965 (Republic 1970)
Ghana	Member State 1957 (Republic 1960; formerly Gold Coast)
Gold Coast	now Ghana
Grenada	Member State 1974
Guyana	Member State 1966 (Republic 1970; formerly British Guiana)
*India	Member State 1948 (Republic 1950)
*Jamaica	Member State 1962

Kenya	Member State 1963 (Republic 1964)
Lesotho	Member State 1966 (formerly Basutoland)
Malawi	Member State 1964 (Republic 1966; formerly Nyasaland Protectorate)
Malaya	now Malaysia
Malaysia	Member State 1957 (elective monarchy; formerly Malaya)
*Malta	Member State 1964
*Mauritius	Member State 1968
Nauru	Special Membership 1968
Nigeria	Member State 1960 (Republic 1963)
Northern Rhodesia	now Zambia
Nyasaland	now Malawi
Pakistan, East	now Bangladesh
Sarawak	now Malaysia
Sierra Leone	Member State 1961 (Republic 1971)
Singapore	Member State 1963 (state in Federation of Malaysia, seceded as Republic 1965)
*Sri Lanka	Member State 1948 (Republic 1972; originally as Republic of Ceylon 1970)
Straits Settlements	now Malaysia
Swaziland	Member State 1968
Tanganyika	now Tanzania
Tanzania	Member State 1961 (Republic of Tanganyika 1962; union of Tanganyika and Zanzibar 1964)
Tonga	Member State 1970
Trinidad and Tobago	Member State 1962
Uganda	Member State 1962 (Republic 1963)
Western Samoa	Member State 1962
Zambia	Member State 1964 (Republic; formerly Northern Rhodesia)
Zanzibar	now Tanzania

Countries no longer in the Commonwealth:

Aden	now Yemen
British Somaliland	Joined Italian Somaliland as Somali 1960
*Burma	left Commonwealth 1948
*Ireland, Republic of	left Commonwealth 1948
*Palestine	end of Mandate 1948 (State of Israel constituted)
*Pakistan	left Commonwealth 1972
*South Africa	Republic 1961
Southern Cameroons	joined French Cameroons as Cameroon Republic 1961
*Sudan	left Commonwealth 1956
Yemen	formerly Aden; left Commonwealth 1967

Colonies, Protectorates, Associated States, etc:

Anguilla, Belize (formerly British Honduras), *Bermuda, British Honduras (now Belize), Brunei, Cayman Islands, Dominica, Falkland Islands, *Gibraltar, *Hong Kong, Montserrat, New Hebrides, *Rhodesia, Seychelles, Turks and Caicos Islands, Virgin Islands.

Currency Boards

Were (or still are) active for groups of states or islands which have not been mentioned separately in this chapter.

The British West African Currency Board

It was appointed by the British Government in 1912, in order to study and decide the best monetary policy for West Africa. The earliest issues of the Board were put into circulation in Nigeria in 1916. The Board continued its paper issues in Ghana, Nigeria, Sierra Leone and Gambia, until independence was declared by each of these countries. The Board gradually ceased its monetary functions and continued in existence for redemption purposes only.

The last issue on record is dated 17 April 1962. The early issues of the Board are highly desirable items, particularly the higher denominations (100 shillings/1 pound) which have been sold for well over £80 ($160) each.

Easier to come by are the issues of the 1950s. Some from the previous decades can be purchased at about £8 ($16) each, and some of the smaller denominations, the 1 and 2 shillings, are even more reasonable in the £4 ($8) range, but are not always in crisp condition.

British East African Currency Board

The notes were circulated by this Board in Aden, British Somaliland, Kenya, Tanganyika, Uganda and Zanzibar from 1920 until 1964. The portrait of the

Currency boards function on behalf of colonial powers in specified territories. Many attractive issues continue to circulate in the Caribbean islands by authority of French and British monetary authorities. The early East African Currency Board 1 florin is much sought after by Commonwealth collectors. (78 × 135 mm)

reigning British monarch appeared on all the notes, an added attraction to the Commonwealth and colonies collector. The high denomination notes – a 10,000 shilling note dated 1947 has been recorded – are extremely rare. The lower denominations, particularly if acceptable in only fair condition, can be purchased at a few pounds.

The East African Currency Board, when it was first established, issued notes in rupee and florin denominations. Rather surprisingly these issues of the 1920s are not as expensive as one would expect. They can be obtained for about £35 ($70) in reasonable condition.

The Republic of Kenya was declared by Jomo Kenyatta in 1964. For a couple of years thereafter the East African Currency Board notes continued in general circulation. They were replaced in 1966 by issues of the Central Bank of Kenya. These issues are all legal tender and when in mint condition priced just above face value.

The British Caribbean Currency Board

This was formed in 1950 and issued notes under the heading of 'British Caribbean Territories – Eastern Group' until 1965. In that year the heading was changed to 'East Caribbean Currency Authority'. The notes of this new Currency Board are currently in circulation.

The two Boards (whose issues did not, however, circulate simultaneously) issued between them paper money for Anguilla, Antigua, Barbados, British Guiana (now Guyana) Trinidad and Tobago and numerous additional islands in the area. All the issues of the East Caribbean Currency Authority are relatively common and very attractive, bearing Annigoni's portrait of Queen Elizabeth II. The notes for the period between 1950 and 1965 are harder to come by.

Many of the Caribbean islands issued their paper money on the Royal Bank of Canada notes. The denominations related to the local currency and the pound sterling as well. This note was issued in Trinidad with a value of 5 dollars or £1.0.10d. (88 × 185 mm)

Higher denominations of the first issues of the British Caribbean Territories – Eastern Group in 1950 are priced at over double their face value. All the denominations bear the portrait of King George VI. The 1 to 10 dollar notes should be available, in very good condition, at not more than £20 ($40) each.

The Pan-Malayan Currency Commission

It was set up in 1938 to provide notes for the area. The notes issued for British Malaya from 1940 were circulated by the Board. Those dated 1941 and 1942 were only issued in 1945 by the British Military Administration for Malaya. These are common issues and complete sets of the fractional notes can be obtained for about £6 ($12). Earlier issues are rarer and as much as £100 ($200) has been paid for the overprinted 5 dollar denomination.

The British civil administration took over in 1946 and new notes were issued under the title of 'The Board's Commissioners of Currency for Malaya and British Borneo'. These issues are also fairly easy to come by. As in the case of Sarawak, there were a number of inter-bank issues of very high denominations of up to 10,000 dollars. If obtainable these would certainly be among the most expensive issues of the Commonwealth. In 1963, by agreement, the State of Sarawak, British North Borneo (Sabah and Singapore) and Malaya united to form Malaysia.

Malaysia remained a British Protectorate, while Singapore left the Union, becoming an independent republic in 1965. The Currency Commission continued until 1967 while each of the countries involved began to issue their own notes. Malaya and Singapore formed their own Currency Board. A Central Bank of

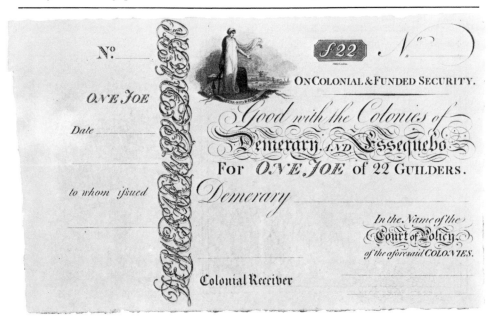

Guyana was known as Demerara and Essequibo when the 'JOE' denomination notes were issued in 1830. The colony was founded by the Dutch in 1620 and its notes are sought by many collectors. (150 × 250 mm)

Malaysia, with authority to issue its own paper currency, began circulating its issues in 1969 – headed 'Bank Negare Malaya'. They are all currently in circulation.

Bibliography:

> Chalmers, R. *A History of Currency in the British Colonies*, London, 1893 (reprint 1972)
>
> Shaw, W. & Haji Ali, K. *Paper Currency of Malaysia, Singapore and Brunei (1849–1970)*, Malaysia, 1971

BULGARIA

Currency: 100 stotinki = 1 lev(a) (silver or gold)

Bulgaria has come into notaphilic prominence due to the huge hoard of 1951 issues which appeared on the market in recent years. In 1973 they were being sold at approximately 50p ($1) per set of seven notes. The colourful, well-printed issues depict on the obverse, Georgi Dimitrov, the Prime Minister, who drew up the communist constitution for the Republic. The 1 lev note is invariably missing from the set, and alone it will fetch ten times the value of the whole set.

All Bulgarian notes have been issued by the National Bank – the Narodna Banka – since the 1870s.

The early vertical notes are priced at above £20 ($40) each. Later issues, with a few exceptions, should not exceed £10 ($20). One example is the 1929 5,000 lev issue, which is the highest denomination issued and is priced at above £85 ($170). Most of the notes issued during and after the Second World War are considerably cheaper, not exceeding £6 ($12) each.

In spite of harsh restrictions relating to the 'export' of circulating Bulgarian notes, a lucrative black market is active and current Bulgarian notes can be purchased in some Middle Eastern countries without difficulty.

BURMA

Currency: 100 pyas = 1 kyat (= 1 rupee until 1948)

There are no issues of Burmese notes of the last century. The earliest issues are the Treasury notes of the Government of India for circulation in Burma. These date from the beginning of the century up to 1935 and are all rare. They may fetch prices of up to £150 ($300) for the high denominations (a 10,000 rupee has been reported).

The British took over Burma from the Japanese and a Currency Board was established in 1945. Note issues ran parallel with those circulated by the Military Administration of Burma. The Currency Board and Military Administration notes are all Government of India issues overprinted for circulation in Burma. The notes are not scarce and can be obtained at less than £10 ($20) on average.

Notes were also issued in the Second World War by the puppet state under the Japanese. These are worth about £5 ($10) each. The modern issues of the Union of Burma since 1948 are quite common and worth not more than £8 ($16) for the more expensive issues.

C

CAMBODIA – see KHMER REPUBLIC

CANADA (also Newfoundland)

Currency: 100 cents = 1 dollar

The first recorded notaphilic issues of Canada are the 1685 'playing card' notes. Jacques de Meulles, the intendant of the French Quebec garrison, paid his troops with these improvised issues (see Chapter IV), which were used as currency until 1759. Notes printed on playing cards are unknown, but later issues on cardboard are highly priced. They have fetched £400 ($800) in some sales. Displeased with the chaotic circumstances of their circulation, the British military government at Quebec prohibited their use in the 1790s and issued copper coins, on which the words 'pure copper – preferable to paper' were inscribed.

In August 1821 an 'Act to facilitate the circulation of Army Bills' was passed by the provincial Parliament of lower Canada in order to finance Canada's defences against the Americans. A total of £3,441,993 had already been issued by February 1815. The denominations for 1, 2, 3, 5 and 10 dollars were payable . . . in bills of exchange on London and not bearing interest, and . . . holders of such bills shall be entitled to demand Army Bills of $50 and upwards bearing interest'. Many considered these to be the first proper paper money issues of Canada. The *Quebec Gazette* in March 1815 published a short but illuminating article on the subject which stated

> 'The introduction of paper currency was certainly a reasonable and judicious experiment, and its unprecedented success has not only been a great pecuniary saving to Great Britain, but it has also contributed in no small degree to the preservation of these Provinces.'

The Army Bill Office was closed in December 1820, although there were no issues after 1815. The notes were redeemed in gold and silver and circulated widely. The Army Bills fully replaced earlier playing card monies and at the end of the hostilities they had restored the public's confidence in paper issues. They are all highly priced.

Notes were now being issued by the Treasury covering many of the new dependencies of Canada. These included British Columbia and Newfoundland and these issues dated in the 1850s are fetching prices well over £100 ($200).

In the mid-nineteenth century a great number of private banks issued a wide range of notes in odd and small denominations. It is these private banks that form the backbone of Canadian notaphilic history. Some of the banks were unscrupulously set up and issued notes well in excess of the funds backing the issues. They soon foundered. The majority, however, traded successfully, and many remained active up to the period of the Confederation.

With its establishment in 1867, the Dominion Government circulated her paper money issues simultaneously with the private banks still in business. This period also saw the first of the fractional currencies, the 'shinplasters' of 1870, put into circulation because of the shortage of silver coins. They continued in use until 1923.

The most expensive of the 'shinplasters' is the 25 cents issue of 1 March 1870 priced at approximately £50 ($100) in crisp condition; the cheapest one is that of 2 July 1923 which, in fine condition, can be obtained at under £10 ($20).

The earlier high denomination notes are more expensive still. They are still redeemable and will fetch over double their face values in direct relation to their antiquity. The 1871 100, 500 and 1,000 notes of the Dominion of Canada are the rarest of all these issues and it would be impossible to value them accurately.

From the date of the Confederation in 1870 and up to the 1920s there was a profuse issue of notes by private banks. These varied in size and colour. Exceptionally rare issues of the private banks are, for example, the 6 dollar notes of 1870 issued by La Banque du Peuple. These were unpopular issues, being quite unattractive, and had a very short career indeed. They and those of similar rarity come onto the market infrequently. When they do, they fetch extraordinary prices.

The law of 1895 established that funds were to be made available for the redemption of all issued notes in case of failure of the bank in question.

A Central Bank of the Dominion Government was created in 1934 and named the Bank of Canada. The first issues of the Central Bank, dated 1935, were reduced in size, and can be obtained at about £10 ($20) in very good condition. The issues with a face value in excess of 25 dollars become expensive. The year 1935 marked the twenty-fifth anniversary of the accession to the throne of King George V and Silver Jubilee issues for 25 dollars portraying King George and Queen Mary in coronation robes were circulated. They are naturally very desirable notes and many were kept back as souvenirs. Consequently they are now scarce and valued at about £75 ($150) even in moderate condition. The 1,000 dollar note of this series fetches only about fifteen per cent over face value.

The 1954 series of the Bank of Canada have become known and popularised as the Devil's Face notes (see Chapter V). Although the notes were withdrawn from

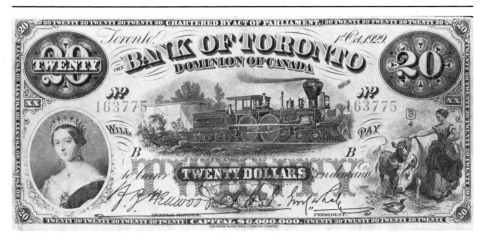

Canadian notes are among the most popular in the world. The Bank of Toronto has particular significance to the residents of that town. Some Canadian notes have been sold in auction at prices exceeding $4,000. (85 × 190 mm)

circulation and replaced, the issues are not rare. Depending on the combination of signatures they are not usually priced at over twenty per cent of the value of the ordinary issue. The issues since 1954 are common, priced at just over face value.

In 1967 a commemorative centennial 1 dollar note was issued and is worth about £1 ($2) in crisp condition.

By January 1950 the chartered banks had dwindled to ten in number. They incorporated thirty-six other banks, including private institutions, which they had previously absorbed. The Bank of Canada had the sole responsibility for the redemption of all notes still outstanding, having received from the existing charter banks the total outstanding amounts.

A collection of the private and chartered bank note issues of Canada gives an opportunity for very wide diversification. The notes are all colourful, intricately designed and trace back to the beginning of the nineteenth century. Prices range from just £4 ($8) for a moderately good issue of the 1930s to possibly several hundred pounds for some of the earlier, and now extremely rare, issues.

One must bear in mind that all notes dated after 1895 are redeemable at face value by the Central Bank.

Newfoundland

Newfoundland was the last of the colonies to join the Confederation in 1949 but the earliest dominion to achieve self-government, in 1885, and has been named 'the Oldest Colony'. As a result of financial difficulties early this century, Newfoundland surrendered her autonomy to the United Kingdom in 1934. In 1949 the colony became Canada's tenth province following a plebiscite and thereafter the economy began to recover. The issues of Newfoundland date back from 1901 to 1920. Most of the notes are expensive, particularly when in perfect condition. The odd denomination fractional issues of 40, 50 and 80 cents dated pre-1909 fetch as much as £75 ($150) each in crisp condition. From that date up to the First World War the only fractional issues were the 25 and 50 cents which can be purchased at about £25 ($50). There were no high denomination notes of Newfoundland, the highest being the 5 dollar bill which is the most expensive. The 1910 issue in crisp condition will fetch over £100 ($200).

Bibliography:

Allen, H. P. *Canada's Counterfeit Originals*, IBNS, March 1971
Bank of Canada *The Story of Canada's Currency*, Montreal, 1966 (2nd Edition)
Carroll, S. *The Bank of Canada Numismatic Collection*, IBNS, 1969
Charlton, J. E. *Standard Catalogue of Canadian Coins, Tokens and Paper Money*, (21st edition) Toronto 1972
   Canada and Newfoundland Paper Money, 1866–1935, Ontario 1966
Elliot, J. A. *Canadian and Newfoundland Currency*, Ontario, 1967
Walker, B. E. *The History of Banking in Canada*, Toronto, 1909
Willey, R. C. *A History of Paper Money in Canada*, WNJ, July/September/October 1968
Zigler, M. *Bank of Canada 1954 Series*, IBNS, March 1971

The collector's attention is also drawn to the Canadian Paper Money Society Journal (see Appendix V).

CEYLON see SRI LANKA

CHANNEL ISLANDS (Guernsey and Jersey)

Currency: 12 pence = 1 shilling
20 shillings = 1 pound sterling
100 pence = 1 pound sterling (since February 1971)
(Also livre tournoise)

Of the current notes of Jersey the unsigned £1 note recently attracted attention. This is a curiosity item, for which £10 ($20) was paid in 1973. The 10 shilling notes in both islands are now out of circulation and those in crisp condition are selling at double their face value. Guernsey notes in general are more expensive because fewer were put into circulation.

The issues under German occupation are popular collectors' items. The two high value notes of this period, the 10 shillings and £1 of both Guernsey and Jersey, are now fetching over £50 ($100) each. Guernsey had two £1 notes and the 1943 issue is a little scarcer than the 1945. The £5 note, issued in Guernsey alone, is a very rare item and when it appears on the market costs over £100 ($200). The lower denominations, from 6d. to 5s. are in the £5 ($10) to £25 ($50) range. The Guernsey notes, particularly the odd denominations of 1s. 3d. and 2s. 6d. which were not issued for Jersey, are the more expensive.

Private bank notes of the Channel Islands have also been fetching high prices. Following rumours that some plates had been found in Jersey from which early notes

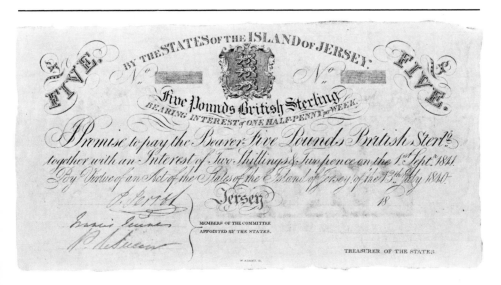

'I promise to pay the bearer' indicates that this Jersey State bond circulated as currency in the 1850s. It is interesting to note that the £5 issue has an interest rate of 'one half-penny per week'. (115 × 220 mm)

were being reprinted, there is some reluctance to purchase unissued and unsigned notes. Notes that did go into circulation, evidenced by signatures and dates, have been changing hands for £25 ($50) to £55 ($110). Unsigned issues fetch lower prices. The great abundance of Jersey provincial note issues did not apply to Guernsey. Only five local Guernsey banks issued notes.

Bibliography:

Banyai, R. *Germany Military Occupation of the Channel Islands*, IBNS, September 1970

Beresiner, Y. L. *An Excursion to Jersey*, IBNS, December 1970

Le Marchant, R. *Paper Treasures of the Channel Islands*, Guernsey, 1970

Marshall-Fraser, W. *A History of Banking in the Channel Islands – And a Record of Bank Note History*, Société Guernesiaise, 1949

CHILE

Currency: 100 centavos = 1 peso

10 pesos = 1 condor

1,000 pesos = 1 escudo = 100 centesimos (1960 monetary reform)

Chile, like many Latin American countries, has produced a wide and colourful range of bank notes in the last 125 years. Many private banks issued notes in the second half of the last century and some are extremely rare. The specialised collector may be prepared to pay up to £75 ($150) for some of the early notes. Many, of course, appear on the market at much lower prices. Some private bank notes can be purchased at under £10 ($20). Of interest among private issues of the last century are the Railroad Issues. These were placed into circulation at times of crisis, promising redemption 'as soon as the banks resume normal operations'. They were dated between 1890 and 1897, and are fetching up to £7 ($14) irrespective, within reason, of condition.

The only official issuing authorities in Chile have been the Treasury Department of the Republic and the Central Bank. The former circulated notes in the 1890s. These are equal in rarity to the private bank issues and fetch over £10 ($20). The issues of the Central Bank, which began in 1927, are somewhat similar to Brazil's. Specialisation based on signatures and dates would allow for a sizeable collection to be assembled comparatively cheaply. The lower values of the common notes from 1935 to date can be purchased at under £3 ($6) each.

Continuous search for different dates, overprints and signatures can often be rewarding. The higher denominations are more expensive, but they can also be purchased on average for £7 ($14) with higher prices for the rarer issues.

Bibliography:

Banyai, R. *Bank Combats Inflation*, NSB, December 1970

Cuba, J. *Origin of Latin American Units*, LANSA, June 1973

Galetovic, J. & Benavides, H. R. *El Billete Chileno*, Santiago, 1973

Matz, A. C. *Overprints on South American Notes*, CC, Summer 1966

Philipson, F. *Banknotes of Bolivia and Chile*, CM, August 1971

Pick, A. *Currency Changes in Chile 1860–1914*, LANSA, February 1974

Watermann, E. *Die Wirtschaftsverfassung der Republik Chile*, (in German), Berlin, 1913

CHINA (also Formosa)

Currency: 100 cents = 1 dollar (big money/small money)

100 jiaro or chiao (= 100 fen) = 1 yuan (= 1 renminbi)

1,000 wen = 1 kwan = 1,000 cash (1368–1858)

(Also tiao copper cent notes which replaced the kwan; yin piao silver

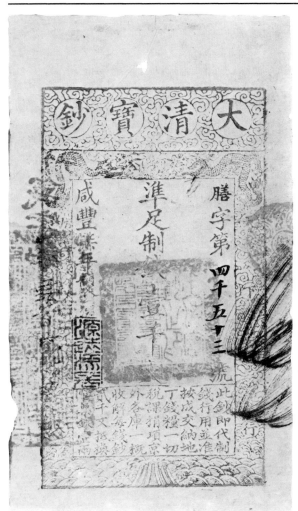

Left: **This Chinese issue in the seventh year of the reign of Emperor Hsien Feng (1857) states 'This paper money is to be used as cash and is also acceptable for all kinds of internal duties and taxes at a set discount. All treasuries outside of the capital must accept this note at its face value.' (160 × 250 mm)**

Right: **China issued currency in great quantities, but possibly no issues were as colourful as the semi-official notes of the Peiyang Tientsin Bank. The vertical notes appeared in 1902 and circulated until 1910 when the Bank was absorbed by the Provincial Bank of Chihli. (99 × 185 mm)**

notes; customs gold units; gold and silver yuan; silver dollars; and string cash – kwan, tiao and ch'uan.)

Collecting Chinese paper currency is a vast field indeed. It has been estimated that well over ten thousand different notes, by types alone, have been issued. Modern catalogues have certainly listed as many as five thousand such issues.

It has already been mentioned in Chapter II that China was the first country in the world to have issued bank notes as we understand them. These date back to the ninth century, but a collector can hope for nothing better than to place a Ming Dynasty note in his collection. An increasing number of these fourteenth- and fifteenth-century notes began to appear on the market in 1973 and although one sold at an auction in October 1972 at over £1,000 ($2,000), they can still be obtained for less. There is, of course, only a limited number of these notes in existence and the price is bound to increase in the future.

Apart from these very early issues, the classification of Chinese notes is a difficult task. It would be futile to undertake a pricing system in the framework of the general guidelines intended here. Let it be said that there are twelve specified and accepted groups of notes in Chinese notaphilic history. Many of these categories are themselves sub-divided and realistic values can only be given to Chinese notes within the context of such sub-divisions. The interested collector must, therefore, refer to the specialised catalogues listed at the end of this section for more detailed price valuations.

China experienced inflation, through the over-issue of paper money, at a time when the very existence of printing was still unknown to the European financial communities. This inflation finally led in the fifteenth century to the suspension of paper money issues which, with only one exception, lasted till the nineteenth century. Thus the notaphilic history of China is relatively new.

The first early issues, more or less readily available, are those dating up to the T'ai P'ing Rebellion which erupted in 1850 and lasted for fourteen years. The notes are rare and expensive but can still be encountered at about £70 ($140). These were printed from wood-blocks up to the end of that century and circulated simultaneously with increasing issues from private sources.

China's history, from the turn of this century until the communist take-over in 1941, traces through catastrophic civil wars and disastrous international conflicts. This has led to a diverse range of paper money issues which can form the basis of a Chinese collection at a price acceptable to every collector.

The republican regime, which came into power in 1911, allowed individual War Lords to usurp militarily the eighteen provinces of China, as well as the three

Opposite. Top: **Barclays Bank DCO issued notes in many Commonwealth countries. Even the modern notes are highly prized by collectors. This South West African issue is priced at about £70 ($140). (76 × 165 mm)**

Bottom: **The colourful designs on Costa Rican paper money make the issues attractive and forgery difficult. The issues of this particular bank are very popular because of its association with England. (84 × 184 mm)**

Manchurian ones, and appoint themselves as governors. The Tuchuns, as they were called, were egotistic individuals, their main aim being self-remuneration. They issued paper currencies but were invariably unable to redeem them because of the lack of backing. They therefore closed down the issuing body, changed their name, and began again! The notes constantly depreciated before eventual repudiation.

The War Lord was clearly uninterested in what would benefit the populace and used every possible trick to force acceptance of his paper money. In 1925 Feng Yuhsiang issued several notes without any backing whatsoever depicting a railway station with a dynamited railroad. A well-known War Lord was Marshal Chang Tso-lin who crowned himself king of the three eastern provinces in Manchuria. He was active during the 1904–05 Russo-Japanese War and financed his own Chinese 'soldiers' and although he issued the notes under the names of official banks these too remained unredeemed. Many of these fascinating issues are valued at no more than £5 ($10) each.

There was great reluctance on the part of the government in power throughout this century to give exclusive note-issuing authority to any one bank. This was mainly owing to the previous experiences encountered in which the Chinese people had completely lost faith in paper money issues.

From 1908 to 1949 four government banks were issuing their notes simultaneously. The most powerful was the Central Bank of China established in 1926. The Bank of Communications, however, formed in 1913, had issued its own notes, as had the Bank of China, which was formed in 1912. The last of the four banks was established in 1934 and named the Farmers Bank of China. In 1942 the Central Bank was nominated by the national government to be the sole issuing authority in the country, and the other three banks ceased their issues. These are the notes that are easiest to come by; most of them can be purchased at under £2 ($4) each, in excellent condition.

The first issues of the Central Bank were in the yuan or dollar denominations, but in 1933 the 'customs gold unit' was introduced by Dr Edwin W. Kemmerer. These vertical notes are extremely attractive issues and can also be obtained for just £2 ($4) each. The portrait of Dr Sun Yat-Sen, the medical surgeon revolutionary who led the revolt in 1912, appears on most of the notes.

In order to avoid panic among the population attempts were made to hide inflation at this time by issuing pre-dated notes. Thus notes being issued up to 1949 were dated as early as 1940. These can be obtained cheaply because, as a result of the inflation, huge quantities were printed and circulated. The gold yuan notes were issued for values of up to 60 million yuan in 1948! When the government fell in 1949 the Nationalists fled to the south. Additional notes, in the silver dollar denomination, were then issued.

Opposite. Top left: **Cuba was the last Latin American country to gain independence from Spain, in 1898. The issues of El Banco Español de la Isla de Cuba continued to circulate for several years thereafter. (115 × 202 mm)**

Bottom left: **Paper money issues of the French colonies were always similar to the notes of the Banque of France, as this Madagascar note shows. (121 × 174 mm)**

Thus in this century alone there have been three distinct note issues: those of the Empire, dating to 1911; of the Republic, from 1912 to 1949 and the nationalist notes from 1949 which circulated simultaneously with the communist issues. At the same time a large number of commercial banks, such as the Imperial Bank of China which began its issues in 1898, were circulating their own issues. The notes of the commercial banks can also be obtained for about £8 ($16) each.

The most attractive Chinese notes are those issued by the foreign banks in China and they are among the more popular issues. Several American-owned companies such as the Asia Banking Corporation, as well as European ones, established their own banks at the end of the last century. These included, among many others, the Charter Bank of India, Australia and China, the Mercantile Bank of India, the Hong Kong and Shanghai Banking Corporation, and the Belgian and British Bank. They all had their own bank notes. Many continued with their issues up to the middle of this century. Some of the notes are highly valued, but most can be obtained at between £15 ($30) and £40 ($80).

The communist issues after 1949 are quite easy to come by. They do not fetch more than a few pence or cents each.

Formosa

Collectors of the paper money of China will also be interested in the issues of Formosa (formerly Taiwan). The Bank of Taiwan issued both vertical and horizontal notes between 1949 and 1955. These are quite easy to encounter and portray either Dr Sun Yat-Sen or General Chiang Kai-Shek. They do not fetch prices much in excess of £2 ($4). The more interesting notes are those of the Japanese administration of Taiwan. This lasted from 1899 to 1901 and notes were issued for values from 1 to 50 silver yens. The notes are vertical and were later issued in the gold denomination, between 1904 and 1906. They are much rarer, and fetch prices close to £30 ($60) each. The few silver denomination notes of the very short-lived Republic of Taiwan under President Soong in 1894 and 1895 depict the tiger symbol and are the most expensive of the Formosa issues, fetching up to £50 ($100) each.

Bibliography:

Hage, S. *About the Tan Notes*, IBNS, Christmas 1962

Hill, A. B. *Central Bank of China Gold Yuan Issues*, CC, Autumn 1964

Kann, E. *Chinese War-Lords and their Paper Money*, IBNS, March 1962
  Ming Notes, IBNS, Summer 1962
  Paper Money in Modern China, IBNS, September 1964
  Foreign Bank Notes in Manchuria, IBNS, Christmas 1964
  The Provincial Banks of China and Foreign Note Issuing Banks in China, (reprint), London, 1967
  Government Banks of China, (reprint), England, 1967

Loeb, W. M. *New Issues of Communist China*, WNJ, September 1966

Loy, S. *The Gem of Chinese Paper Money (First Silver Dollar)*, IBNS, Autumn, 1962

History of the National Bank of China, CC, Summer 1965

Mao, K. O. *First Issues of Government Bank of China*, IBNS, Summer 1963

   Collecting Chinese Banknotes, IBNS, Christmas 1963

   Paper Money in China's Dependencies (Sinkiang), IBNS, Christmas 1965

   Chinese Currency throughout the Ages, IBNS, Christmas 1966

   Farmers Bank of China, IBNS, Spring 1969

   Rise of Foreign Banks in China, IBNS, June 1970

   Central Bank of China Silver Dollar Issues, IBNS, June 1971

   The Paper Money of Communist China, IBNS, March 1973

   Some Ch'ing Dynasty Banknotes, CC, Spring 1965

   The Sino-Scandinavian Bank in China, IBNS, Summer 1964

   History of Chinese Paper Currency, Vol. I, Hong Kong, 1968

   History of Paper Currency as Issued by the People's Republic of China from 1921–1965, Hong Kong, 1972

Narbeth, C. C. *Bank that Backed a New Dynasty (Bank of Communications)*, C & M, January 1973

Randall, R. E. *Chinese Notes Reflect Fiscal History*, WC, June 1970

Smith, W. D. *Communist Chinese Banknotes*, CC, May 1972

Smith, W. D. & McTravers, B. *Chinese Bank Notes*, California, 1970

Sullivan, A. J. *Gold Yuan Issues of the Central Bank of China*, IBNS, Christmas 1965

Sullivan, R. R. *Fantasy on the Ten Yen Note*, CC, Winter 1964

COLOMBIA

Currency: 100 centavos = 1 peso

   1 peso = 10 reales

The vast area that Colombia has covered through her history led to many interesting

Since independence in 1819, Colombia has changed its political standing several times. The United States of Colombia (1862–86) issued attractive notes which are now valued by collectors. (194 × 85 mm)

issues. One might have hoped to encounter notes of Venezuela and Ecuador issued when they belonged to 'Gran Colombia' but only those of the Banco de Panama dated in the 1860s belong to the period. Very few notes have been issued by Panama in her history and these are particularly rare issues priced at over £150 ($300).

The 1,000 Day War, at the turn of this century, led to a huge quantity of locally printed private issues and a large number of contemporary forgeries. The national government took over private banks, overstamping their issues, and adding to the overall confusion.

There are some very attractive early notes issued by private banks and dated in the 1860s; some of these, without overprints, will fetch prices up to £50 ($100). The overprinted issues date to the end of the last century and can be obtained at between £5 ($10) and £25 ($50). As is the case with many Latin American countries, local notes were printed on very frail paper and such notes in extremely fine or better condition are infrequent.

The modern issues of Colombia began with the 1886 series of a non-convertible currency and were followed by a number of distinct government issues. Among these notes the higher denominations are the more expensive. Very few 1,000 pesos notes were issued and one dated 1908 sold in 1972 for $2,000. Most collectors are unlikely to encounter these high values.

Most low denomination notes, until the formation of the Banco de la Republica, in 1923, are reasonably priced within the £5 ($10) to £20 ($40) range. Banco de la Republica issues are now circulating and the replacement notes of the 1, 5 and 10 pesos issued from 1959 – when the notes were first printed locally – appear with a tiny 'r' or an encircled star in between the signatures on the obverse. They have fetched very high prices in Colombia and one should keep a look out for them.

Bibliography:

Banco de la Republica *Billetes del Banco,* Bogotá, 1949

Catalogo de Billettes 1923–1973, Bogotá, 1973

Barriga Villalba, A. M. *Historia de la Casa de Moneda, Vol. III,* Bogotá, 1969

Beresiner, Y. L. *Catalogue of the Paper Money of Colombia,* London, 1973

Gould, M. *Colombian Currency of the Insurrection,* IBNS, March 1962

Matz, A. C. *Popayan Varieties,* CC, Vol. IV, No. 3, 1963

CONGO BRAZZAVILLE – see FRANCE

CONGO FREE STATE – see BELGIUM

CONGO KINSHASA – see BELGIUM

COSTA RICA

Currency: 100 centavos = 1 peso

100 centavos = 1 colon since 1895

Costa Rica is among the profuse paper money issuers in Central America.

Columbus's name in Spanish, 'Colon', was adopted for her monetary unit during the currency reform of 1896 and his portrait is frequently represented on the notes.

The peso notes of the Republic from 1835 to the end of the century fetch high prices above £50 ($100). In 1885 a number of emergency peso notes were issued to finance the war against her neighbouring countries; these denominations are exceedingly rare and valued above £100 ($200).

A number of private banks issued their own notes in both pesos and colones. The private issues began with the Banco Anglo-Costa Ricense in 1863 and fetch prices above £10 ($20) but become more common in relation to the private issues of the 1910s; later notes are even more common. Many were overprinted by the National Bank of Costa Rica which itself was formed in 1939. The overprinted notes, as well as the original National Bank of Costa Rica issues, can be obtained for about £10 ($20), but many of the issues since 1939 are still legal tender.

Bibliography:

Burstyn, L. *Paper Money of Costa Rica*, Parts I and II, CC, Spring and Summer 1966

Soley, T. *Historia Monetaria de Costa Rica*, (in Spanish), San José, 1926

CRETE – see GREECE

CROATIA – see YUGOSLAVIA

CUBA

Currency: 100 centavos = 1 peso

The Banco Nacional de Cuba issues from 1949 to date have now become well known to collectors throughout the world. Since 1972 the National Bank has been releasing these notes, overprinted 'Specimen', to collectors and dealers at a very low price. The issues bearing Ernesto 'Ché' Guevera's signature as President of the Central Bank are well worth purchasing. In view of the vast quantity of these notes now on the market, prices are unlikely to exceed £5 ($10) per note. The sets being sold by the Central Bank include the 1897 fractional notes of El Banco Español de la Isla de Cuba and these too, when overprinted 'specimen', are in the £8 ($16) price range.

The issues of El Banco Español de la Habana, which date back to 1857 and continued to be issued up to 1888, are extremely rare. They fetch prices of £80 ($160) or more when in good condition.

The Spaniards circulated a large number of issues for Cuba. Of interest among these are the 1869 notes which appertain both to Cuba and Puerto Rico. The 20 peso note of this issue is very rare and was recently sold for £225 ($450) in an auction. The lower denominations are a little cheaper, but still in the £30 ($60) range.

All the Banco Español de la Isla de Cuba notes of 1896 were overprinted on the reverse in large red letters with the word 'Plata', silver. These issues are only a little dearer than those without overprint. The whole series of the Banco Nacional de Cuba issues are easy to come by, except for the very high denominations issued in

1950. The 1,000 and 10,000 pesos can fetch up to £30 ($60) each. The remainder are priced under £10 ($20) with the exception of the original, as opposed to specimen, notes with the 'Ché' signature; the higher denominations of these could fetch up to £25 ($50) each.

Bibliography:

Matz, A. C. *Cuba – A Check List 1869–1961*, CC, Summer 1965
Morton, L. E. *Cuban Rebel Currency of 1870*, WC, August 1967
Philipson, F. *Paper Money of Cuba*, CM, August 1972
Shafer, N. *Commemorative Paper Money*, WNJ, June 1965
    Silver Certificates of Cuba, WNJ, February 1965
    Cuban Paper Money, IBNS, December 1970
    Unknown Cuban Issues, WNJ, January 1968
    Death of a President, WNJ, March 1968
    A Specimen Book of Unknown Cuban Issues, WNJ, November 1967

CURAÇAO – see NETHERLANDS

CYPRUS (Kibris)

Currency: 1,000 mils = 1 pound sterling = 20 shillings until 1943

The earliest notes to circulate in Cyprus were those of the Ottoman Empire, which lasted until the introduction of the pound sterling by the British at the outbreak of the First World War. The notes dating back to 1916 with the portraits of the reigning British monarch are popular and fetch up to £80 ($160) each. The Republic of

Cyprus still uses the pound denomination. The text is in Turkish and Greek on the face and English on the reverse. (95 × 166 mm)

Cyprus issues, which bear the legend in Greek and Turkish only, are more common notes within the £5 ($10) range. They are all redeemable and will fetch prices over their face values when in crisp condition.

The most interesting Cyprus issues are those of the American Joint Distribution Committee. These notes were used in the British internment camps for Jews fleeing from Europe on their way to Palestine. They are dated 1946 and are priced over the £75 ($150) range.

Bibliography:

> Atsmony, D. *Notes of the Joint Distribution Committee of Cyprus,* IBNS, Autumn 1962
> Haffner, S. *Palestine Notes fill Cyprus Void,* WC, June 1972
> Pridmore, F. *Cyprus Currency Turkish Treasury Notes,* IBNS, August 1969
> Shafer, N. *Emergency Money of Cyprus,* WNJ, January 1968
> *Emergency Small Change of Cyprus,* WNJ, April 1967

CZECHOSLOVAKIA (Československa)

Currency: 100 haleru (heller) = 1 koruna (crown)

The first Czech issues are the 1919 overprints on Austro-Hungarian bank notes dating from 1902 to 1915. These notes are priced around £2 ($4) each. Czechoslovak notes have been released by the National Bank of Czechoslovakia to collectors after being perforated with the word 'specimen'. All the specimen issues are easy to come by. None should fetch over £5 ($10). The exception, possibly, is the 1945 2,000 kronen of the Czechoslovak Socialist Republic.

There are in fact very few rare notes of either Czechoslovakia, Bohemia and Moravia (issues of the Second World War) or of the issues of the Socialist Republic of Czechoslovakia. With the exception of the 500 koruna 1923 and the 100 to 5,000 koruna issues of 1919, which may cost up to £50 ($100), all issues are in the £4 ($8) range, but the 20 and 100 kronen Socialist Republic of Czechoslovakia notes dated 1953 and 1951 respectively, may be worth up to £150 ($300) each as they were never issued. The Czechoslovak Republic issues of 1919 have the text of the denomination of the note in six languages: Czechoslovak, Slovak, Russian, German, Polish and Hungarian.

Bibliography:

> Jonfrst, J. *Pairove Penizenia Uzemi,* (Paper Money of Czechoslovakia), (in Czech), CSR, Prague, 1960
> Kovarick, J. *First Czechoslovak Notes,* IBNS, Christmas 1963
> Kupa, M. *Currencies of Czechoslovakia,* BCTEM, Second Semester 1953
> Philipson, F. *Bank Notes of Czechoslovakia,* CM, November 1970
> Rauch, B. *An Introduction to Czech Relief Notes,* CC, May 1972

D

DANZIG – see POLAND

DENMARK (also Faroe Islands, Lesser Antilles or Virgin Islands [Danish West Indies] and Greenland)

Currency: 100 øre = 1 krone

Collecting Danish paper money is ideal for those who enjoy diversification in signatures and dates. The early National Bank issues from 1875 and up to the present time are dated for every year and have been issued with a wide range of signatures. The early issues are, of course, more expensive. The 500 krone of 1875 will fetch up to £100 ($200), but other issues are cheaper and should not exceed about £45 ($90). Many Danish issues of this century can be obtained at prices in the £5 ($10) range. Rare notes are the exception rather than the rule.

The best known are the Treasury issues of 1914, which were placed in circulation with five per cent interest and declared to be legal tender. The 500 krone of this issue will fetch up to £30 ($60). An additional attraction in Danish notes is the great number of famous portraits appearing on them, including that of Hans Andersen.

The modern bank notes of Denmark do not appear to have a date on the face of the note, but there is one in fact: the prefix coding on the left hand side of these notes has a serial letter followed by four numbers and another serial letter. The two middle numbers indicate the year of issue; thus a serial coding A4684B is for a 1968 issue.

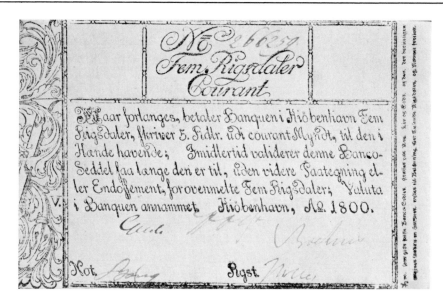

Denmark issued its earliest notes in the rigsdaler denomination and the Copenhagen Bank was one of the earliest financial institutions in the country. (110 × 175 mm)

Faroe Islands

The Faroe Islands, situated in the north Atlantic Ocean have been part of the Danish kingdom since the fifteenth century. A number of issues in the kroner denomination were put into circulation by the government council, at the very beginning of this century. These circulated up to 1939. With the outbreak of war, Danish notes overprinted for the Faroe Islands came into circulation. Simultaneously a new issue of the island notes was made in 1940. Most are priced at over £10 ($20) except for the current ones. Since 1950 Danish currency has circulated in the Islands.

Danish West Indies

The group of islands in the Lesser Antilles, now designated the Virgin Islands, were sold by the Kingdom of Denmark to the United States in 1917. The currency denominations are expressed in francs as well as dollars and some of the early issues, dating as far back as 1849, are within the collector's reach. The state notes of 1849, for two dalere (5 francs = 1 daler) can still be obtained for about £30 ($60). Later issues of the last century are much rarer. Some of them fetch as much as £90 ($180). Since 1917 the islands have used the dollar as the monetary unit.

Greenland

Greenland, since 1924, has been a Crown colony of Denmark and an integral part of the Danish Kingdom. From early times the note issues in the skilling and ragstaler denominations have circulated. The earliest of these are dated 1803. These extremely

This Greenland issue circulated between 1913 and 1926 and is particularly rare when in perfect condition. (90 × 192 mm)

rare issues are valued at about £120 ($240) each. Later denominations were in the øre and krone units but most notes of Greenland since 1880 can be obtained at reasonable prices. The note issues of the government of Greenland from 1912 onwards are priced at about £18 ($36) each, while some of the later issues (excluding the high denomination notes of 100 kroner) are within the £10 ($20) range.

Bibliography:

Flensborg, P. *Paper Money of Greenland 1803–1967*, Copenhagen, 1971
  Paper Money of Greenland, IBNS, September 1972
Higgie, L. W. *The Colonial Coinage of the U.S. Virgin Islands*, Wisconsin, 1965
Loeb, W. M. *The Danish West Indies*, IBNS, Autumn 1961
McKercher, M. *Denmark's First Currency Notes*, IBNS, December 1971
Meyer, R. *The Greenland Trade Certificates*, CC, Spring 1965
Philipson, F. *Early Paper Currency of Denmark*, IBNS, March 1971
  Danish Isles Currency, CM, June 1973
Rixen, J. *Danish 'S.O.C. Bank' 1927 Notes*, WC, September 1966
Sieg, F. *Sieg's Settelkatalog 1874–1970*, (in Danish), Norden – Ulbjerg, 1971

DOMINICAN REPUBLIC (Republica Dominicana)

Currency: 100 centavos = 1 peso

After the capital city of Santo Domingo was destroyed by a cyclone in 1933, it was rebuilt under the stern dictatorship of President Trujillo who named it after himself in 1936. The name of Santo Domingo was restored in 1961, after the assassination of the President.

The Banco de la Republica Dominicana notes show the legend of the city in which they were issued. As a general rule the notes bearing the name 'Cuidad Trujillo' are a little scarcer and may fetch up to fifty per cent more than the 'Santo Domingo' issues. The Banco Central issues are all quite easy to come by and they do not fetch more than £3 ($6) or £4 ($8) each for the low denominations. The portrait of Trujillo appears on some of the issues.

Local private issues circulated at the end of the last century are more expensive, many reaching the £45 ($90) level.

One of the attractive series that can be made into a collection are the notes of the 'Banco de la Compañia de Credito de Puerto Plata'. Denominations from 25 centavos to 10 pesos were issued in 1889; the six notes together in a set could be purchased at about £85 ($170). Among the most interesting early notes are those of the Republica Dominicana dating back to the 1840s. These have reached prices as high as £70 ($140) each at some auctions.

Bibliography:

Matz, A. C. *Politics and Paper Money*, WNJ, October 1964
  Varieties of Latin American Paper Money, CC, Summer 1963
Remick, J. H. *Coinage of the Dominican Republic*, C & M, December 1972
Rudman, I. *The Paper Money of the Dominican Republic* (in preparation)

E

ECUADOR

Currency: 100 centavos = 1 sucre (also pesos)

The issues of the Central Bank of Ecuador since 1928 are fairly easy to come by. Those dating back to the formation of the bank are scarcer and as they are all still legal tender they fetch prices over their face value. Denominations up to 100 sucres are priced from £6 ($12) to £10 ($20) each when in good condition. The 500 and 1,000 sucres, placed into circulation after 1948, are considerably more expensive. The 500 sucres is among the rarest of these because very few were originally printed.

The private bank notes, which are the only additional type in Ecuador, are suprisingly difficult to find. The commonest are the unissued notes of the Banco Sur-Americano – a bank that never came into being. The notes were printed in huge quantities and reached the market in sets of three denominations of 1, 20 and 100 sucres all dated 1920. A set of these should not fetch more than £4 ($8). The 5 sucre issue of the series is a more difficult denomination to encounter, and is priced at £10 ($20) on its own.

Private bank issues are not found in good condition but still fetch prices above £15 ($30). At least twelve banks have issued notes and the easiest to come by are those of 'El Banco Comercial y Agrícola' dated from 1904 to 1925. The 1 sucre note may be purchased at about £8 ($16). After 1927 many of the private banks continued to issue their own notes until the Central Bank of Ecuador obtained the sole right in 1928. The Banco del Azuay 20 sucre note dated 1927 in mint condition sold at £75 ($150) in an auction in 1973. A set of notes recently making an appearance on the market are tax receipts dated about 1774 and issued by the Spaniards in different cities in Ecuador. These extremely early 'notes' are rare items and fetch about £75 ($150).

The note issues of Central Banks are a feature of this century alone and thus usually depict relatively modern vignettes. This high denomination Ecuador note shows the Central Bank of Ecuador. (186 × 90 mm)

Bibliography:

    Alberto Carbo, L. *Historia Monetaria y Cambiaria del Ecuador*, (in Spanish),
    Quito, 1953

EGYPT

Currency: 100 piastres = 1 Egyptian pound

The first Egyptian note this century was not placed into circulation until 1916 when the Egyptian currency notes were issued by the government. The text appeared in Arabic and English. These fractional notes for 5 and 10 piastres continued to be issued until the Second World War and can be purchased at a few pence each.

Until 1882 Egypt was part of the Ottoman Empire. From that time until 1922 British authorised issues appeared, including Currency Board notes, and were issued in conjunction with the Ottoman bank notes.

The National Bank of Egypt notes were issued under a British decree dated 1898. Ten types circulated simultaneously up to 1961 for denominations from 25 piastres to 100 Egyptian pounds. The 50 and 100 Egyptian pound issues are very scarce and some dates fetch over £150 ($300). The 1899 notes, if they could be encountered, would be considerably more expensive. The smaller denominations are much cheaper and can be obtained for about £10 ($20).

The Central Bank of Egypt began its own note issues in 1961; the notes are still legal tender and are therefore priced at over face value when in mint condition.

Egypt and Syria formed the United Arab Republic in 1958 and despite the fact

The National Bank of Egypt first released its notes early last century and continued to do so until the Second World War. All the notes dated before 1926 are particularly rare because of demand by collectors of Israel and Palestine notes, where they circulated until that date. In spite of its mediocre condition, the 1918 note illustrated is valued at several hundred pounds. (100 × 186 mm)

that Syria left the alliance later, Egypt continued to use the joint name, the United Arab Republic; some of the currency notes for 5 and 10 piastres still bear this legend in Arabic and English.

The notes issued by the Italians for the intended occupation of Egypt are now valued collector's items (see Chapter IV).

Bibliography:

Philipson, F. *The Paper Money of Egypt*, CM, November 1971

EL SALVADOR

Currency: 100 centavos = 1 peso = 1 colon since 1924

All the notes issued since the formation of the Banco Central de Reserva de El Salvador after 1934 are common and can be obtained for about £5 ($10). The exception is the 100 colon issue. Even recent issues are scarce, having been issued in small quantities, and may be priced at above £25 ($50). 'Banco Central de Reserva' was a change of name from the earlier 'Banco Agricola Comercial'. The latter issued its notes from the end of the last century right through to 1934 in conjunction with several other private banking institutes. Most issues of the private banks, particularly high denomination notes, are rare and priced over £35 ($70). One series which can be obtained fairly easily is that of the 'Banco Nacional del Salvador'. These notes of 1, 5 and 10 pesos, as a set, should not cost more than about £20 ($40). A similar set of the 'Banco Salvadoreno' is priced around £15 ($30). In both cases, the denominations from 25 to 1,000 pesos are much rarer.

Some six hundred private banks in England issued their own paper currencies. The Workington Bank was set up by the partners Bowes, Hodgsons, Falcon and Key in 1801, but like many other banks of that period, it had to cease its activities, in 1810. The 1 guinea note is particularly attractive. (106 × 188 mm)

Many private bank issues were circulated almost yearly from 1885 to 1934.

Bibliography:
Almanzar, A. F. & Stickney, B. R. *The Coins and Paper Money of El Salvador*, Texas, 1973

Calales, J. M. *Evolución Bancaria en El Salvador*, (in Spanish), San Salvador, 1942

Matz, A. C. *Varieties of Latin American Paper Money*, CC, Summer and Fall 1962, 1st Semester 1964
El Salvador Varieties, CC, Fall 1964

Stickney, B. *El Salvador's Papel Moneda*, LANSA, October 1973

Young, J. B. *Central American Currency and Finance*, Brinston, 1925

ENGLAND

Currency: 12d. = 1 shilling
20s. = 1 pound
100 pence = 1 pound since February 1971

Bank of England

The general rule applicable to Bank of England notes is that the earlier the date the higher the value. All Bank of England and Treasury notes (issued since the reign of Charles I) are redeemable and consequently their market value exceeds face value. A £1,000 note of the Bank fetched £3,500 in an auction held in 1973. The successful bidder later revealed that he would have been prepared to pay as much as £5,000 for the note. These high denomination notes are unlikely to come onto the market frequently as there are only thirty-six recorded notes in existence. The denominations that are easily available are the issues of this century: the 10 shillings, £1, £5 and £10 issued in London can be obtained without difficulty. The £20, £50 and £100 also come onto the market, but less frequently.

Until 1939 Bank of England notes of denominations higher than £1 were issued by branches of the Bank throughout the country. The name of the city of issue appears on the notes. All issues by the branches are popular and fetch prices often double that of the London issue.

This century's Bank of England notes were only introduced after the Great War. The 10 shilling note in 1928 was the first fractional issue in British banking history and the £1 issues released at the same time had not appeared for over a century.

The signatures on Bank of England notes (see Appendix II) determine the issue, and collectors use the names of the chief cashiers as reference. Thus notes prior to Harvey (E. M. Harvey, 1918–25) would be expensive. The 'white fiver' of J. G. Nairne's period of office, dating from 1902 to 1918, might well fetch a price above £120 ($240) in crisp condition.

Many Nazi forgeries of the 'white fiver' and £10 notes still appear. A collector in the UK is forbidden by law from possessing a forged note. One can only verify whether a Bank of England note is genuine by presenting it to the Bank. Counterfeit notes are confiscated against a receipt. There is, therefore, no market for German forgeries in the UK. But a lucrative market exists in Europe and in the United States. In the States in particular forgeries of the Second World War at times fetch prices higher than the original notes.

A very thorough study of the serial numbers of the modern Bank of England notes was undertaken, and written up, by Vincent Duggleby. This has been based on the principle that the exact priority of issue and the frequency of replacement notes can be determined from the serial lettering and numbering. This has caused the respective prices of two apparently identical notes to vary in price by maybe two hundred per cent; the highly priced note may have a serial letter, for example, preceding the number – and vice-versa for the 'worthless' one.

The rarest Bank of England notes of this century are the war-time 2s. 6d. and 5s. notes, which were never issued. The very few that appear on the market fetch prices in excess of £300 ($600). They were not given serial numbers and were signed by K. O. Peppiatt in the 1940s.

Treasury Issues

There are only six Treasury issues – excluding the 'Gallipoli Overprints' – and the only denominations are the 10 shilling and £1 notes. They circulated between 1914 and 1927. Once again a great deal of emphasis has been placed by some on the variation of the serial letters and numbers. Some of the issues are priced at sixty per cent more than an identical note because of the size or shape of a full stop!

The John Bradbury issues printed in 1914 – the first and second issues – are being traded in crisp condition at prices in excess of £50 ($100). The third Bradbury issue and the three Warren Fisher issues vary in price between £6 ($12) and £25 ($50). As a general rule the 10 shilling note is more highly valued. The Turkish overprint on the 10 shilling and £1 Bradbury second issue has become an extremely rare find, particularly the £1. The 10 shilling is worth £70 ($140) in almost any condition; the £1 was changing hands at £420 in 1973!

Provincial Banks

Provincial banks in England date back to before the formation of the Bank of England. The more recent banks have been listed in the *Banker's Almanack* for many years, and the notes of private and provincial banks are popular collectors' items. Their prices have gone up considerably in recent years and now that they are in demand they sell at prices from £10 ($20) to £30 ($60).

Issues in the late 1700s, and before are more expensive. The odd denomination notes, such as the 6 guinea and £2 10s. issues, are also highly valued. There are a large number of provincial banks that were amalgamated with each other and later taken over by one of the well-established banks still in existence today. These notes, irrespective of how far back they are traced, are still redeemable at the main office of the bank, unless they were cancelled. They are therefore even rarer. The last private English bank to issue notes was Fox Fowler Ltd of Wellington, Somerset, which closed its doors in 1923, and note issues of this bank are highly sought after.

Bibliography: *Books*

Bank notes

Bevan, D. *A Guide to Collecting English Banknotes*, Surrey, 1970

Duggleby, V. *English Paper Money*, London, 1975

Graham, W. *The One Pound Note in the History of Banking in Great Britain*, Edinburgh, 1911

Institute of Bankers *Catalogue of the Maberley Phillips' Collection of Old Bank Notes, Drafts, etc.*, London, 1906

Mackenzie, A. D. *The Bank of England Note; A History of its Printing*, Cambridge, 1953

Miller, D. M. *Bank of England and Treasury Notes 1694–1970*, Newcastle-upon-Tyne, 1970

Narbeth, C. C. (Editor) *Collect British Bank Notes*, London, 1970

Banking

Easton, H. T. *History and Principles of Banking*, London, 1924

Pressnell, L. S. *Country Banking in the Industrial Revolution*, London 1956

Richards, R. D. *The Early History of Banking in England*, London, 1929

Bank of England

Acres, W. M. *The Bank of England from Within: 1694–1900*, London, 1931

Andreades, A. *History of the Bank of England*, London, 1944

Bowman, W. D. *The Story of the Bank of England*, London, 1937

Clapham, Sir J. *The Bank of England*, London 1944

Guiseppi, J. *The Bank of England, A History from its foundation*, London, 1966

Warren, H. *The Story of the Bank of England*, London, 1903

Individual Banks

Chandler, G. *Four Centuries of Banking (Martins Bank)*, London, 1964

Crick, W. F. and Wandsworth, J. E. *Hundred Years of Joint Stock Banking, (Midland Bank)*, London, 1936

Fulford, R. *Glyn's 1753–1953*, London, 1953

Gregory, T. E. *The Westminster Bank Through A Century*, Plaistow, 1936

Hoare, C. and Co. *Hoare's Bank – A Record*, London, 1955

Howarth, W. *Barclays and Co. Limited*, London, 1901

Lloyds, S. *The Lloyds of Birmingham*, Birmingham, 1907

Matthews, P. W. and Tuke, A. W. *A History of Barclays Bank*, London, 1926

Sayers, R. S. *Lloyds Bank in the History of English Banking*, Oxford, 1957

Withers, H. *National Provincial Bank 1833–1933*, London, 1933

Articles

Anthony, I. *Background to Bank of England Issues 1928*, IBNS, Summer 1968

Opposite. Top: **Greenland notes are not easy to find. This specimen note (also overstamped 'cancelled') has a map of the island on its reverse and is a scarce and interesting item. (90 × 165 mm)**

Bottom: **The Haiti notes of this century are extremely colourful compared to the early issues of the 1850s which were simply black on white. (85 × 166 mm)**

Behind the British Note issues, IBNS, Christmas 1968

Links with British Banknotes, IBNS, Summer 1968

Beresiner, Y. *Bank of England's Note Printing Craftsmen*, WC, January 1973

Bressett, K. E. *Sterling Commentary (Treasury)*, WNJ, March 1967

Loeb, W. *New English £10*, WNJ, 1964

MacWhirter, N. *The One Pound Note*, IBNS, Summer 1965

Morgan, L. *Bank Charter Act of 1844*, IBNS, Summer 1963

Bank Charter Act of 1844, II, IBNS, Autumn 1963

Bank Charter Act of 1844, III, IBNS, Christmas 1963

Yorkshire District Bank, IBNS, Christmas 1965

Narbeth, C. C. *Bank of England Notes*, IBNS, January 1961

Treasury and Bank of England Notes, IBNS, Summer 1963

The Story of the Bradbury, IBNS, Summer 1965

The History of the Anti-Forgery Notes, IBNS, March 1971

Child and Co. Britain's First Bankers?, C & M, December 1972

Obojski, R. *Institute of Bankers Collection*, WNJ, May 1966

Philipson, F. *Some Early Banknote Designs*, IBNS, Summer 1969

Research on the Gallipoli Notes, IBNS, June 1971

ETHIOPIA (Abyssinia)

Currency: 100 cents = 1 Ethiopian dollar since 1946 (also thalers)

Formerly part of the Sudan, Ethiopia's first notes were issued by the Bank of Abyssinia in 1905, which changed its name in 1932 to the Bank of Ethiopia and continued its bank note issues under the new title. The notes of the Bank of Abyssinia fetch as much as £100 ($200) each but those of the latter bank, some depicting the Emperor and Empress Haile Selassie, are more common and can be obtained at between £12 ($24) and £70 ($140).

When the State bank of Ethiopia was established in 1946 the Ethiopian dollar was adopted as the currency unit. With the exception of the 100 and 500 denomination notes, all issues of the State Bank can be obtained at about £5 ($10). The higher denominations mentioned are far more expensive, sometimes appearing on the market at around £40 ($80).

Ethiopia was under Italian rule from 1935 until its liberation by the British in 1941. This period saw a number of issues of the Banca d'Italia overprinted for special use in Italian East Africa. They are valued at about £30 ($60) each. It has been reported that some bank notes have been overprinted with the words 'British Occupation' and dated 1941, but there is no record of these issues.

Opposite. Top: **The Hong Kong and Shanghai Banking Corporation has issued a wide range of colourful notes in several Far Eastern countries. This 10 dollar note circulated in Hong Kong in the 1940s and 50s and is indicative of the beauty of some of the issues. (107 × 185 mm)**

Bottom: **The Toman notes of the Bank of Persia were payable in different towns and cities to avoid large quantities of notes being presented for redemption at one branch, thus causing serious disruption to the economy. (90 × 155 mm)**

F

FIJI

Currency: 100 cents = 1 dollar (also UK currency)

There are very few recorded Fiji issues before this century. Nearly all existing issues reported are those of the government, depicting the coat of arms of the island or a portrait of the reigning monarch of the Commonwealth. The issues prior to 1925 are among the scarcer notes, fetching as much as £75 ($150). Of the remaining issues, most can be obtained at reasonable prices but above their face values.

One of the most interesting series of notes of the Fiji Islands are those sent to the islands at the beginning of the Second World War to alleviate the critical shortage of currency. Issues of the Reserve Bank of New Zealand were overprinted in Australia and stated 'Government of Fiji only'. The issues were for denominations of £1 and £5 sterling. It is possible that 10 shilling notes were also printed but never issued; they would be very highly priced. The overprinted £1 and £5 notes are within the £40 ($80) range. Most of these issues were withdrawn by 1950.

Bibliography:

> Lorimer, H. J. *New Zealand's Bank Notes Overprinted for Fiji*, IBNS, May 1971
>
> Sprake, A. *Fiji's Wartime Emergency Notes*, IBNS, December 1972

FINLAND

Currency: 100 pennia = 1 markkaa
100 kopek = 1 rouble
(Silver roubles since 1840; silver marks since 1860; gold marks since 1897)

The earliest issues recorded for Finland date back to the beginning of the nineteenth century. They were put into circulation in Helsinki by a loan and deposit agency, which in the 1820s became the Loan and Deposit Bank. Until the formation of the Bank of Finland in 1840, note issues of Finland were erratic and are now extremely scarce.

The first issues of the Bank of Finland are in silver roubles. A wide range of dates covers all the denominations from 3 to 25 roubles. These are scarce and can be purchased for £45 ($90) or so. The first issues are more easily available and date from the 1860s. The Bank of Finland silver mark and, later, the gold mark denominations of this century can be purchased for about £16 ($32).

The Republic of Finland has among its most interesting notes those known as the 'cultural series' depicting nude effigies against a factory in the background. The series were issued from 1918. The higher denominations of that date, and those of 1922, are extremely attractive large notes fetching up to £35 ($70) each. Later issues gradually become more available, except for high denomination notes. Issues up to 100 markkaa are within the £12 ($24) range, but those of 500 markkaa (which date back to 1909, although not issued until 1918) and the 100 markkaa issued in the same year are priced at over £40 ($80). The 5,000 markkaa of 1922 is within the £75 ($150) range.

Bibliography:

Bank of Finland *The Paper Money of Finland, 1809–1951*, (in Finnish), Helsinki, 1952

Borge, E. *Suomi-Finland 1860–1972*, (in Finnish), Helsinki, 1972

Siege, F. *Sieg's Settelkatalog 1874–1970*, (summary in English), Ulbjerg, 1971

FRANCE (also Monaco and French overseas provinces)

Currency: 100 centimes = 1 franc

100 francs = 1 nouveau franc (monetary reform 1960)

(sous; sols; livres – denominations relating to state notes and assignats.)

With the exception of the British Commonwealth, France has the largest number of overseas territories in which notes were issued while under her sovereignty. When considering the paper money of France there are two issues which come immediately to mind. The 'assignats' issued during and following the French Revolution, and the Banque de France issues which have dominated paper money emissions in France since the early nineteenth century.

The inflationary French assignats of the late eighteenth century had become completely worthless by 1786 (see Chapter IV). With regard to the Banque de France, a superb and huge collection can be formed by dates, signatures, and the 'portrait gallery' of famous Frenchmen who appear on these issues. Mr Roger Outing of England has done much research here and some publications are now awaited. The early notes, from 1803 to the 1880s, printed black on white for denominations from 25 to 5,000 francs, are among the rarest issues. At Napoleon's instigation the 'caisse des comptes courants' in the early months of 1800 raised 30 million francs by sale of

The French Revolution period saw many kinds of paper money issues. This note for 20 francs was circulated in Rouen and cancelled after redemption. It is dated 1794. (90 × 192 mm)

shares to the public and changed its name to 'Banque de France'. These only appear very infrequently on the market and, irrespective of date, fetch prices above £75 ($150).

The series that followed were, in some instances, re-issues of early denominations and continued to circulate well into the Second World War. The notes from 5 to 5,000 francs have a wide variety of both dates and signatures relating to each denomination. Most of these are priced within the £10 ($20) range. The war issues of the Banque are, with a few exceptions, probably the most common and easiest to come by, and include the 100 franc denomination. They can be obtained at less than £5 ($10) in excellent condition.

The third series of the Banque de France consisted of note issues from the end of the War to 1950. These are mostly of a high denomination and include, for the first time, the 10,000 franc note. This was followed by the fourth series from 1953 to 1965. All these notes are quite common. The exception is the 1938 5,000 franc issue which is priced at about £30 ($60). The issues from 1953 to 1965 were overprinted as a result of the monetary reform of 1960 when the *nouveau franc* was introduced. The overprint applied to the large denomination notes of 500, 1,000, 5,000 and 10,000 francs with an effective devaluation to one per cent of the original value. These issues are rarer than the originals but can still be obtained for about £25 ($50). The currency reform of 1960 was abandoned in 1963 and the word 'new' was dropped from the monetary unit.

There were several private entities which issued notes in France, notably the banks at Rouen, Bordeaux and Nantes, which were granted government charters in 1818 – but only lasted until the Banque de France's more aggressive policy of opening branches in the provinces. All private bank note issues are rare. There are also some issues relating to the railroad administration during the French and

French colonial notes can form a very attractive collection and the war-time issues of some of the colonies are particularly interesting. This 100 franc note was circulated in 1942 in Martinique. (89 × 182 mm)

Belgian occupation of Germany, after the First World War. These extremely interesting notes have a locomotive on the obverse and are sought after by many collectors who specialise in the railroad theme. They were placed in circulation in 1923 and all the denominations are easily obtainable; a set of ten notes from .05 franc to 100 francs can be purchased for about £60 ($120).

Monaco

Monaco was originally annexed by France in 1793 and entered into a customs union with her in 1869 after restoration of the ruling family in 1814. All note issues have been French with the exception of a series of city notes following the First World War. The 25 and 50 centime and 1 franc notes were the only issues. They were dated 1920 and headed 'Municipalité de Monaco'. They are obtainable at £5 ($10) each in reasonable condition.

French overseas provinces

As with some of the British Commonwealth countries, a currency board of the French Republic functioned in many islands. The Bank of Indo-China was a French overseas institution whose notes circulated in Tahiti, New Caledonia, the New Hebrides and French Somaliland. The lower denominations are not very rare and are priced between £10 ($20) and £45 ($90).

The Caisse Centrale de la France Libre also issued currency board notes from 1941 to 1944 in French Equatorial Africa and the Cameroons. A common currency board for these countries was set up in 1947 and functioned until 1960, when the Cameroons became independent. The same currency board also circulated notes during the war in French Guiana, Oceania, West Africa and Guadeloupe, Martinique and Réunion. Most of these notes are within the £10 ($20) range. They are often referred to as the 'Free French issues' in view of the overprint on many of the notes, which states 'France Libre'. This was a slogan of support for France in the Second World War. The Institut d'Emission, another pseudo-currency board, circulated its own notes in French West Africa and Togo, as well as the Indo-China Federation. The latter was formed in 1946 and included the Khmer Republic (Cambodia), Laos and Vietnam within the French Union. Many of these issues, too, are easily encountered as large quantities were put into circulation, albeit for limited periods of time. They do not fetch prices above £20 ($40). From 1955 onwards, the West African Monetary Union was responsible for paper money issues in the Ivory Coast, Dahomey, Mali, Niger, Togo, Senegal and the Republic of Upper Volta. These issues have a letter under the serial number which relates them to one of the mentioned territories. The majority of the note issues of the West African Monetary Union are comparatively common and within the £15 ($30) range.

French Equatorial Africa, which was known as the French Congo until 1910, declared its independence from the Vichy government in 1940. The general government of French Equatorial Africa began its note issues in 1917 and these were continued by 'Free France' from 1941. All the notes of the French colonies are interesting and colourful issues. They often depict local rustic scenes.

It must be remembered that identical notes headed, for example, 'Central Bank for the Overseas Territories' were overprinted for individual countries and circulated in many of them simultaneously. Thus the same type of note issue of French Equatorial Africa was separately overprinted for use in Guyana, Guadaloupe, Martinique, Réunion, St Pierre-Miquelon and a few more territories.

Although the general indication given is that most of these issues are within the £10 ($20) to £45 ($90) range, there are many exceptions. These relate to the higher denominations (up to 5,000 francs) which are often priced at £100 ($200) and over, as well as certain overprints and dates.

Bibliography: Assignats – see Chapter IV

Banque de France *Les Billets de la Banque de France*, 'Change' No. 85, 1972
Dickerson, R. E. *Style and Design on French Bank Notes*, CC, March 1973
Guitard, H. *Vos Billets de Banque*, (in French), Paris, 1963
Lafaurie, J. & Habrekorn, R. *Les Billets de la Banque de France et du Trésor 1800–1952*, BCD, Nos. 8 and 9, 1953
Mazar, D. J. *Histoire Monetaire et Numismatique Contemporaine 1790–1967*, (in French), Paris, 1967
Muszynski, M. *100 Franc Bank of France 1906 Type*, CC, Winter 1964 and Summer 1965
   Celebrated Frenchmen on the Banque de France Notes, IBNS, December 1971
   Les Billets de la Banque de France, (in French), Paris, 1975
Philipson, F. *Paper Currency of Laos, Cambodia and Vietnam*, CM, May 1971
Pick, A. *French Colonial Bank of the Sugar Colonies*, LANSA, October 1974

G

GERMANY

Currency: 100 pfennig = 1 mark since 1874–1923
   30 groschen = 1 taler
   60 kreuzer = 1 gulden
   100 Rentenpfennig = 1 Rentenmark (1923–48)
   100 Reichspfennig = 1 Reichsmark (1924–48)
   100 Pfennig = 1 Deutsche mark (West and East since 1948)

Newcomers to notaphily immediately associate Germany with the disastrous hyperinflation of 1919–23. This subject is considered in detail in Chapter IV.

Germany has a vast notaphilic history. The earliest issues are of the Empire following the unification sponsored by Prussia in 1866. These notes, dating from 1874, were issued only in the 5, 10, 20 and 50 mark denominations. They continued in circulation until 1906; and although some of the earlier issues are in the £80 ($160) range or over, the ones of the 1900s are well below that price and many can be obtained at less than £10 ($20) each.

The bank note issues of the German Reich, which began in 1893, are also quite common, particularly the issues of this century. The 100 mark note of 1908, known as the 'Long 100', is one of the better known issues, obtainable at about £2 ($4). The next series of issues were those of the 'Darlehenskassenschein' from 1914 up to the inflation period. These issues are obtainable at under £1 ($2). From there the Reich Bank notes followed until 1924. Some of the interesting issues which made an appearance immediately after the inflation period were the fixed-value loan notes, from October 1923. These were soon declared by the German government to be legal tender issues. Some of the fractional notes were in the gold mark denomination, with the equivalent values given in dollars. The issues are certainly collectors' pieces and obtainable for under £8 ($16).

The rentenmark has been called 'the Saviour of Germany'. In 1923 it was declared equal to one billiard marks and indicated the end of the inflation period. Issues of 100 and 500 rentenmarks are extremely rare pieces priced at above £70 ($140). This applies even more so to the 1,000 rentenmark, but issues thereafter become common. The notes from 1925 to the Second World War are almost all within the £10 ($20) range. The German Federal Republic introduced a currency reform in 1948 and with it the 'Deutsche mark' as a new currency unit. Denominations from ½ to 100 Deutsche marks were issued. Most of these notes are obtainable for just £5 ($10). The 1948 50 Deutsche mark, in green and blue as opposed to the normal violet colour, was in circulation for only a few days. It fetches a very high price, near the £120 ($240) level. The year 1948 also saw the issues of the 'Bank of German Lands' with notes from 5 pfennig to 100 Deutsche mark. The German government bank, the Bundesbank, was established in 1960 and has since been issuing bank notes.

The German Democratic Republic (East Germany), simultaneously introduced its

In spite of the abundance of German inflation notes of the post-First World War period, several of the very high denomination notes are scarce. The 2 billion mark was issued in November 1923 when inflation reached its peak, and is consequently a rare note. (68 × 118 mm)

own currency in the 'Ost-Deutsche mark' denomination. Its issues from 1948 to 1964, when the present notes were put into circulation, are all common; the higher denominations are not priced at over £10 ($20).

Germany saw a large number of regional bank notes. Many of the issues were not considered legal tender but circulated as notgeld during the period leading up to the First World War, and for a short time thereafter. Some of the issues were accepted by the central government and these included the notes of the Bank of Baden, the earliest of which dated back to 1817. They are fetching prices of up to £50 ($100) with the higher denomination of 100 marks of 1890 having fetched £68 ($136) in an auction in 1972. Notes after that date, including inflation issues, cost just a few pence each. Bavaria also issued its own notes. The 1875 notes and the 1924 1,500 reichmark issues are very rare. The rest are inflation notes usually priced under £2 ($4) each.

The scarce notes of the regional banks are the gulden and taler issues of, among others, the Banks of Frankfurt, South Germany (in Hesse) and Saxony (in Dresden). These issues, dated during the last quarter of the nineteenth century are priced above the £35 ($70) level.

Germany, like the British Commonwealth and France, also issued a great number of notes for her overseas territories. The ones that immediately come to mind are the German East Africa issues. Notes depicting the young Kaiser are dated 1905 and are well known in the notaphilic world. They fetch prices upwards of £10 ($20) and £100 ($200) or more for the 500 rupien note.

The other interesting currencies circulated by the Germans in East Africa were provisional issues; but the Great War continued for longer than expected. These

The German East African notes of the pre-First World War period depict the Kaiser. They were issued in the rupiah denomination. (102 × 159 mm)

'interim' notes, issued in March 1915, are crudely printed on coarse coloured paper. They circulated for almost two years until they were finally replaced with a second series. The first interim issues, printed in Africa, are priced at about £2 ($4). The second series are now known as 'bush notes'. They were even more primitive in appearance than the first, having been printed on a children's rubber-stamp printing set while the Germans were retreating through the bush! They became available to the collector in large quantities and appear on the market usually below £4 ($8).

The collector of German notes must also consider the issues of the German occupation of Poland, Rumania and the USSR.

The 1920s inflation issues of Baden and Bavaria come under a general German collection and the prices are in line with the German inflation notes. An interesting series in 1948 were the bank notes issued for Berlin. They merely carry the overprint of a large encircled 'B' in the centre of the ordinary issues. Denominations run from $\frac{1}{2}$ to 100 Deutsche marks and the prices they fetch are between £4 ($8) for the smaller denominations and up to £20 ($40) for the higher.

The reader is referred to Chapter IV where concentration camp issues, inflation notes (notgeld) and military issues have been separately considered.

Bibliography:

Banyai, R. *Dr. Hjalmar Schacht: Central Banker and Financial Wizard*, WNJ, December 1968

Bramwell, D. *The Small Prussian Bank that Gave Germany her Notes*, CCW, October 1971

Dickerson, R. E. *Banknotes of the German Empire 1871–1888*, CC, December 1972

Jaeger K. & Haevesker, U. *Die Deutschen Banknoten seit 1871–1963*, (in German), Engelberg, 1963

Keller, A. *Paper Money of the Old German States from 17th Century until 1914*, Berlin, 1921

Keller, A. *German Banknotes*, Part II, IBNS, Christmas 1962

Lawrence, J. *German East African Currency, 1885–1917*, IBNS, June 1970

Loeb, W. M. *West Germany New Issues*, WNJ, September 1964
East Germany New Issues, WNJ, July 1965

Lotkik, V. *Occupation Money of Hitler in the USSR*, IBNS, Christmas 1964

Milich, A. *Who's Who on German Reich Notes*, IBNS, March 1970

Morse, L. W. *The Rosemar Reichsbanknote Reserve 1874–1920*, CC, May 1972
From Numismatic Mine Tailings to Gold Scales, IBNS, September 1972

Pick, A. *Allied Military Currency in Germany*, CC, Summer 1963
German Gold Discount Bank Signatures on Some German Bank Notes, IBNS, June 1972

Rosenberg, H. *Die Banknoten des Deutschen Reiches ab 1871*, (in German), Hamburg, 1970

Schaaf, B. *The Paper Money of German East Africa*, I, CC, May 1972

Talisman, M. R. *Investing in the German Mark*, CC, May 1973

GIBRALTAR

Currency: (As in UK)

Gibraltar's paper money has been popularised by the fact that a complete collection can conceivably be formed of all the paper money issues of the peninsula.

The earliest recorded notes bear an embossed stamp stating 'Anglo Egyptian Bank Limited, Gibraltar'. Two series of these notes were issued in 1914 for denominations from 2 shillings to £50. The 2 shillings, 10 shillings and £1 have appeared in the market for values of up to £100 ($200). The £5 and £50 notes are very rare indeed.

In 1938 a 2 shilling note was also apparently issued, but never placed into circulation, the notes having been destroyed in 1968 with only a few exceptions, which have found their way into collectors' hands.

Two currency laws passed in 1927 and 1934 relate to the Government of Gibraltar notes. The latter dealt with notes still in circulation, and all issues authorised by the 1934 ordinance continued as legal tender notes. They are valued, when crisp, at a premium over their face value. Those dated 1927 and issued by authority of the ordinance of the same year are fetching prices of between £10 and £35 ($20 and $70) for the 10 shillings, and £40 and £60 ($80 and $120) for £1 and £5 notes respectively. Later notes, dated up to 1965, are cheaper.

Bibliography:

Devenish, D. C. *Currency Notes of Gibraltar*, IBNS, Christmas 1969
Malcolm. J. C. *Gibraltar Paper Money*, C & M, 1970

GREECE

Currency: 100 lepta = 1 drachma(i)

The National Bank of Greece was established in 1841 and issued notes from that year until 1867 for values from 5 to 500 drachmai. These issues, and later ones from 1860

The notes of the Ionian Islands are keenly collected by those interested in Greek notes. The small size 2 drachma 1885 issues are particularly rare. (48 × 82 mm, actual size)

to 1897 are extremely rare, as are all issues of the last century. They usually fetch above £100 ($200) subject to their condition, date of issue and denomination.

There has been a total of twelve series (*ekdosis* in Greek) of the National Bank of Greece notes issued this century; the series is indicated on the reverse.

A large collection of the National Bank of Greece issues can be accumulated by dates, signatures and series. With the exception of the 1,000 drachma issues of the eighth series, all the issues can be obtained for under £15 ($30), in good condition. These issues almost invariably portray Georgios Stavros, the first Governor of the Bank of Greece. The National Bank continued with its own issues until the Bank of Greece was formed in 1928.

The first notes of the Bank of Greece were the issues of the National Bank overprinted with the words 'Bank of Greece'; these circulated from 1928 to 1935 and are not priced above £12 ($24). Some of the issues mentioned have been overprinted with the word *neon* (new). But the value of the notes is only slightly affected by either of the overprints. The regular issues of the Bank of Greece began in 1932 and long before the disastrous 1944 inflation that raged through the country, high denomination notes of up to 5,000 drachmai were being circulated. They are very common and generally fall within the price range of the inflation issues. (See Chapter IV).

Small notes, somewhat similar to the Treasury issues of 1917, were issued by the Treasury of the Kingdom of Greece from 1944. These were placed in circulation by the Ministry of Finance. They can be easily encountered and are valued at under £3 ($6). The rarer Treasury issues are the very high denomination bills of 100 and 200 million drachmai issued in Karkyras and dated 1944. These fetch over £30 ($60) each.

Modern bank note issues of Greece, since 1954, have been issued by the Bank of Greece. Of the issues of Greece that very rarely appear in the market, and consequently would fetch prices in the range of £100 ($200) or more, are the regional bank issues at the beginning of this century, issues of the Epirus-Thessaly, Ionion and Crete Banks. The first of these was taken over by the National Bank in 1899, but issued notes until 1905. The Bank of Crete, founded in 1899, only ceased to issue notes in 1929.

Some legal tender notes issued by the Italians (Casa Mediterranea di Credito per la Grecia) in 1940 and the Germans ('Saloniki–Agais' overstamp on German army notes) in 1942, were placed into circulation on the Islands during the Second World War and are among the sought-after military occupation issues.

Bibliography:
Barbary, C. H. B. *Bank Notes of Greece*, MCBN, November/December 1970

Papadopoulos, C. A. *The First Modern Greek Paper Money*, IBNS, December 1972

Philipson, F. *The Glory That Was Greece*, CM, January 1971

GREENLAND – see DENMARK

GUATEMALA

Currency: 100 centavos = 1 peso = 1 quetzal (since 1925)

The first recorded issues of Guatemala date back to 1804. Notes for obligatory circulation were accepted by the government in payment for customs duties and taxes. There are no surviving specimens of these issues. The first notes of which specimens are available are those of the Banco Nacional de Guatemala, founded in 1874. The Bank was defunct by 1876, but sufficient bank notes were issued in the brief period of its existence to ensure survival to this day. Even in very good condition these early issues do not fetch more than £15 ($30). The Bank was the first of the seven private Guatemalan banks formed during the last century. All the issues are extremely attractive and many are large in size. All have denominations of up to 100 pesos. In good condition any of the 100 peso notes will fetch £45 ($90). The banks issued almost identical notes, the only difference being in dates.

Treasury issues dating back to 1887 were also circulated in large quantities and still fetch only £15 ($30) as do most government issues.

The modern notes of the Banco Central de Guatemala, established in 1925, are common. The issues, now demonetised, are priced at about £2 ($4).

Bibliography:

Clark, O. H. *Paper Money of Guatemala 1834 to 1946*, Texas, 1971
Keller, A. *Guatemalan Local Emergency Issues*, CC, 1970
Musser, D. *The Paper Money of Guatemala*, WPMJ, December 1959
Prober, K. *Numismatic History of Guatemala*, (in Spanish), Guatemala, 1957
Shafer, N. *The Early Paper Money of Guatemala*, IBNS, Winter 1969

H

HAITI

Currency: 100 centimes = 1 gourde

The important point about Haiti notes is that the date appearing on the earlier notes is not the date of issue but of the law by which the issue was authorised. Not until 1919 did Haiti notes indicate a date of issue. These are headed 'Republic of Haiti' and priced at about £20 ($40).

The short-lived Empire of Haiti (1850–57) was created by the former negro slave, Faustino Soulouque, who declared himself Emperor. A few notes for denominations between 1 and 50 gourdes were issued under him by a law of 1851. These are rare, priced in the £65 ($130) region.

After the re-establishment of the Republic in 1857 the earlier notes of the Empire continued in circulation and notes authorised by even earlier laws (after the official declaration of the Republic in 1820) were now issued and began to circulate in denominations from 1 to 100 gourdes, with a great number of odd denominations in between (such as the 2 and 8 gourde notes). All the issues, including those authorised by the decrees of 1851 and 1871, are difficult to find in good condition because of the flimsy paper on which they were printed. When they appear on the market in crisp condition they fetch prices often over £45 ($90). New currency laws were enacted in 1892, 1903, 1904, 1908, 1914, 1915, 1916 and 1919.

With the formation of the National Bank of Haiti in 1916, the notes of the Republic were overprinted with the new heading. The old issues that were re-dated are a little scarcer and fetch up to £35 ($70). The rest of the issues of the National Bank are undated, but several types of each denomination exist. The notes are currently in circulation and the 1 to 500 gourde issue can be purchased at a premium over their face value.

A study has recently been undertaken by Leon Burstyn of the Argentine in order to establish whether the exact date of early Haiti issues, as well as those of the Republic, can be determined by the signatories to the notes.

In this context, reference should be made to the Latin American Paper Money Society (see Appendix IV).

Bibliography:
Benson, W. L. *Corrections to Sten Catalogue*, IBNS, Summer 1968
Gunn, J. W. *Notes, Marks and Prints of Time*, WC, June 1972

HOLLAND see NETHERLANDS

HONDURAS

Currency: 100 centavos = 1 peso = 1 lempira (since 1932)

With the exception of the modern notes of El Banco Central de Honduras, the only popular and easily found early note is the 50 centavos of the private American enterprise, *Aguan Navigation and Improvement Company* issued in Trujillo (not to be

The bank notes of Honduras are very difficult to find. The Aguan Navigation and Improvement Company was American-based and received authorisation to issue its own currency by the Honduras Government in the 1880s. (82 × 191 mm)

confused with the temporary capital of the Dominican Republic) in 1886. The note is worth about £6 ($12). The higher denominations fetch as much as £75 ($150) each.

A number of additional private banks issued notes in the peso and lempira denominations up to 1951. All are within the £50 ($100) range.

The Central Bank, formed in 1951 as a result of the amalgamation of the private banks the Banco de Atlantida and the Banco de Honduras, has been the sole issuer of notes since her foundation and there have been no scarce notes circulated by the bank in the last twenty years or so.

Bibliography:

Holsen, P. J. *Honduras Notes*, LANSA, October 1974, Vol. II, No. 3

HONG KONG

Currency: 100 cents = 1 Hong Kong dollar

The Government of Hong Kong issues since 1941, portraying the reigning British monarch, are well known to most collectors as they are among the cheapest notes one can find. The very small denominations are sold in quantities for a few pence and one need pay no more than £1 ($2) for the 1 dollar note in good condition, this being the highest denomination of these issues, which were discontinued in 1959.

Some of the bank issues are also well known. The Hong Kong and Shanghai Banking Corporation and the Chartered Bank of India, Australia and China notes have attracted the attention of many collectors. These issues are all beautifully designed and available at reasonable prices for denominations of under 50 dollars. A multitude of these notes was issued, bearing different dates. The earliest recorded is the 100 dollar note of 'Hong Kong and Shanghai' dated 1872 which is priced at over £300 ($600) when in very good condition.

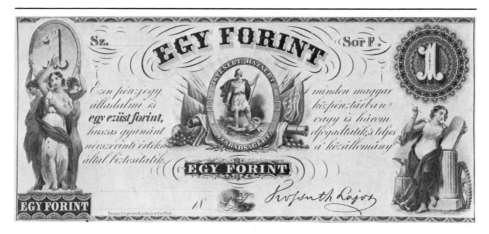

Lajos Kossuth was the first President of the Hungarian Republic of 1848. He issued notes of the Hungarian Finance Department in 1848 in order to raise funds for his fight for freedom. These bore his signature and are important historic items today. (80 × 188 mm)

Chinese, Dutch, German and many other foreign banks issued notes in Hong Kong; all are attractive items and mostly within financial reach of the collector. As a general guide one would expect an average price of £10 ($20) per note for the low denomination notes of the second quarter of this century.

The Hong Kong Government also issued over printed Chinese and Japanese notes during this century. All the well-known Chinese issues, those of the Bank of Communications, the Bank of China and the Farmers Bank, have been overprinted in English and Chinese and all are scarce, priced above £25 ($50). The propaganda notes of Hong Kong have been considered separately in Chapter IV.

Bibliography:

de Sousa, A. B. *Standard Catalogue – Hong Kong*, CCW, Hong Kong, 1967

Hauck, R. *The Hong Kong Annual Report 1963*, CC, Winter 1964

Mao, K. O. *A Short Account of Hong Kong Currency*, IBNS, Christmas 1963

Narbeth, C. C. *Hong Kong Duress Notes*, IBNS, Christmas 1964

HUNGARY (Magyar)

Currency: 100 filler = 1 korona until 1926 = 1 pengo
12,500 korona until 1946 = 1 forint

Disastrous inflation in Hungary resulted in a wide range and variety of notes being issued after the Second World War. They are easy to find and cheap to buy. (See Chapter IV – 'Inflation'.) (The notes that circulated in the Austro-Hungarian Empire have been considered in this chapter under 'Austria'.)

Hungarian notes proper, with one exception, did not come into being until the beginning of this century. The earliest of these are the Kingdom of Hungary notes issued in 1914 in order to finance the Great War. These high denomination notes of 250, 2,000 and 10,000 korona are only known in specimen and are worth about £40 ($80) each.

The Austro-Hungarian Bank continued to issue notes until 1919; the rarest of these are the non-interest-bearing high denomination notes with various dates up to 1918 and which are valued upward of £30 ($60). The 1902 and later issues of this Bank were overprinted in Hungarian and circulated in 1920. They are common notes fetching a little over £5 ($10) in crisp condition. Before the Hungarian National Bank came into being in 1919 paper money was issued by the postal authorities of the country. Most of these are common, with the exception of the 10 million korona specimen note of 1921 which may be valued at £35 ($70). These were followed by notes of the Ministry of Finance until the Currency Reform Act of 1925 led to the circulation of the overprinted issues with the new denomination, the pengo (= 12,500 old korona). The rarest of these is the highest denomination note of 80 pengo overprinted on 1 million korona.

The National Bank issued its first notes in 1926, although specimen notes dated 1919 have made their appearance on the market at prices above £20 ($40). These issues continued until the end of the War, and again, most are within easy reach at £3 ($6) or less.

The forint notes that followed the catastrophic inflation of 1946 have continued in circulation.

The only pre-twentieth-century notes are those issued by the first President of the first Hungarian Republic, which came into being as a result of the War of Independence of Hungary in 1848–49. Lajos Kossuth, greatest of the Hungarian patriots and champion of freedom and liberty, had led the national renaissance since 1815. Altogether there have been a total of eleven distinct Kossuth issues, six of which were circulated in 1848 and 1849 within Hungary's borders. The notes of the Hungarian Finance Department dated 1848, and signed by Kossuth as Minister of Finance are fetching about £20 ($40) today in crisp condition. Earlier issues both of the Hungarian Commercial Bank (undated notes) and the Hungarian National Bank, which were the first ones to circulate in mid-1848, are much rarer. The latter were issued for a total of 5 million forints, in denominations from 50 to 500 forints. The majority are extremely rare issues. The fractional notes of both the banks are also exceedingly rare, having issued denominations as low as one-sixteenth of a forint!

Additional issues by Kossuth still hard to obtain are the Treasury notes for small denominations issued in January 1849. In July of the same year a new set of Treasury notes was issued by Kossuth as Governor of Hungary in Debrecen. He had escaped there from the Allied army of the Emperors of Austria, after the declaration of independence of Hungary on 14 April. In the last month in which Hungary was free, August 1849, 2 and 10 pengo-forint notes were issued by Kossuth. Only twelve known specimens exist of this note. The remaining Kossuth notes are the better known ones issued after his departure from Hungary and during his continued struggle for revolution. He issued dollar 'Hungarian Fund' notes in the United States with the permission of the US Government in order to support the independent Hungarian government. The notes are dated 1852 and have fetched fairly high prices for the 1,500 dollar denominations. Much more common are the Philadelphia issues of the same year. The legend is in the Hungarian language and there is no printed date, although Kossuth's facsimile signature appears on the notes. These are priced at about £3 ($6) each; uncut sheets of 1, 2 and 5 forints have been fetching around £15 ($30).

Finally, Kossuth's most unsuccessful efforts on behalf of his revolution were the 1860–61 notes issued in London: the text was in Hungarian but the notes were unfortunately confiscated in a hurry by the British government who wished to avoid embarrassing the Austrian Emperor. All notes issued in England have been destroyed. Very few have recently been reported to be owned by the son of one of the engravers now living in Australia. The notes would undoubtedly fetch startling prices if genuine.

Finally, two additional issues relating to Kossuth are the 1866 notes from plates engraved by Kossuth's son, but never issued, and the 1948 and 1949 notes for the celebration of the centenary of the Hungarian War of Independence. The National Museum exhibited and reprinted some pieces from the original plate of the two pengo-forints. These are not real issues and therefore have merely a curiosity value.

Bibliography:

Kupa, M. *Paper Money of Hungary*, (2 volumes), Budapest, 1964
Kossuth's State and Bank Notes, IBNS, Summer 1962
Narbeth, C. C. *Early Paper Money*, IBNS, Summer 1969
Weissbuch, T. N. *Louis Kossuth: Moneys of a Rebel in America*, WNJ, March and April 1965

I

ICELAND (Island)

Currency: 100 aurar = 1 krona

The eighteenth-century special issue for Iceland in the regstaler denomination were Danish notes overprinted on the reverse 'For Circulation in Iceland'. They are priced at about £80 ($160). The earliest notes of the island are the 1 krona issues with the legend 'Fyrir Landssjodur Islands', and are dated up to 1900, obtainable at about £125 ($250).

The notes at the beginning of this century issued by the Bank of Iceland are easier to come by. The exception is the 100 krona notes which are priced at £30 ($60). Government notes issued during the same period and headed 'Rikissjotur' are far more common. Those which circulated from 1941 can be obtained at under £5 ($10) and the same issues, that is government notes authorised by decrees dating to the end of the last century, are also within the £10 ($20) range.

The National Bank of Iceland was not formed until 1928. With the exception of the first series, its notes, including the higher denominations, are priced at under £10 ($20). Issues since 1957 are in circulation and still legal tender.

Bibliography:

Philipson, F. *Danish Isles Currency*, CM, March 1973
Sieg, F. *Sieg's Settelkatalog 1874–1970*, (with English summaries), Ulbjerg, 1971

INDIA (also Hyderabad)

Currency: 100 paise = 1 rupee

A number of early Indian banks issued their paper money in the last two decades of the eighteenth century. Examples of such issues are those of the Bengal Bank and the Bank of Hindustan. These notes are exceedingly rare.

At the turn of the nineteenth century the government issued its own Treasury bills which were redeemable *in specie* (metallic coin) after a determined period of time. The bills never appeared on the market but would certainly be valued at over £150 ($300) should they do so. The serious loss in the value of the Treasury bills led the government to establish the Bank of Calcutta, which became the first note-issuing government-backed bank in India. This Bank issued its own notes for the first time in 1806 for denominations of 10 to 10,000 rupees. It changed its name to the

Bank of Bengal only a few years after it was established and continued its paper money issues up to 1858, when the British Crown took over the East India Company's administrative responsibilities.

By this time other banks had made their brief appearance and all their issues are exceedingly rare. Occasionally proofs are to be found, but it would seem that these are not as scarce as the original issues. The specimens are all valued in the £100 ($200) range.

In 1861 the Paper Currency Act was established. It specified that the Government of India was to be the sole administrative authority to issue notes. The first of these, printed in England, began circulating in 1862. The three banks that had left their mark on the Indian economy so far, the Bank of Bombay, the Bank of Madras and the Bank of Bengal, were nominated agents for the management of issues under the control of the government commissioners. These notes are known as the 'British colonial type issues' and were put into circulation in different parts of the country. The name of the city of issue is indicated on the face of the note. Those from 1860 to 1899 appear with different dates and for a wide range of denominations. They are all very hard to come by, priced at over £75 ($150).

The Government of India issues of this century gradually become more easy to encounter. Those dating between 1903 and 1920, which do not bear the portrait of the monarch, are obtainable at about £35 ($70) each. Later notes, issued up to 1935 with the portrait of King George V, are in the £22 ($44) range. From that date, the issues both of the Government of India during the War and the Reserve Bank of

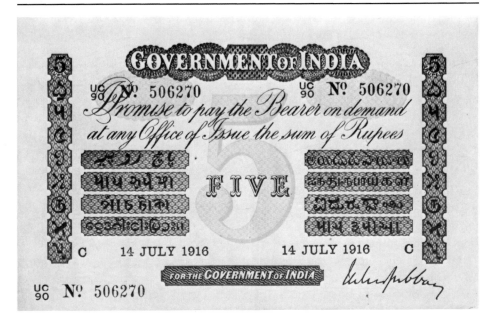

The colonial notes of India were typical in their design. The denomination, in numerals as well as letters, appears on the notes in eight local dialects. (109 × 176 mm)

India, which followed these issues, are fairly common. The exceptions are the 100 rupee notes valued at over £15 ($30). The more common notes of India are those of the Republic dated after 1949; most depict the famous Ashoka Column. One of the most interesting features of these issues is the multilingual text relating to the denominations. On some notes this is in no less than thirteen local dialects.

A great number of Indian notes circulated in the Persian Gulf area. Some were overprinted for use in Bahrain, Kuwait and the Yemen, but the most interesting among these are the note issues which were officially overprinted for pilgrims travelling to Mecca in Saudi Arabia. The notes were issued by the Reserve Bank of India and are not particularly difficult to obtain. (Pakistan also overprinted such notes.) They are mostly priced at under £15 ($30).

Hyderabad

Completely separate from the Indian government notes are those of Hyderabad, the central Indian state which refused to enter the Union in 1947. It was nevertheless forced to do so a year later. This state had its own paper money issues in rupee denominations dating back to 1918. The notes continued to circulate until 1942. The original notes, in Urdu, are scarce and are certainly priced at over £25 ($50) each, the 1,000 rupee being worth £100 ($200). The later issues of 1938, for 1, 5 and 10 rupees, are much more common. The collector must beware not to confuse these issues with any from Arab states as the texts look very similar.

The paper money of Hyderabad came into the limelight some years ago when the S.S. *Egypt*, an Italian vessel, hit a mine off the coast of France and sank. When it was retrieved some ten years later, a large quantity of British gold sovereigns were discovered in the strong room. With it were large packets of bank notes which had been printed for Hyderabad. Most of these notes had been destroyed by the damp but some, which remained in reasonable condition, were subsequently overstamped in English, stating that the notes had no monetary value and specifying the facts of the incident. Very few of these notes have survived in perfect condition, and those in reasonable condition (with the overstamp) appear on the market regularly and can be purchased at about £35 ($70).

A number of state notes from Kashmir have also been reported as being issued in 1936. There is no record of these notes as far as price values go.

The Reserve Bank of India had some of its notes overprinted with the word 'Karachi', and these issues are, of course, interesting to collectors of Pakistani notes too. They are undated and valued at about £35 ($70).

Bibliography:

Fonner, G. *An Underwater Bank Note*, CC, Autumn 1969
Kraus, H. *Emergency Currency of Hyderabad*, IBNS, July 1970
Kumar, S. *New Note Issues of India*, IBNS, Summer 1967
Leader, R. *Some Notes of India*, IBNS, March 1972
Mao, K. O. *'Egypt' Yields Hyderabad Rupees*, WC, August 1971
Rao, R. *Hyderabad Error Note*, WC, November 1966
Robin, P. *More on Current Indian Notes*, CC, Summer 1966

INDONESIA

Currency: 100 sen = 1 rupiah

Until 1945 special note issues of the Netherlands for the Netherlands Indies circulated in the area. The earliest of these notes, of the De Javasche Bank, were issued in 1828 and they are extremely rare. Later notes of the same Bank are more easily obtainable.

The year 1940 saw issues of the 'Muntbiljet' again especially emitted for the Netherlands Indies. The notes are comparatively common, worth about £8 ($16). The De Javasche Bank also issued notes in the Celebes'Islands in 1948. These are obtainable within the £5 ($10) range.

The modern notes of Indonesia began with the state issues of 1945 in both the sen and rupiah denominations. These issues, which continued until 1948, have the city of issue indicated on the obverse. The notes, including the odd 400 denomination issue, are quite common. They appear on the notaphilic market in large quantities. They are extremely attractive in colouring and design and very popular, at prices not exceeding £1 ($2). The same applies to the issues of the United Republic of Indonesia, which began to circulate in 1950. Almost simultaneously, in 1952, the Bank of Indonesia (formerly De Javasche Bank) issued its own notes. All are very common.

The struggle of the Indonesians to free themselves from the Dutch brought about a number of emergency notes. Odd denomination emergency notes are not particularly common and there are many interesting items among them, some priced at no more than £5 ($10), but others much more highly.

Bibliography:

Loeb, W. M. *New Issues*, WNJ, July 1964

Philipson, F. *Paper Money of Indonesia*, CM, December 1971

Pick, A. *Interim and Inflation Notes*, CC, Summer 1964

IRAN (Persia)

Currency: 100 dinars = 1 rial
100 kran = 1 toman until 1918

Extensive research on the early notes of Persia has been undertaken by John Wall. While stationed in Iran he discovered records of a paper money issue dating back to AD 1294 issued during the Il-Khan Dynasty under Emperor Kai Khatu Khan. The notes from $\frac{1}{2}$ to 10 dinars, printed in Persian, Arabic and Mongolian, were apparently oblong in shape and withdrawn from circulation shortly after being issued. There are no known specimens of this note in existence.

The next recorded issues are dated 1888 and were circulated in Teheran and the provinces by the New Oriental Banking Corporation. This Bank was taken over by the Imperial Bank of Persia, which began its own issues in 1889 until 1931. The latter was a private British-owned banking corporation with concessions from the Shah of Iran. It issued two series of notes, the first from 1890 to 1924 and the second from the latter date up to 1931. The issues from 1 to 1,000 toman printed by Waterlow and Sons and Bradbury and Wilkinson are rare. All are overprinted with the city of issue and the date on which they were put into circulation. The portrait of Nasir-ed-din-Shah Oajar appears on all the notes. When they appear on the market the lower denominations fetch between £25 ($50) and £60 ($120) depending on their date and condition.

The beginning of this century saw additional issues of Iranian local commercial companies, many difficult to come by, but because of lack of demand they can be purchased cheaply. The notes were illegal issues since only the Imperial Bank of Persia and the Russian State Bank had the authority to issue notes.

The Bank Melli, national bank of Iran, was established in 1932 and has issued notes since then. Most of these are fairly easy to come by. The earlier issues fetch higher prices, particularly the 500 and 1,000 rials which in crisp condition are valued at over £40 ($80). Later issues, small denominations especially, are common. The National Bank of Iran was followed in 1961 by the formation of the Central Bank of Iran, which has issued the notes for the country since then. All issues are redeemable. Some of the interesting notes of Iran are the German Reichsbank note issues overprinted in red Persian numerals and in the mark denomination. These notes in denominations up to 20 marks can be purchased at about £120 ($240). The year 1946 also saw special communist issues by the Government which was set up in Iranian Azerbaijan for several months. The notes issued in the kran and toman denominations are rare and now fetch over £25 ($50) each.

Iranian overprints on German notes circulated up to the First World War and are highly prized. (88 × 135 mm)

Bibliography:

Berman, R. *Paper Money Rates Most Unstable Money Forms*, CW, March 1969
Jahanshahi, A. A. *Thirty Years of Bank Melli Iran 1928–1958*, Teheran, 1958
Keller, A. *Das Papiergeld der Deutschen Kolonien*, Munster, 1967
Philipson, F. *The Paper Currency of Persia*, CM, March 1973
Rabino, J. Banking in Persia, *Journal of the Institute of Bankers*, London, 1892
Wall, J. *The Paper Money of Persia*, CC, June 1973

IRAQ

Currency: 1,000 fils = 1 Iraqi dinar

Notes of the Ottoman Empire circulated in Iraq up to the end of the Great War. These were followed by the British currency issues for the area which was under the mandate of the League of Nations.

The first notes of Iraq were circulated in 1931 by the government in denominations of $\frac{1}{4}$ to 100 dinars, convertible into sterling. The notes are among Iraq's rare issues priced above £50 ($100). The second series, dated the same year continued until the formation of the National Bank of Iraq in 1947. They are a little more common, but most of the notes still fetch upwards of £15 ($30) each.

The National Bank of Iraq, which changed its name to the Central Bank of Iraq a few years later, continued to issue notes for denominations from $\frac{1}{4}$ dinar but ceased issues of the 100 dinar note. All the notes until the establishment of the Republic in 1948 portray the reigning monarch. A collection of the notes after 1931 could form a picture gallery of King Faisal II from the age of twelve or so, to manhood.

The Currency Commission in Ireland issued under its own authority consolidated banknotes for eight 'shareholding banks' of which the Bank of Ireland was one. The notes were issued between 1927 and 1942 and continued in circulation until 1953. (82 × 149 mm)

The Central Bank of Iraq continued to issue the $\frac{1}{4}$ to 10 dinar notes after 1958, but the portrait of the king was removed and the coat of arms of the Republic took its place. These issues are easily encountered and priced at just over face value when in perfect condition.

Bibliography:

Al-Azzawi, A. *History of Iraqian Currency, for the Post-Abbasid Period, 1258–1917*, Baghdad, 1958

IRELAND (Eire)

Currency: (as in England)

The Banking Commission, set up in 1926, led to the Currency Act of 1927. This established a Currency Commission empowered to issue legal tender notes for denominations of 10 shillings to £100. The notes were to be made available to any bank within the Free State of Ireland against normal commercial securities. The first notes of the Currency Commission were dated September 1928. All notes of the Irish Republic are still redeemable at face value. The Currency Commission continued its issues until 1942 with the names of individual banks appearing overprinted on the notes. These were known as 'Consolidated Bank notes' and involved some eight distinct note issues by different banks.

The Central Bank of Ireland replaced the Currency Board in 1948, and took over the note-issuing authority of the eight major Irish banks which had issued notes since 1928. The Consolidated Bank notes, which incidentally were also payable in London, have all been redeemed by the Central Bank of Ireland.

As all notes are redeemable there are no modern Irish notes, irrespective of condition, priced below their face value. On the other hand, no 10 shilling note, for example, is valued at over £10 ($20) in crisp condition and the most expensive £1 note will not fetch more than £35 ($70).

The earliest Irish private bank issues date to the end of the eighteenth century but are extremely hard to obtain; some nineteenth-century notes make an occasional appearance on the market with, often, odd denominations such as that of the Cork Bank in the 1820s, for 30 shillings. In good condition it is priced at about £50 ($100).

Bibliography:

Nellor, G. *North Irish Volunteers*, CM, November 1972

Young, D. *Guide to the Currency of Ireland Legal Tender Notes 1928–1972*, Dublin, 1972

Legal Tender Notes of the Currency Commission and Central Bank of Ireland, CC, February 1972

ISLE OF MAN (Manx)

Currency: (as in England)

There seem to be no easily obtainable notes of the Isle of Man. Issues go back to 1788 when the Isle of Man Bank was founded, but even recent notes of 1969 are scarce in uncirculated condition. Early notes are of such rarity that one cannot price them.

With the exception of the current issue, an Isle of Man note purchased at £5 ($10) would be a bargain. Some of the notes, of which a recorded number of specimens exist, are priced in the £50 ($100) to £75 ($150) bracket. The 'card money' issues are also items that are fetching prices above £50 ($100).

In the Second World War, internment camps on the island issued a wide variety of denominations and some of the very low values, such as the half penny, are extremely scarce. More common ones may be obtained at between £6 ($12) and £25 ($50).

Bibliography:

Gibb, A. *Meet John Nicholson*, cm, January 1970
Quarmby, E. *Bank Notes and Banking in the Isle of Man 1788–1970*, London 1971.

ISRAEL (also Palestine)

Currency: 1,000 mils = 1 Palestine pound
100 prutah = 100 agurot = 1 Israeli pound

The earliest notes in circulation in Palestine included not only bank notes of the Ottoman Empire but also many of the Egyptian currencies which were declared legal tender in the area after 1920 and remained so until the establishment of the Palestine Currency Board in 1927. These pre-Currency Board issues do not have any indication on them that they were intended for circulation in Palestine, and so they cannot be exclusively associated with the area.

The Palestine Currency Board notes are among the rare issues of Israel. The 100 Palestine pound note is particularly scarce, with only seven still unredeemed. In a London auction in 1973 one of these notes fetched £1,100 ($2,200) and another was sold for no less than 10,000 dollars in the United States in January 1974. The Palestine Currency Board issues consist of six different denominations with four distinct dates to each denomination. Single control of these issues is in the hands of

The Isle of Man had many emergency issues for small denominations. This tiny note for 1 shilling, dated September 1815, is highly sought after by collectors. The reverse shows the Manx coat of arms. (38 × 56 mm actual size)

the Crown agents, who still redeem the notes at face value.

The first issues of the State of Israel were the fractional currencies of the Treasury. The 50 and 100 mils issued in 1948 were vertical notes which became known as 'carpets' because of their mosaic design. They are priced at about £40 ($80).

The only other Treasury issues were circulated in 1952 for values of 50, 100 and 250 pruta. These carry a variety of signatures, which affect the price. The most common are priced within the £4 ($8) range and the most expensive at £35 ($70). The Anglo-Palestine Bank, which had already been active for many years, became the sole issuing authority when the state of Israel was created in 1948. The notes fetch about £5 ($10) to £15 ($30) for the denominations up to 10 pounds. The 50 Israeli pound note is extremely rare and has been priced at £150 ($300). The Anglo-Palestine Bank evolved into the Bank Leumi Le-Israel in 1951 and this was replaced by the Bank of Israel in 1955, which is today the sole issuing authority and has had a number of issues, most of which are easily obtainable. The portraits appearing on the last series of 1970 make an interesting collection of Judaic personalities through the ages.

Emergency notes of Israel and some issues reported for circulation in the occupied territories after the 1967 war are considered separately in Chapter IV.

Bibliography:
Atsmony, D. *The Chaos Pound*, IBNS, March 1962

Beresiner, Y. L. *Bank Notes of Israel*, MCBN, February 1972

Bertram, F. & Webber, R. *Israel's Twenty Year Catalogue of Coins and Currency*, New York, 1968

Fisher, J. H. *Palestine Currency Board Coins and Paper Money*, IBNS, Christmas 1968

Collector Nets Elusive 100 Pound Note, WC, July 1971

Palestine 10 Pound Note Survey, WC, July 1972

Palestine Currency Board 50 Pound Note, IBNS, June 1972

Haffner, S. *The History of Modern Israel's Money*, California, June 1970, (2nd edition)

Palestine Notes Fill Cyprus Void, WC, June 1972

Kadman, L. *Israel's Money*, Tel Aviv, 1963, (2nd edition)

Matalon, S. *A Unique Token of Palestine*, IBNS, March 1972

Narbeth, C. C. *Bank Notes for a State With No Name*, CCW, April 1972

Russell, M. *Israel Study Tour*, WC, July 1971

Serxner, S. J. *As Long as you Can Spend It . . .*, CC, Summer 1965

Siemsen, C. *Emergency Currency in Israel and Palestine*, CC, Winter 1968

ITALY

Currency: 100 centisimi = 1 lira (also paoli, scudi and bajocchi)

The main problem the collector of Italian issues has to face is the several distinct dates appearing on individual notes. These include various dates of decrees and laws authorising note issues as well as the dates of issue. Classification problems relating to the dates have been solved in different manners. The general and logical rule is that the later date on the note is the date of issue. The confusion only arises in relation to the notes of the Banca d'Italia, which took over the activities of the National Bank of the Kingdom of Italy in 1893, and in 1926 became the sole note-issuing authority in the country.

There were a number of private banks in Italy which operated during the last three decades of the nineteenth century. Many, such as the Banco di Napoli and Banco di Sicilia issued extremely interesting notes, some of which are priced at several hundred pounds.

A large quantity of state notes were issued when Italy was a kingdom and before and just after the Banca d'Italia became the sole note-issuing authority. These are headed 'Regno d'Italia'. Although some date back to the end of the last century, there are very few rare issues, most fetching no more than £10 ($20) at the most. Banca d'Italia notes were for denominations from 25 to 10,000 lire and have a wide variety of dates and signatures. These were issued from the turn of the century until the currency reform of 1962. Most can be obtained for under £5 with a few exceptions, such as the last 10,000 lira issue which circulated from 1948. In crisp condition the note fetches well over £20 ($40).

Papal State notes and those of the Roman republic dated between 1780 and 1850 form an independent group of Italian currencies.

Some high denomination Italian notes were called 'Bed Sheets' because of their extravagant size. They were issued in the 1940s and 50s and are extremely attractive in appearance. (145 × 241 mm, notes not in proportion to each other)

Overseas territories

Italian overseas territories are mostly recent and short lived. Italian East Africa, which included Ethiopia and Italian Somaliland, was established as an Italian territory in 1936 and lasted until the British took over in 1941. The Banco d'Italia had some of its 50, 100 and 1,000 lira notes overprinted for circulation in the territories. The words 'special series for Italian East Africa' appear in Italian in the margin of the notes. These issues are quite difficult to obtain and fetch close to £25 ($50) for the 50 lira and over £100 ($200) for the 1,000. It is also reported that when the British occupied the area, the same Banca d'Italia notes were overprinted by them with the words 'British Occupation'. However, there is no record of the existence of such notes; if they were to make an appearance, demand for them would be high.

Italy also issued notes for Italian Somaliland authorised by a special law of 13 May 1920. These notes from 1 to 100 rupias are exceedingly rare, priced above £70 ($140). In 1960 Italian Somaliland became the Somali Republic.

There were many other Italian issues, including notes for Albania, Libya and the Sudan, and a number of issues circulated in specific geographical areas such as Venice, Trieste and Sicily.

Bibliography: *Books*
Banco Popolare di Novara, *La Moneta Italiana, Un Secolo dal 1870*, Navarra, 1971

Bobba, C. *Cartamoneta Italiana del 1746 ai Giorni Nostri*, Asti, 1971

Capozzi, F. *Prezziario della Carta-Moneta Italiana*, Padova, 1965

de Fanti, M. *Cartamoneta Italiana (1745–1961)*, Forli, 1966

The modest overprint on these Italian notes does not do credit to their value. These were issued by the Banca d'Italia for circulation in Italian East Africa, and the 50 lire note is worth approximately £30 ($60) (58 × 136 mm)

Gamberini, C. *Raccolta delle Principali Leggi, Ordinanze, Decreti e Manifesti Relativi alla Cartamoneta in Italia (dal 1746)* Vols. I and II, Bologna, 1965, Vol. III, Bologna, 1969
Descrizione della Carta-Moneta in Italia, Vols I and II, Bologna, 1967, Vol. III, Bologna, 1968
Giuffrida, R. *Il Banco de Sicilia,* Vols. I and II, Palermo 1971 and 1973
Mancini, L. *Catalogo Italiano della Cartamoneta 1746–1966,* Imola, 1966
La Cartamoneta Italiana Antica e Fuori Corso (1746–1951), Bologna, 1965
Marcon, P. *La Cartamoneta nello Stato Pontificio,* Roma, 1965
Mini, A. *La Carta Moneta Italiana 1746–1960,* Palermo, 1967
Cenni Storici Sui Banchi di Napoli e Sicilia e Loro Titoli, Palermo, 1965
Sollner, G. *Catalogo della Cartamoneta d'Occupazione e di Liberazione della II Guerra Mondiale,* Castel d'Ario, 1965

Articles

de Caro, L. *The 50 Lira Banknotes of the Bank of Italy 1915–1920,* IBNS, Spring 1969
Garbarino, T. *The first 'Cedole' notes of Rome,* IBNS, 2nd quarter, 1973
Pick, A. *Casa Veneta dei Prestititi,* IBNS, Autumn 1969
Rullau, R. *Florentine Parchment Promissory Notes in the Early 18th Century,* IBNS, 1st quarter, 1973
Sollner, G. *Italian Notes from 1848,* IBNS, March 1965

J

JAMAICA

Currency: 100 cents = 1 dollar
12 pence = 1 shilling
20 shillings = 1 pound

Bank of Jamaica dollar notes are the common issues, and have been circulating since 1970. The earlier notes of the Government of Jamaica, authorised by laws of 1904 and 1918, are also fairly easy to obtain. The 2 and 5 shilling notes are not as scarce as the £10. The latter may fetch up to £70 ($140) sterling. All lower denominations too, dated from 1940 up to and including 1960, are expensive. They can be encountered on the market at between £15 ($30) and £45 ($90). The issues are collectable by dates.

The currency notes which preceded the government issues, but were authorised by the same laws, are much rarer notes. They normally portray King George V and are priced at or above £50 ($100).

With regard to private banks, Barclays Bank, the Canadian Bank of Commerce, the Colonial Bank (which was later absorbed by Barclays Bank), the Bank of Nova Scotia and the Royal Bank of Canada all issued their own notes through their offices in Kingston. No issues of these banks are easy to obtain or cheap. The least expensive will certainly fetch as much as £150 ($300). It is also an interesting point that none of these banks issued their notes in any denomination other than pounds (shilling notes were never printed) and that the highest denomination was the £10 note.

Bibliography:

Byrne, R. & Remick, J. H. *The Coinage of Jamaica*, Montreal, 1966

Crawford, D. A. *Jamaica's 2/6d notes*, IBNS, December 1970
A Couple of Early Jamaican Bank Notes, IBNS, Summer 1971
The Political Notes of Jamaica, IBNS, December 1971
Government of Jamaica Issues 1920–1960, IBNS, March 1972
Bank of Jamaica Issues 1961 to Date, IBNS, June 1972
Jamaica – Other Bank Issues, IBNS, September 1972
Jamaican Islands Certificates, IBNS, March 1973

Renick, J. *Jamaican Bank Notes*, IBNS, Christmas 1968

JAPAN (Nippon)

Currency: 100 sen = 1 yen

(Also rupees, dollars, pounds, pesos, gulden, rupiah, cents and shillings for Second World War occupation notes, and chiao, fen, yuan, cents and dollars for Japanese puppet banks in China 1938–44.)

Japan has had profuse issues of military notes, those used by her own forces and others used in occupied territories. These have been separately considered in Chapter IV.

Mention must be made here of the Japanese puppet banks in China after the Japanese occupation of the mainland in 1938. The Federal and Central Reserve Banks of China (the former established in 1938 and the latter in 1941) were the two major Japanese puppet banks, and each fell into embarrassing difficulties when anti-Japanese engravers decided to 'sabotage' some of the notes. The best known example is that of the Federal Reserve Bank of China; the dollar issue of 1938 shows a Japanese sage on the obverse of the notes signalling with his fingers in what is easily interpretable as a rude sign directed at the Japanese occupying forces. A less drastic but equally obvious incident involved the common 200 yuan note of the Central Reserve Bank of China, issued in 1944. The engraver cleverly included four letters in the general background of the notes; these were U, S, A and C. Chungking radio later announced that these letters stood for 'United States Army Coming'. The engraver, under threat of decapitation, succeeded in escaping to Hong Kong. He returned to Shanghai in the summer of 1945 after his prediction became a reality. This is a common note and it can be obtained for as little as £1 ($2). Only two other puppet banks issued their notes but the public was overtly reluctant to use these issues.

The earliest recorded notes of Japan are the popular 'Hansatsu' notes dating back to the mid-seventeenth century. The ones which appear on the market at about £5 ($10) each are usually dated after 1868, when the government, under Emperor Mutsuhito (1852–1912) ended the clan and feudal system. It was under this system that the long vertical carboard notes were first put into circulation as long ago as 1661. The very early issues are almost impossible to obtain. They fetch prices in the £250 ($500) region.

With the advent of the Meigi Restoration in 1867, Japan became a world power and the government took over the original 'Hansatsu' issues. The notes which can still be collected are those which date from 1730 to 1867. The earlier dates are priced at over £25 ($50) each.

The initial note issues of the government of Japan were poorly printed late in the nineteenth century and appeared as imitations of the earlier 'Hansatsu' notes. They were replaced by a series of vertical notes in the sen and yen denominations printed in Germany in 1872, and headed 'Imperial Japanese Government'. From 1873 onwards regular horizontal notes were printed, and although the 1 yen denomination of these is comparatively common, the 10 and 20 yen denominations are far more difficult to obtain and are priced at £45 ($90) or more.

During this same period, from 1869 to 1873, a large number of private trading companies issued their own paper money. Many were printed on cardboard, others circulated their issues as regular paper money. The denominations of these were in rios (1,000 mon = 1 rio) and some, such as the Yokohama Trading Company, issued its notes in dollars. Many are obtainable for about £20 ($40).

The Imperial Japanese Government continued its issues from 1881 with the notes dated according to the dynasty of the ruling Emperor. The issues continued until 1945. The dates relating to 'Meiji 17' or 'Taisho 3' indicate the Emperor under whom the issues were made, and the year of his reign.

Japan has printed portraits on many of its bank notes. The 100 yen issue of the Bank of Chosen portrays Daikokuten, the god of wealth – one of the seven gods of good fortune. He is seated on bags of straw rice, carrying on his shoulders a bag of money. (95 × 160 mm)

From 1945 the Bank of Japan, which had already been established and had been issuing its own bank notes since 1885, took over the sole right of note issuing in the country and continues to do so. The issues since 1948 are easy to come by, with the exception of the 10,000 yen of 1958, priced at about £40 ($80). Earlier notes, particularly the first issues which were redeemable in silver, are more expensive. The 10 yen and higher denominations dated from 1885 to 1927 are rare issues, although the lower denomination notes may be obtained for as little as £10 ($20) each.

Bibliography:

Atterton, D. *A History of Japanese Paper Currency*, IBNS, September 1973

Daiwa Bank Ltd *A Glimpse of Money in Japan*, Osaka, 1960

Loeb, W. *New 100 Yen Issue 1963*, WNJ, March 1964

Mao, K. O. *About Japanese Notes*, IBNS, Christmas 1967

Narbeth, C. C. *Early Paper Money*, IBNS, Summer 1969

Philipson, F. *Japanese Paper Currency*, Parts I and II, CM, August and September 1973

Tokai Bank Ltd *A Story of Japanese Currency from Olden Times to the Meiji Era*, Tokyo, 1965

JORDAN

Currency: 1,000 fils = 1 Jordanian dinar

The first paper money issues of the Hashamite Kingdom of Jordan did not circulate until 1949, when notes with the portrait of King Abdullah appeared. These issues, for denominations of up to 5 dinars are easy to come by, and priced at about £12 ($24). The high denominations of 10 and 50 dinars, however, are much scarcer in view of the smaller quantities that were circulated. They are priced at over £45 ($90).

Cambodia chose to represent military action on the reverse of its modern bank notes. Government forces are repelling the enemy. (85 × 168 mm)

From 1952 the issues portray King Hussein. These notes, many of them still redeemable and others in circulation, fetch prices at a premium over face value.

The Central Bank of Jordan was formed in 1965 and continues with note issues portraying King Hussein. Most of the notes state that they were issued by authority of a decree dated either 1949 or 1959.

Before the 1949 issues, paper money circulating in Jordan was authorised by the British Mandate Administration.

Bibliography:

Hauck, R. *Jordan Currency Notes*, CC, Spring 1965

K

KATANGA – see BELGIUM

KHMER REPUBLIC (Formerly Cambodia)

Currency: 100 sen = 1 riel

Issues of Cambodia before its independence in 1955 were the notes which circulated within the Federation of Indo-China. The issuing authority, controlled by France, was responsible for issue in the states of Cambodia, Laos and Vietnam (see 'France').

Since 1955 a number of undated bilingual notes have been issued by the National Bank for denominations from 1 to 500 riels. These are all easily obtainable. Many of the lower denominations can be purchased for just a few pence.

Bibliography:

News Item *Cambodia Demonetises 500 Riel Note*, WC, July 1970
Philipson, F. *Paper Money of Laos, Cambodia and Vietnam*, CM, May 1971

KOREA

Currency: 100 sen = 1 yen
            100 jeon (chon) = 1 won after 1947

Because of continuous Japanese influence in Korea, the currencies (and indeed traditions) of the two countries are closely associated. Korean notes can be distinctly divided into four groups. The earliest issues, in the 1890s, circulated under the Li Dynasty, which lasted until the Japanese annexation of Korea early this century. The first notes of the Kingdom of Korea were issued in the yang denomination and are extremely difficult to obtain. Later issues were those of the First Bank of Korea; the vertical notes are nearly identical to the Japanese issues of the same year – 1904. The notes are common and can be differentiated from the Japanese issues by the presence of two superimposed five-pointed stars at the head of the note. The later horizontal notes are also quite common. These can be obtained at about £5 ($10).

In 1907 Japan declared Korea to be a Japanese protectorate. Three years later it was formally annexed. The note issues that followed the annexation, those of the Bank of Korea formed in 1910 and the Bank of Chosen in 1914, continued to circulate up to the time of the Japanese surrender to the Allies in 1945. The notes of the former Bank are more difficult to come by and fetch comparatively high prices. The issues of the Bank of Chosen can be obtained easily.

After the War, while Korea was under Russian and American administration, notes of the Bank of Chosen continued to circulate in the whole country. The high denomination note of 1,000 won can be obtained at under £10 ($20).

After the division of the area into South and North Korea notes for the North were the first to be issued by the Russians from the headquarters of the Red Army. They were followed by the notes of the Central Bank of North Korea in 1947. This Bank has continued to issue notes and all are quite common.

The first issues of the Central Bank of North Korea coincided with the issues of the Bank of Chosen in the South. These were followed, from 1949, by the issues of the Bank of Korea, the values of which are in the same bracket as those of the Central Bank of North Korea.

Bibiliography:

Chashi, Y. *Nippon Shihei Taikei Zukan*, (in Japanese), Tokyo, 1957
Loeb, W. M. *New Issues of North Korea*, WNJ, September 1966

KUWAIT

Currency: 1,000 fils = 1 Kuwait dinar

Because of the strong involvement of Iraq and Saudi Arabia in Kuwait's affairs the currencies of these two countries were initially prevalent in Kuwait. Up to recent years the special Indian issues for the Persian Gulf also circulated in Kuwait.

The earliest issues of the Sheikdom itself are very recent indeed, circulated by the Currency Board of Kuwait since 1960. The 10 dinar note is the most expensive and the highest denomination issued. It would cost about £20 ($40).

L

LEBANON (Liban)

Currency: 100 piastres = 1 Lebanese pound (livre)

Lebanese bank notes date back to 1942. Those of that date and later, headed 'République Libanaise', are easily come by. All previous circulating notes since the Great War have been the issues of the Banque de Syrie et du Liban (or du Grand Liban). The notes intended for use in the Lebanon were overprinted 'Liban'. These issues, which naturally fit into a Syrian collection as well as a Lebanese one, have an interesting characteristic in that the series to which each issue belongs is indicated by one of five different geometric designs on the face of the note; these appear as thick and clearly visible lines in the form of a single line, double parallel lines, a single V, a double V, or a rhombus shape.

The higher denominations, from 10 livre upwards, are fetching above £30 ($60) but the lower ones can be obtained in good condition for £5 ($10).

Prior to the Second World War the notes that circulated in Lebanon, as in the whole Middle East area, were the note issues of the Ottoman Empire.

Bibliography:

Tessier, M. *Catalogue des Papiers Monnaie de la Syrie et du Liban*, BCD, 1953

LIBERIA

Currency: 100 cents = 1 Liberian dollar

The fascinating issues of the Republic of Liberia are the only notes ever issued by the Republic. They date from 1863 to 1880 and they occasionally appear on the market in the cent and dollar denominations for about £45 ($90) each. The 3 dollar note – the series runs from 50 cents to 10 dollars – is the rarest issue. From the 1880s paper money issues were replaced by the United States dollar notes. Today only these and Liberian coins are legal tender.

LIBYA

Currency: 100 piastres = 1 Libyan pound
1,000 dirhams = 1 Libyan dinar

Turkish currency was prevalent in the area until the introduction of the Italian currency for the colonies in Africa. This probably included Banca d'Italia notes overprinted 'British Occupation' after Britain occupied the territory during the Second World War.

From 1951 notes of the United Kingdom of Libya were put into circulation and

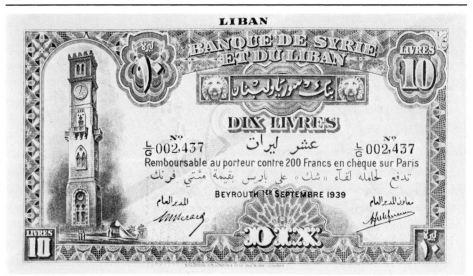

The Bank of Syria and Lebanon was formed under the French Mandate before the partition of Lebanon and Syria. When the bank was dissolved, in order to differentiate between the two countries the note issues were overprinted with either 'Liban' or 'Syrie'. (111 × 190 mm)

the heading was replaced by the words 'Kingdom of Libya' in 1952. Both series of these issues are very common and do not fetch over £4 ($8) each. In 1955 the Bank of Libya began its paper money issues; notes headed 'National Bank of Libya' were circulated simultaneously and continue to be legal tender.

LIECHTENSTEIN – see SWITZERLAND

LUXEMBOURG

Currency: 30 groschen = 1 taler (1856 only)
100 pfennig = 1 mark (since 1865)
20 mark = 25 francs (since 1876)
100 centimes = 1 franc

Without exception the Luxembourg notes of the last century are rare items, valued at over £90 ($180). The first of these were the notes of the International Bank in Luxembourg dated 1856 in the 10 taler denomination. These early yellow notes continued in circulation until 1873. From 1876 a new currency unit, the mark, was established at a rate of 25 francs to 20 marks. The current franc notes were over-printed. The only recorded issues are for 25 and 200 francs with the overprint. The dates of issue are also overstamped on the note, although the printed date is 1856.

There are some rare issues of this century too, outstanding among them the 10 franc issue of April 1940 placed in circulation by the Grand-Duchy of Luxembourg and the 20 and 50 mark issues of the International Bank of Luxembourg dated 1 July 1900. These notes are priced at about £50 ($100) each when in EF condition. Other issues include the Treasury notes from 1914 to 1923, of which the high denomination 500 francs issued in 1919 is the rarest, valued at about £35 ($70). The remainder of these issues, and most other Luxembourg notes, can be purchased at between £4 ($8) and £10 ($20).

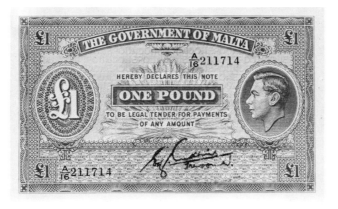

Maltese bank notes still portray the reigning British Monarch. Some issues also show the George Cross. (88 × 145 mm)

Bibliography:

Banque Internationale *Les Cahiers Luxembourgeois*, (in French), Luxembourg, 1956

Swailes, A. J. *'New' Luxembourg 50 Franc Note*, CC, Summer 1965

Victor, F. *Monnaies et Essais-Monetaires du Grand-Duché de Luxembourg 1795–1965*, (French, English and German), Luxembourg, 1965

M

MALTA

Currency: 12 pence = 1 shilling
20 shillings = 1 pound
100 cents = 1 pound

There are records of Maltese coins in scudi denominations having been withdrawn from circulation in 1886, and it is therefore concluded that the Banca di Malta issued its paper monies for scudi notes. These and the Banco Anglo-Maltese issues, again of the last century, are extremely rare, fetching prices above £75 ($150). The latter Bank is recorded to have issued notes in pounds; undated notes were still in circulation in 1926.

From 1914, and up to the establishment of the Central Bank of Malta in 1967, the only issues are those of the Government of Malta. The £5 and £10 denominations of 1914 are rare issues fetching about £35 ($70) and £55 ($110) respectively. One note that is quite common is the 2 shilling King George V issue dated 20 November 1918 which was reissued during the Second World War after being overprinted with half the value. The 'red overprint', as it became known, is valued at about £12 ($24) when in excellent condition, but a note in moderate state can be obtained for as little as £4 ($8). Notes without the overprint were never placed into circulation but occasionally appear in the collectors' market priced at more than £45 ($90).

From 1919, and up to the outbreak of the Second World War, regular currency of the United Kingdom circulated on the Islands. Thereafter, government notes for denominations from 1 shilling to £1, with one issue only of the £5 note, circulated until 1967. All issues are redeemable and fetch prices over their face value, but they are easily obtainable.

The notes of the Central Bank of Malta continue to circulate.

Bibliography:

Keller, A. *Check list of Malta and Andorra*, CC, Summer 1965

George VI Varieties, CC, Spring 1965

MAURITIUS

Currency: 100 cents = 1 rupee

There is a record of paper money issues circulated by the French dating back to the 1780s. The British are also on record with an issue in 1810, but none of these notes have come to light and they would be very highly priced if any examples were to make an appearance.

The earliest notes available are those of the Mauritius Commercial Bank for denominations from 10 to 1,000 dollars. The 10, 20 and 50 dollar notes have appeared in sets on the market for prices of approximately £25 ($100) each. Of the remaining four denominations the 500 and 1,000 dollar notes are the rarest and these are valued at about £400 ($800). All of the notes are dated between 1839 and 1843. The first government notes, those of the Government of Mauritius, appeared in 1848. All of the series of the last century are very rare. The Government also issued notes from 1914 to about 1930; these are highly priced, valued at over £50 ($100).

The notes available to the collector today are the King George V notes of the 1930s and those of King George VI from 1937 to 1954 when the first of the Queen Elizabeth notes made an appearance. The last of these can be encountered at prices in the £5 ($10) range. The King George V notes are valued at about £20 ($40).

The Bank of Mauritius came into being in 1968 and all the note issues portray Queen Elizabeth. There are no fractional notes, but for the first time since the 1920s a 50 rupee note was issued.

Bibliography:
News Item *Mauritius New Currency*, WC, September 1973
Philipson, F. *Paper Money of Islands*, CM, July 1972

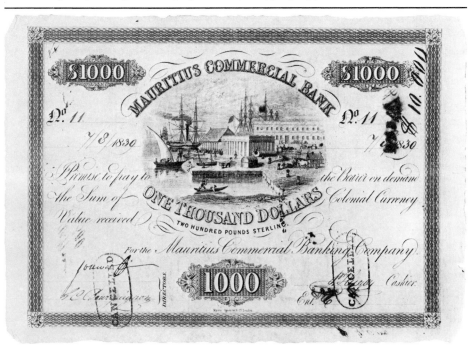

The 1,000 dollar note issued by the Mauritius Commercial Bank is an extremely rare item and only ten are known to exist. The figure $10,000 written down the right side is probably the marking of a bundle. (180 × 255 mm)

MEXICO

Currency: 100 centavos = 1 peso

Mexico, with the possible exception of China, is the most profuse paper money issuing country. An immense collection of delightfully colourful and interesting issues can be assembled at comparatively little cost. Mexico, like China, has been very well covered by catalogues and literature – a factor that adds to the pleasure of the collector. The easily obtainable notes of Mexico are in the main the revolutionary issues dating between 1910 and 1917. The year 1913 is the key to cheap notes. Nearly fifty distinct banks issued their notes during this period and that excludes the numerous regional and military authorities many of whose notes appeared on cardboard as emergency money. The field is vast and still open to discovery. The notes of the revolutionary period are mostly in the £2 ($4) range or below. There are, of course, exceptions, outstanding among them the German-South American Bank overprints on the Banco de Mexico and Banco de Londres y Mexico notes. These, dated October 1913, are priced at over £15 ($30).

The cardboard issues of the revolutionary period form a class all their own. There are some expensive and rare issues which are greatly appreciated, and well paid for, by specialised collectors. The beginner should concentrate on the ordinary issues before seeking out emergency notes.

Mention must be made of the popular 'Sabanas de Villa' notes. These have been separately considered in Chapter IV under 'Inflation'.

The more expensive and rarer Mexican issues are those of private banks of the last century and some in the first decade of this century, referred to by the collector as 'bancos'.

The first issues of Mexico are the notes circulated by Emperor Augustin Iterbide in 1823, shortly after Mexico's freedom from Spanish rule. The 1 and 2 peso notes of the Imperio Mexicano are valued at around £90 ($180) and £125 ($250) respectively. They are among the rare Mexican issues.

The founding of the first 'banco', the Banco de Londres, Mexico y Sudamerica, was authorised in 1864, while Mexico was ruled by the Austrian Emperor Maximilian. The first issues date back to the first year of the Bank's existence.

All denominations including the 1,000 peso note of this Bank are rarities, priced at over £500 ($1,000). Notes of earlier dates fetch relatively higher prices; the ones mentioned here are the 1887 issue.

Most of the banks began their paper money issues during the last thirty years of the nineteenth century. Many with early dates are obtainable at reasonable prices. Banco Mexicano notes dated 1888 can be purchased at £20 ($40) each, Banco de Cohuila notes of 1898 at less than £5 ($10), and so forth.

Some of the notes issued early this century are also exceedingly rare. The 50 centavos issue of September 1900 of the Banco de Nuevo Leon is worth well over £50 ($100), although the unsigned (therefore uncirculated) notes of the bank are worth only about £5 ($10). There are many thousands of different Mexican issues that have been recorded, and it is a good field upon which to base a collection.

Bibliography:

Bank of London & Mexico *100 Years of Banking in Mexico*, London, 1964

Benson, W. E. *Some History Brought to Life*, IBNS, March 1970

Beresiner, Y. L. *Banknotes Without a Source*, COINS, December 1972

Brown, M. R. *Mexican Revolutionary Bills*, ANA, reprint 1950

Byers, C. *Mexico's Revolutionary Currency*, CC, Summer 1969

Deloe, V. *General Francisco Villa*, IBNS, Christmas 1966

Gaytan, C. *Paper Currency of Mexico 1822–1971*, California, 1971

Gaytan, C. & Utberg, N. W. *Paper Money of Mexico*, Mexico, 1961

Long, R. *Mexican Illustrated 'Banco' Catalogue*, WC, series since June 1966

Moreno, A. J. *La Efigie de las Damas en los Billetes Mexicanos*, Mexico, 1971

Rosovsky, E. *Mexico's Paper Money*, IBNS, March 1970 and Summer 1970

Siggers, P. *Background to Collecting Mexican Notes*, CPMJ, June 1971

Slabaugh, A. *Paper Money of the Mexican Revolution*, NS, reprint Illinois, 1956

Shafer, N. *The Story of Mexican Paper Money*, WNJ, June 1965

    Some Important Varieties of Mexican Paper Money, WNJ, August 1966

MONGOLIA (Inner and Outer)

Currency: 100 cents = 1 dollar

           Outer: 100 mongo (mung) = 1 tugrik

           Inner: Chinese currency

Up to the foundation of the National and Commercial Bank of Mongolia in 1924, all

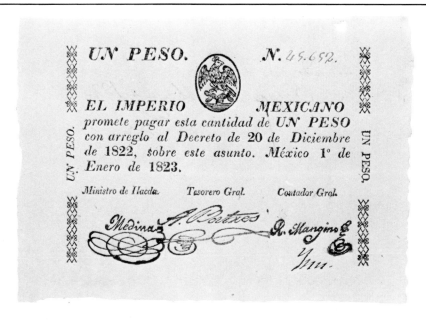

The first issues in Mexico were those of the Empire in 1822 for denominations of 1 and 2 pesos, and they were not signed. They are popular because they are among the earliest of Latin American issues. (109 × 157 mm)

sorts of currencies had circulated in the area since the formation of the People's Government, proclaimed in the old capital, Urga, three years earlier. Currency substitutes in the form of livestock, pressed tea and other commodities were being used; and silver bars circulated in conjunction with paper money issues of the USSR, China, Japan, the United States, Britain, and others.

The first paper issues of this period are now quite rare. They were the 10 to 100 dollar short-term (from August to November 1921) bonds which circulated for three months as currency, and bore an annual interest of six per cent. These are fetching prices of £12 ($24) on the open market.

Orders for dollar denomination notes were placed by the government with the Russians, and bore the legend 'Mongolian State Treasury Notes' but were never issued because of the change in denomination from the dollar to the tugrik. The first colourfully ornamented tugrik notes issued by the Commercial and Industrial Bank made their appearance in 1925. They are difficult to come by, fetching prices in the £25 ($50) range.

By 1928 Mongolian currency had replaced all other issues and the notes continued in circulation until 1939 when a new set of notes was issued, headed 'Mongolian People's Republic'.

From 1941, when the Cyrillic alphabet was adopted, the notes appeared in both languages.

A new series was issued in 1955 and a further one in 1966. The last two series are obtainable and the complete sets fetch prices of £50 ($100) and £30 ($60) respectively.

The State Bank of the Riff came into being in 1920 under Abd El-Krim, chief of the Riffian tribes in north Morocco, who tried to gain independence from the Spaniards in the 1920s. The tribes were defeated and their chief exiled in 1926. (81 × 127 mm)

Bibliography:

    Gribanov, E. D. *The Currency of the Mongolian People's Republic*, IBNS, March 1965

    Kann, E. *Foreign Banknotes in Manchuria and Inner Mongolia*, IBNS, Christmas 1964

    Loeb, W. M. *New Issues of Outer Mongolia*, WNJ, May 1967

    Shafer, N. *Specimen Paper Currency of Outer Mongolia*, WNJ, March 1964

MOROCCO (Maroc)

Currency: 100 centimes = 1 franc

             100 francs = 1 dirham

The Spaniards and French are recorded as having issued the first paper currencies of Morocco at the end of the last century. The first issues proved to have been circulated are those of the Banco de España, dated 1906, overprinted for use in Africa. These peseta issues are rare, fetching prices in excess of £100 ($200).

The Central Banking Institution in Morocco was created in 1917 and in that year the first notes of the Banque d'Etat du Maroc made their appearance. Many new series came out in the following years. All the later issues (the 1928 type and onwards) are fairly common and fetch prices of about £6 ($12).

One of the interesting war-time issues were the 'De Gaulle' notes printed in 1943 in Casablanca, for values from 5 to 100 francs. These, too, are rare, valued at over £50 ($100).

Other emergency issues, some on cardboard, made their appearance during the War years. In spite of the rough printing, the issues are quite attractive, having an elaborate background and well-executed designs. The best known of these are the 50 centime, 1 and 2 franc issues of the 'Empire Cherifien', dated 6 April 1944. They are not particularly common and would cost about £10 ($20) each.

One other interesting and comparatively expensive note is the 1959 franc issue overprinted with the dirham denomination. This was a temporary measure and the issue lasted a short time, so the notes still encountered with the overprint are fetching prices of over £25 ($50).

Bibliography:

    Philipson, F. *Paper Currency of Morocco*, CM, October and November 1973

N

NEPAL

Currency: 100 paisa = 1 rupee

Nepal is one of the countries of the world whose whole range of paper money can be assembled as a complete collection. Its first currency did not appear until 1945. Before that date paper money issues of India and Tibet circulated in the Kingdom and the currency was based on coins. Note issues of denominations from 1 to 100 rupees of the Kingdom of Nepal, under King Tribhubana Biarabikram, are priced at about £12 ($24).

The coronation of King Shri Mahendra Vira Vikrama Birabikram in 1956 was marked by a new issue of notes of the same denominations as the previous ones, with the legend of the denominations in English on the reverse. For thematic collectors the appearance of a coin on the left of the 1 rupee note may be of added interest. Although the 100 rupee would today fetch as much as £20 ($40) the lower denominations can be purchased at less than £5 ($10) each. The current series was issued in 1942.

Bibliography:
> News Item *Notes re Note*, CC, May 1972
> Philipson, F. *The Paper Money of Tibet and Nepal*, CM, June 1972

NETHERLANDS (Holland) (also Netherlands Indies, Curaçao and Netherland New Guinea)

Currency: 100 cents = 1 gulden

The earliest reported government issue of the Netherlands dates back to the period following the French Revolution, when the revolutionary army was welcomed. These issues, dated 1795, although reported, have not been recorded, and they are extremely rare, if they exist at all.

The Nederlandsche Bank was established in 1814 with the accession of William I to the throne. Although the Bank did not have sole issuing authority until almost fifty years later, the reported issues of the national emergency notes in 1814 are attributed to this Bank. The notes are rare and valued above £100 ($200). The Bank continues to be the sole note-issuing body of the government in the Netherlands and all its issues of the last century are expensive; the ones that are more likely to be

The Netherlands, like Spain, chose to depict famous paintings on some of its currency. On the right of this 10 gulden note is a reproduction of one of Rembrandt's masterpieces. (67 × 143 mm)

encountered date back to 1890 and fetch prices in the £60 ($120) range. The 1,000 gulden black and blue note is among the most difficult to come across.

New series were issued in 1914 and at regular intervals thereafter. Many of these notes after 1924 are obtainable at prices below £5 ($10) but the exceptional rare note is there, too, such as the 'Amsterdam' overprint issue dated March 1941 portraying Queen Emma on the 20 gulden note, which is worth about £20 ($40), and the 1930 500 gulden valued at £70 ($140).

During the War years from 1943, state notes for values of 1 to 100 gulden were put into circulation and continued until 1949. These well known 'Muntbiljets' increase in notaphilic value in relation to the denomination. The $2\frac{1}{2}$ gulden notes are very common, obtainable at under £1 ($2) but the 100 gulden is priced at over £100 ($200). The Muntbiljet was preceded by the silver certificates (Zilverbons) issued in 1938 and up to 1944, which are among the cheapest Dutch issues one can encounter. They do not fetch more than a few pence each.

State notes and Zilverbons had also been issued at the outbreak of the Great War and these are rarer items. Some of the state notes dated 1914 are fetching prices in the £20 ($40) range and the Zilverbons dated between 1914 and 1927 are valued at between £5 ($10) and £25 ($50).

Netherland notes in general appear with a wide range of different dates and signatures which can be a solid basis for a large collection.

Netherlands Indies

The Netherlands is probably among the earliest issuers of paper currency for any overseas territories. The Javasche Bank had its notes first circulated in the Netherlands Indies (now Indonesia) as early as 1828. These have the name of the capital, Batavia (Djakarta), on them and are very rare issues valued at over £100

Many of the notes issued by the Dutch colonies are almost exact reproductions of those issued by the Netherlands. The portrait of Queen Juliana usually appears on the Treasury notes and those of the war period show scenes of the armed forces on the reverse. (72 × 151 mm)

($200). Other notes dated 1850 are also recorded for the area, headed 'Nederlandsch-Indie'. They circulated up to 1943 in the form of state notes, with the word 'Muntbiljet' on them. The notes with later dates are easier to come by, but still expensive, valued in the £20 ($40) range.

Curaçao

Curaçao had its name changed to the Netherlands Antilles in 1954, from which date it was raised to an equal standing with the home country. When it was still a colony several note issues had been circulated, those of the Bank of Curaçao outstanding among them. Notes from this Bank, for values of 25 centen to 500 gulden, were issued over a period of some fifty years. The notes up to 1925 are very rare and although later notes may be valued at about £10 ($20), the earlier ones are at least double that figure.

Netherland New Guinea (West Irian)

From 1602 to 1963, West Irian was a Dutch colony and until Indonesia took over the administration and control of the area in 1963, Dutch notes for overseas areas circulated there. The more modern low denominations of West Irian – as it is known now – are not valued at more than £5 ($10).

Bibliography:

Bakker, A. *Old Denominations in Netherland Banknotes*, IBNS, March 1970
Pick, A. *Printing Error on the Dutch 10 Guilder 1945*, IBNS, Christmas 1969
Philipson, F. *Paper Money of the Netherlands*, CM, June 1971
Sasbourg, C. P. *Dutch Bank Notes*, IBNS, September 1969

NEW ZEALAND

Currency: 12 pennies = 1 shilling ⎫
20 shillings = 1 pound ⎬ until 1967
100 cents = 1 New Zealand dollar ⎭

The vast range of New Zealand notes that have been issued in the past one hundred and thirty years are keenly sought after by many collectors. Their values, particularly those of the last century, are increasing rapidly.

Up to the time of the first issues of official paper money by the Union Bank of Australia, through her Wellington Branch in 1840, coins and currencies from a wide range of countries were accepted as circulating media on the islands. These were brought to New Zealand by travelling businessmen and sailors. From 1840 British currency was considered legal tender in the area, but a shortage of coins forced the issuing, by private tradesmen, of small denomination notes that are extremely rare items today. Cheques were also issued in New Plymouth in 1841 and circulated as notes because of the loss in value when redeemed. These items fetch prices well over £100 ($200) on the rare occasions that they make an appearance on the market. The Union Bank of Australia issued legal tender notes for New Zealand from 1840 to 1923.

The issues of this century can still be obtained at £25 ($50) or so for the £1 denomination but earlier notes, particularly the first ones, are considered to be great rarities. It is therefore startling to read in R. P. Hargreave's book *From Beads to Banknotes* that a note of this Bank dated 24 March 1840 was honoured for its face value in 1934 when it was returned from the United States. A number of other banks opened their doors and issued notes in the 1840s. All are extremely rare, as are the Fitzroy debenture and promissory notes, intended by the then Governor as a last measure to save the island from bankruptcy.

These issues circulated as currency despite being inconvertible. When permission by the Crown to declare them legal tender was denied they depreciated in value but remained redeemable, and a very small quantity survived. They are prized items today. Following the government example with these issues, a number of businesses issued their own notes in 1844 and 1845 and many of these are common, fetching prices not in excess of £32 ($64). With the British Government at last helping out in the supply of coins, economic problems were alleviated and non-convertible notes as well as foreign currency gradually disappeared.

In 1850, for the first time, an official bank was formed: the Colonial Bank of Issue. The notes circulated for only six years. Most were redeemed, and are therefore very scarce. Commercial banks were thereafter established and many continued to issue their notes well into this century.

The year 1860 is considered the turning point in New Zealand's economic history. With the discovery of gold mines in different parts of the islands, people began to arrive from everywhere, strengthening the economy and banking enterprises. Outstanding among the banks of this period is the Bank of New Zealand, established in 1861, which continued in existence until 1933. This Bank's notes, again, are easier

The Reserve Bank of New Zealand portrayed on its notes a Maori chief and a kiwi, the national emblem. This series was in circulation between 1934 and 1940. The higher denominations are harder to find, especially in good condition. (89 × 177 mm)

to come by if one's search is directed at issues of this century. They fetch prices upwards of £45 ($90) but earlier notes are priced at double or triple that amount. Notes of this century, invariably printed by British bank note companies, are very attractively designed. But up to the formation of the Reserve Bank of New Zealand in 1933 the notes were often in an extremely bad state.

The Reserve Bank of New Zealand first issued a set of notes of 10 shillings to £50 (excluding the £10 denomination) in 1934 as a temporary measure while awaiting the printing of official notes. These issues, which are still redeemable, fetch prices at a premium over their face value because, although temporary, they circulated for four years. In 1940 the second and permanent series of the Bank – including the £10 denomination – were placed into circulation. New Zealand converted to decimal currency in 1967 and the notes since that date bear the dollar denomination. The portrait of the reigning monarch also appeared on New Zealand notes for the first time in 1967.

Bibliography:

Butlin, S. J. *Australia and New Zealand Bank*, London, 1961
Chappell, N. W. *New Zealand Banker's Hundred*, Wellington, 1961
Hargreaves, R. P. *From Beads to Banknotes*, Dunedin, 1972
Lorimer, H. J. *New Zealand Banknotes Overprinted for Fiji*, IBNS, March 1971
News Item *New Zealand Recent Banknotes*, CM, February 1973
Sinclair, K. & Mandle, W. F. *Open Account*, Wellington, 1961

NICARAGUA

Currency: 100 centavos = 1 peso until 1912
100 centavos = 1 cordoba

Like those of other Latin American countries, the earliest recorded notes of Nicaragua are those of a private bank, the Banco Agricola Mercantil, whose notes were issued in León in the 1880s. These issues, including the ones overprinted with the words 'Tesoreria General', are rarely encountered, and fetch prices in the £50 ($100) range, particularly for the high denomination 100 pesos notes.

More common are the issues of the Republica de Nicaragua which were first circulated in 1893. These notes are easily obtainable despite the early date. They are not priced above £10 ($20). They continued until 1910, when the National Bank of Nicaragua was established and issued its own currency until 1941. This Bank was followed by the Banco Nacional de Nicaragua established in Managua in 1941. All issues since 1912 are common, priced between £5 ($10) and £15 ($30).

Since 1962 the note-issuing rights have been vested with the Banco Central de Nicaragua whose notes are currently in circulation. The notes of the Banco Nacional de Nicaragua were withdrawn from circulation in 1935, overstamped 'Revalidido', and then reissued. As the practice continued for some years the overprinted notes are not more expensive than those without the overprint.

Bibliography:

Stickney, B. *Coins and Paper Money of Nicaragua*, Texas, 1974

NORWAY (Norge)

Currency: 5 rigsort = 1 speciedaler

100 øre = 1 krone since 1875

The earliest recorded issue of Norway, now a museum piece, was a private note printed on thick rag paper issued in 1695 by the Norwegian merchant Jorgen Thor Mohlen. He received royal consent to issue his own paper currency. All notes were redeemed and presumably most destroyed. There are few surviving specimens.

Although there is a record of the existence of a Danish-Norwegian Bank dating back to the end of the eighteenth century, there is no trace of bank note issues. The Norges Bank was established in 1816, and her paper money issues have been dominant in Norway since that date. Three series of notes were issued; the first from 1817 to 1840, the second from 1841 to 1865 and the last from 1866 to 1877. These notes are priced at over £120 ($240). The series from 1877 to the turn of the century, with denominations from 5 to 1,000 kroner, are quite easy to encounter, although the 500 and 1,000 notes are expensive, valued at about £45 ($90) each. The lower notes can be obtained for under £20 ($40).

Designs were changed with the issue dated 1901. They portray President Christie, and these issues, until 1925, are for the same denominations as the previous series. Their values are only about £5 ($10) below the value of the 1877 to 1900 notes.

The paper money issues since 1925 are comparatively easy to obtain with the exception of the 500 and 1,000 krone notes, which are always more expensive because of the high denomination.

The 1942 War issues are also interesting, and the lower denominations can be obtained for about £10 ($20).

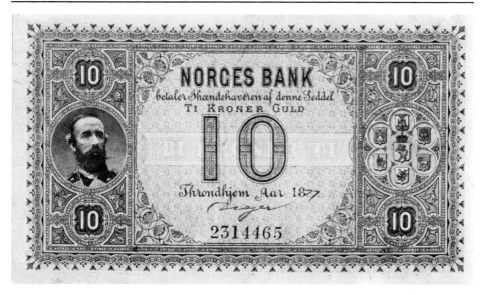

The Bank of Norway has been issuing notes for over a century. The 10 krone note of 1877 is quite scarce, particularly in good condition. (78 × 133 mm)

Bibliography:

Hauck, R. H. *A Century and a Half of the Bank of Norway*, CC, February 1972
Narbeth, C. C. *Early Paper Money*, IBNS, Summer 1969
Ronning, B. R. *Banknotes of Norway 1817–1877*, IBNS, Autumn 1969
Sieg, F. *Sieg's Seddelkatalog 1874–1970*, Ulbjerg, 1971
Skaare, K. *Moneta Norwei*, (in Norwegian), Oslo, 1966

P

PAKISTAN

Currency: 100 paisa = 1 rupee

Until 1947 the paper money issues of India circulated in the area. The first note issues of Pakistan were the Indian Reserve Bank notes overprinted with the words 'Government of Pakistan'. These 1948 notes are among the more difficult to obtain. In good condition they fetch about £20 ($40) each. The official Pakistan notes were issued in 1949 by the government of Pakistan and emphasised the religion of the new country in the national emblem, a crescent and a five-pointed star, symbol of Islam. Government of India paper money issues were also overprinted and are equivalent in rarity to the Reserve Bank of India issues.

The Government of Pakistan notes, circulating simultaneously with the State Bank of Pakistan notes, were first issued in 1947 and continue to circulate, with denominations from 1 to 500 rupees. Most of the notes of both the Government of Pakistan and the State Bank of Pakistan are common, obtainable at under £5 ($10) for the lower denominations. In view of the fact that most of the issues continue to be redeemable, they will always fetch prices over face value.

One interesting set of Pakistani notes is intended for use by Moslem pilgrims visiting Mecca. These are State Bank of Pakistan issues overprinted in English and Arabic stating 'for Haj Pilgrims from Pakistan for use in Saudi Arabia only'. Different denominations of these notes were issued and are still in use. They fetch about £12 ($24) each in good condition.

Bibliography:

News Item *New Signatures on Pakistan Bank Notes*, WC, March 1968
Philipson, F. *Paper Money of India and Pakistan*, CM, January 1972

PALESTINE – see ISRAEL

PANAMA

Currency: 100 centesimos = 1 balboa

The most interesting aspect of Panamanian paper money is the fact that with one exception, and for a period of only one week in October 1941, independent Panama has never issued its own paper currency. The main notes circulated in the country were, and are, the US dollars. But the exception is fascinating. A set of the now well publicised 'Arias' or 'Seven Day' notes of Panama were introduced by Dr Arnulfo Arias, the twice-deposed President of the Nation. He enacted a law allowing for the issue of legal tender notes of the state. Within a week of this paper money issue, on 9 October 1941, President Arias was deposed – and the new regime withdrew all the issues, as well as closing the Central Bank of Issue formed by the President.

The number of notes that were circulated before withdrawal is not exactly known; estimates vary from 14,000 to 400,000 balboa notes. It is confirmed that only the denominations for 1, 5, 10 and 20 balboas were placed into circulation and exact figures of each are unavailable. The largest quantity printed were 720,000 one balboa notes. This denomination, the easiest of the four to encounter, now fetches approximately £50 ($100). A complete specimen set of all balboa notes, including the never-issued 50 and 100 denomination, was put on auction in 1973 and sold for £320 ($760).

Much earlier issues of Panama, such as those of the Estado Soberano de Panama, belong to a Colombian collection and are therefore separately considered. It can, however, be generally stated that all Panamanian notes are exceedingly rare.

Bibliography:

Bunn, T. *Panama's Seven Day Notes*, WC, July 1965

Grigore, J. *Coins and Currency of Panama*, Wisconsin, 1970

News Item *1941 Balboa Notes*, WNJ, March 1964

Remick, J. *The Republic of Panama*, C&M, June 1972

Stickney, B. R. *Numismatic History of Republic of Panama*, Texas, 1971

PARAGUAY

Currency: 100 centavos = 1 peso fuerte

100 centimos = 1 guarani since 1923

The best publicised and most popular notes of Paraguay are those resulting from the

Liberty is depicted on this 5 peso note of Paraguay, wearing the Pharyngian hat. This vignette has been popular in many Latin American countries. (80 × 140 mm)

disastrous Triple Alliance War, when Paraguay faced the military forces of Brazil, Argentina and Uruguay. From the Paraguayan point of view the war was catastrophic. Most of the male population, as well as a large number of women and children who also fought in the war, were annihilated. The bank notes issued during this period, 1864 to 1870, are known as the 'Carlos Antonio Lopes' and 'Francisco Solano Lopez' issues, because they were placed into circulation by these two Presidents, who were father and son. Most of these issues are within the £6 ($12) to £15 ($30) range. The notes are not dated but the issues between the years 1847 and 1862, and 1863 and 1870, can be distinguished because the signature of Carlos Antonio appears on the top right hand corner of the former. These issues are priced a little higher than the notes without the signature.

The rare notes of this period are the issues that circulated after the death of Francisco Lopez in 1870. These are different in design and smaller in size. They are priced well over £25 ($50).

A new series of notes was issued by the Republic, as state notes, following a law of 18 November 1899. This date appears on the notes but is not the date of issue. Denominations from 50 centavos to 500 pesos fuertes were circulated in the first quarter of this century. The rarest of these is the 500 peso priced at £45 ($90). The rest of the notes are worth between £45 ($90) and £20 ($40).

In 1907 the Banco de la Republica was formed, and issued notes until 1936 when it was replaced by the Banco del Paraguay. All notes are fairly easy to obtain. The exception is the 1,000 pesos denomination issued in 1907 and 1920, which is priced at about £60 ($120). The remaining issues are within the £5 ($10) range. The Banco del Paraguay was replaced by the Banco Central del Paraguay in 1952 and her note issues continue to this day.

Only since 1952 have denominations in excess of 1,000 guaranis been printed and these are in the 5,000 and 10,000 denominations. It should also be noted that these last issues are the first to portray the two Presidents, Francisco and Carlos Lopez.

There is very little information available about private issues. Records and specimens exist of the 'Banco Commercial' and 'Banco del Paraguay y Rio de la Plata' which issued their notes in the 1880s, but few details exist.

Bibliography:

Burstyn, L. *Paraguay Paper Money of Ca. and Fs. Lopez*, (in Spanish), Buenos Aires, 1972

Loeb, W. M. *New Issues of Paraguay*, WNJ, June 1964

Matz, A. C. *Varieties of Latin American Paper Money*, CC, Autumn 1963

Seppa, D. A. *Paper Money of Paraguay and Uruguay*, Texas, 1970

PERU

Currency: 100 centavos = 1 sol = 1 real de inca in 1881
10 soles = 1 libra (1914–26)

The earliest paper issues of Peru are the 1822 miniature San Martin notes which are extremely rare, valued at about £100 ($200) each.

The common early Peruvian notes are those of 30 June 1879 headed 'Republica del Peru' and issued by the administrative committee for values from 1 sol to 500 soles. The rarest of these is the 50 sol denomination note valued at about £45 ($90). The 500 sol is valued at about £35 ($70). The remainder can be obtained at under £10 ($20) each.

In 1881 the then President of Peru decided to try a new denomination, the 'real de inca', equivalent to a specific quantity of gold. These issues were originally printed for provisional purposes on the bank notes of the Banca de la Compañia General del Peru and the 1, 5 and 100 real issues are all comparatively common, obtainable at under £20 ($40). The note is not known to exist without the overprint. A few months later original notes in the inca denominations were placed into circulation and the 100 inca of these is the most difficult to obtain at about £45 ($90).

At the beginning of this century, the libra denomination was established, and from 1914 until well into the 1930s it stood at par with the pound sterling.

The Iquitos Revolution notes headed 'provisional cheques' issued on 1 October 1921 for denominations from 10 centavos to 5 libras are the only non-government issues this century. The whole set can still be obtained at about £40 ($80).

The Banco de Reserva del Peru first issued its libra notes in 1922 and has continued to do so, under the name of Banco Central de Reserva del Peru. Many of these notes fetch no more than their face value, as all the issues since are redeemable.

A number of the earlier Banco de Reserva notes were overprinted Banco Central de Reserva and the denomination on the note in libras was changed to soles. These issues with the overprint are worth about £22 ($44) each.

A number of private banks had their own issues in the last century; outstanding among them, and most difficult to come by, are those of the Banco de la Providencia. A series of the notes of this Bank, unissued and unsigned, was sold in auction in 1972 and fetched approximately £400 ($1,000). Another interesting private bank is the Banco del Peru y Londres, its early issues are also valued at about £100 ($200) each.

It is interesting to note that many Peruvian notes are not obtainable in any better than fine condition because of the nature of the climate in that part of the world.

Bibliography:

Beresiner, Y. L. and Dargent, E. *Catalogue of the Paper Money of Colombia and Peru*, London, 1973

Miniature notes were placed in circulation in Peru in 1822 by José San Martin. He was on his way to the north, to the famous meeting with Bolivar in Guayaquil. The tiny notes are the Republic of Peru's first issues. (25 × 56 mm, actual size)

Beresiner, Y. L. *Revolutionary Notes of Peru*, C&M, January 1972
  The Modern Bank Notes of Peru, PM, Summer 1972
  Peru Notes Depict the Death of Atahualpa, NSB, January 1973
Camprubi, A. C. *History of the Banks of Peru*, (in Spanish), Lima, 1957
  El Banco de la Emancipación, (in Spanish), Lima, 1960
Dargent, E. *A Listing of the Banco Central del Reserve*, CC, Summer 1961
  Hacienda and Mine Notes of Peru, CC, Spring 1969
  A History of the Banco de Lima, IBNS, September 1971
  Henry Meiggs and Republic Works Company of Peru, LANSA, June and
  October 1973
Povar de Albertis, A. *The First Notes of Peru*, LANSA, October 1973
Shafer, N. *Interest Bearing Notes*, WNJ, February 1967

PHILIPPINES (Pilipinas)

Currency: 100 centavos = 1 Philippine peso
(pesos fuertes until 1904, and silver pesos 1903 to 1918)

The Banco Español Filipino de Isabel was established in 1852 by the Spaniards in Manila and issued its own currency in that year, considered to be the first of the paper money issues in the islands. These notes in denominations from 5 to 100 pesos fuertes are exceedingly rare and were issued up to 1896. All the notes, but particularly the very early ones, are known only to exist in museums.

This Bank was given the right to issue official notes of the Philippines in 1908. The first notes were entirely in Spanish and circulated until 1912. In that year the name of the Bank was changed to the 'Bank of the Philippine Islands' and all issues from then on were entirely in English up to and including 1933. The high denomination notes of the 1908 series are extremely rare. The smaller denominations in EF condition are priced at about £30 ($60) each. The note issues of the Bank of the Philippine Islands, which were placed into circulation from 1912, are obtainable at prices ranging between £15 ($30) and £50 ($100). It is again the high denomination issues such as the 50 and 100 peso which are very difficult to obtain, as is the 200 peso, known to exist in highly specialised collections.

There were three other paper currencies circulating in the islands from 1903 to 1949. The silver certificates of the United States circulated from 1903 to 1916 in denominations from 2 to 500 pesos; a second set of Treasury certificates was released from 1918 until 1949 (these include the 'victory' overprint) and from 1916 until 1937 a series of notes of the Philippines National Bank also circulated. The smaller denominations in good condition fetch between £5 ($10) and £12 ($24).

The Philippine National Bank issues began with the emergency notes of 1917. For the first time, fractional denominations for 10, 20 and 50 centavos were introduced. All can be obtained at under £8 ($16). The highest denomination of this series was the 100 peso priced at over £50 ($100).

Japan, during the Second World War, occupied the Philippines and a number of guerilla notes were also issued (see Chapter IV).

Right: **The earliest notes of Poland were issued in 1794 in the midst of the revolution in which the peasantry and aristocracy backed Thaddeus Kosciuszko during the Polish War of Independence. (Kosciuszko is also remembered for having fought at George Washington's side in the American War of Independence.) (95 × 173 mm)**

Opposite: **General Douglas MacArthur commanded the forces which landed in the Philippine Islands in 1944, carrying with them Treasury certificates with the word 'Victory' boldly overprinted on the reverse. The series was known as the Victory Series No. 66. It is interesting to note that these notes were again overprinted in 1949, with the name of the Central Bank of the Philippines. (66 × 161 mm)**

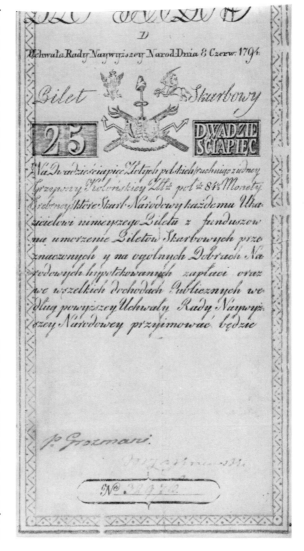

From 1949 to date the Central Bank of the Philippines has been responsible for the issue of bank notes. The smallest fractional currency denomination ever issued, the 5 centavo, was circulated in 1952. All modern notes from the 5 centavo to 500 peso are obtainable in mint condition at a premium over face value.

Bibliography:

News Item *Pesos Fuertes of the Español Philipino*, WC, October 1972

Obojski, R. *Philippine Guerrilla Banks Produce World War II Notes*, BNR, April 1973

War Time Philippine Guerilla Notes, WC, August 1966

Shafer, N. *A Guide Book of Philippine Paper Money*, Wisconsin, 1964

The Philippine Guerilla Currency Board, CC, Spring 1965

Complete Check List of Philippine Guerilla and Emergency Currency of World War II, WNJ, September 1965

A Guide to the Philippine Paper Money Issues, WNJ, August 1965

POLAND (Polska) (also Danzig)

Currency: 100 fenigow = 1 mark
100 groszy = 1 złoty since 1923

The first note issues of Poland came about during the struggles for freedom in the

Even the relatively modern, high denomination notes are very popular with collectors because of their attractive designs and historic illustrations on both sides. This is a Portuguese 1,000 escudo note of 1961. (103 × 164 mm)

1790s when a revolutionary body led by General Kosciuszko issued his own Treasury notes of the supreme National Council, dated 8 June 1794. Denominations issued were from 1 to 1,000 złoty of which the 1, the 500 and the 1,000 are the most difficult to come by, fetching prices in the £85 ($170) region. The 25 złoty is among the common notes of these issues and they fetch about £15 ($30) each. These notes depicted both the white eagle, representing Poland, and the design of the galloping horseman representative of Lithuania.

Another series of early notes of Poland was issued in 1824. They are known as the 'cash' notes which circulated in conjunction with some Treasury assignats for values of between 100 and 1,000 złoty, bearing an interest of six per cent. Although the country did not come into its own until after the Great War of this century, the Bank of Poland had already been established by 1830 and issued its first notes in that year. These are extremely rare, and priced in the £75 ($150) bracket. Polish efforts to free themselves from Russian dominance are well represented by currency issues of the period between 1863 and 1870. These issues of the Bank Polski had the legend both in Russian and Polish on the obverse. English, French and German were used for the reverse and they were issued in the rouble denomination instead of the złoty.

The Bank of Poland lost its right of issue in 1870 and until the Great War the only circulating paper money in Poland was Russian. In 1916 Germany invaded Poland and issued her own currency. These Polish National Land Loan Office notes, for denominations from ½ mark to 100 mark, are dated 1917; they portray the eagle and they are quite easy to come by, most of them obtainable at under £5 ($10).

The Republic of Poland was established in 1918 and issued a number of notes which circulated until the inflationary period of the 1920s. By 1923 notes for 50 and 100 million marks had been issued. These are the rarer of the issues fetching about

With the ever increasing number of Portuguese dependencies gaining independence, their notes are becoming very popular. The Azores will be printing their own notes when they become independent and this illustrated note will then be of far greater interest. (81 × 125 mm)

£40 ($80). The rest of the inflation issues do not exceed £5 ($10) each. From 1924 to 1938 the Ministry of Finance issued notes in the groz (plural 'groszy') denomination; most of these, with the exception of some error notes, are in the same price range.

The Bank of Poland was re-established in 1919 and began issuing its notes in 1924 for denominations from 1 to 5,000 złoty. They are extremely attractive and many of them can be obtained for about £5 ($10).

In the Second World War a number of military notes were circulated by the German government and there were additional issues in conjunction with those of the Bank of Poland during the German Occupation. All are quite common. The National Bank of Poland came into being once more in 1944. The latest issues, which continue to date, are for denominations of up to 1,000 złoty.

Danzig (Gdansk)

Collectors of Polish, German, Baltic and East European notes will find an interest in the issues of Danzig. The city's fascinating history, going back ten centuries, is unfortunately not depicted on its bank notes. The earliest issues are the emergency notes put into circulation by the Municipal Council in 1914 and up to 1919. They fetch under £15 ($30) each. The notes after 1920, when Danzig was within the Polish Union under the League of Nations, were inflationary issues dated 1922 and 1923. The denominations run from 100 to 10 million marks. Like other inflation notes, these are extremely common and are valued at under £4 ($8) each.

The notes that are a little more valuable are those issued in the gulden denomination put into circulation by the Treasury after the inflation period. The Bank of Danzig also issued notes in the 1920s, and both these series fetch over £35 ($70) each. Some of the rarer notes, such as the 25 gulden issue of 1924 are valued at around £90 ($180).

Bibliography:

Chojnaski, P. *Early Polish Bank Notes*, CC, Spring 1969
Jablonski, T. *Tolski Pienadz Papierowy, 1794–1948,* (in Polish), Warsaw, 1964
Kovalski, M. *Katalog Banknotów Polskich 1916–72*, (in Polish), Warsaw, 1972
Kupa, M. *Currency of Poland*, BCD, 2nd Siemester, 1953
News Item *Polish Military Token*, WC, June 1969
Philipson, F. *Paper Currency of Poland*, CM, April 1971

Opposite. Top: Nineteenth-century Mexican notes are almost all rarities, particularly if they are in good condition. The Mercantile Bank of Veracruz was established in 1897 and issued notes from 1 to 1,000 pesos. (80 × 185 mm)

Centre: The balboa is the name of the paper notes circulating in Panama; they are in fact US dollar notes. There was a short-lived issue of Panamanian national currency in October 1941 when the notes were withdrawn after only seven days in circulation. (67 × 151 mm)

Bottom: Danzig, now part of Poland, was a free state following the First World War. Between 1914 and 1937 it issued several notes of which the 1,000 gulden is among the most desirable items. (93 × 172 mm)

1000

EL BANC

PAGARÁ

2,432,434

MIL

MADRID 15

EL INTERVENTOR

1000

Bradbury Wilk

E ESPAÑA 1000

ORTADOR

2.432.434

ETAS

IO DE 1907.

EL CAJERO

OR

pany Ltd Londres

1000

PORTUGAL (also Azores and Macau)

Currency: 100 centavos = 1 escudo

1,000 reis = 1 milreis

The best known Portuguese notes, particularly to collectors who base their interests on historical aspects of notaphily, are the issues put into circulation during the 'War of the Two Brothers' in the first quarter of the nineteenth century. The war involved the two sons of King John VI of Portugal, Dom Pedro IV and Dom Miguel I. Dom Pedro had remained in Brazil when his father returned to Portugal in 1816. Upon his father's death in 1826, instead of returning to the motherland he nominated his seven-year-old daughter Maria as Queen of Portugal, with his brother Miguel as Regent. When Miguel began to gain more power than Pedro had expected, the latter returned to Portugal and the war between the two broke out.

At this time bank note issues in the form of State Treasury notes had already been in circulation from 1798, and before Miguel gave up the struggle for power in Portugal the note issues had been overprinted and officially repaired several times (see Chapter V). They circulated over a period of almost thirty years and it is impossible to find them in very good condition. They can still be obtained at about £14 ($28) each. The early Treasury issues, without the overprints, are far scarcer and when they appear on the market they are priced around £80 ($160).

The Banco de Portugal first issued notes in 1846. Until 1911 the denominations used were the reis and milreis units. These can be fairly easily encountered and the earlier notes fetch up to £35 ($70). The notes after 1900 are cheaper, although some of the higher denominations are priced at over £100 ($200). All of the issues printed locally by the Casa da Moeda are easy to come by. The rarer notes are the 1,000 gold escudo issues dated from 1926 to the 1950s, depicting famous Portuguese heroes. These were gradually withdrawn from circulation and high denominations are valued at about £75 ($150).

Azores

Here we must also briefly consider the many issues of the dependencies of Portugal. The largest of these, Brazil, and Angola, have been separately dealt with. The Bank of Portugal issues, with the overprint 'Moeda Insualana' and 'Açôres', circulated in the Azores in denominations of up to 50,000 reis. These issues can be obtained at about £25 ($50) to £45 ($90) each. The escudo issues, which circulated until 1931, are difficult to obtain and will fetch prices over £45 ($90).

Overleaf: **Bradbury Wilkinson and Co. Ltd are famous for their superlative designs, often portraying figures from Greek mythology. This 1,000 peseta note of the Bank of Spain depicts Commerce surrounded by symbols of Agriculture, Plenty and Industry. This is the highest denomination of the July 1907 series of five notes; it is priced at about £100 ($200). (120 × 158 mm)**

Opposite. Top: **The first notes of the Turkish Republic were still printed in Arabic script, although they were issued after Attaturk took power in 1923. (100 × 175 mm)**

Bottom: **The period just before the Civil War saw many US private bank issues, some extremely colourful and now relatively easy to obtain. (78 × 178 mm)**

Macau

Other Portuguese dependencies had notes headed 'Banco Nacional Ultramarino' which circulated in their territories. In Macau these were the only notes that were handled by the public and the fractional issues in avos (100 avos = 1 escudo) are quite common and fetch prices below £5 ($10). The pataca denomination notes are not as readily available as one would expect. However, the lower denominations do not fetch more than £20 ($40) each.

Portugal was the third largest colonizing country in the nineteenth century after Great Britain and Spain. The Banco Nacional Ultramarino was established and active, issuing local paper currencies in Mozambique, Cape Verde Isles and other territories in Africa as well as in Goa, Damao and Diu in India, and Timor in the Malay Archipelago.

Bibliography:

Narbeth, C. C. *Notes that Must be in Bad Condition*, CM, October 1971
   Macau Notes, CCW September 1971
Philipson, F. *Paper Money of Portugal*, CM, October 1971
Shafer, N. *Commemorative Paper Money*, WN, June 1965

PUERTO RICO

Currency: 100 centavos = 1 peso
           100 cents = 1 dollar since 1902

Since 1922 the currency of the United States has been legal tender in Puerto Rico. In the period between 1902 and 1922 paper money was issued by the First National Bank of Porto Rico. The denominations from 5 to 100 dollars were dated on the

Puerto Rico issued its first notes long before it became a US protectorate in 1898. Among the most valuable notes issued prior to the declaration of self-government in 1952 were the 1909 notes of the Bank of Porto Rico which were punch-holed to indicate cancellation. These cancellation marks do not detract from the value of the note, which is worth several hundred pounds. (188 × 81 mm)

reverse. All of these issues were withdrawn by the United States Authorities and are, therefore, extremely rare, valued at over £200 ($400) each.

The first Puerto Rico currencies are dated 1813 when Treasury notes, for a value of 8 reales, were issued under Spain. There were a number of additional early issues which circulated between 1815 and 1820 under the reign of Fernando VII headed 'Treasury of Puerto Rico'. Circulation of the notes of the Banco Español de Puerto Rico began in 1896. The very attractive notes were followed, at the beginning of this century, by the note issues of El Banco de Puerto Rico. Both these series were overprinted with the words 'moneda Americana' and they are now exceedingly rare, fetching approximately £85 ($170) or more.

The change over from the Puerto Rican money to the American was witnessed by the commonest of the Puerto Rican notes, known as the 'Billete de Canje' (exchange notes) dated 1895. These are encountered both with and without the counterfoil. The notes fetch not more than about £20 ($40) without the counterfoil and about £60 ($120) with.

One additional interesting issue applicable to Puerto Rico was issued in New York dated 1869. The Junta Central Republicana de Cuba y Puerto Rico issued notes headed 'La Republica de Cuba' for denominations of 1, 5, 10 and 20 pesos, valued at between £35 ($70) and £350 ($700) respectively.

Bibliography:
Friedberg, R. *Paper Money of the United States*, (7th edition, part V), New York, 1972
Gould, M. M. & Higgie, L. W. *The Money of Puerto Rico*, Wisconsin, 1962
Loeb, W. *Listing of the Notes of Puerto Rico*, IBNS, Autumn 1961
Remick, J. *Money of Puerto Rico*, C & M, June 1970

R

RHODESIA

Currency: 12 pennies = 1 shilling
20 shillings = 1 pound until 1970
100 cents = 1 dollar

The early Rhodesian notes are in great demand by Commonwealth collectors, particularly because the shilling and pound denominations were first circulated as early as 1895. The Treasury issued the notes in accordance with the Bank Act 1891.

The Boer War of 1899 caused a coin shortage and led to the circulation of notes in Southern Rhodesia, known as the 'Marshall Hole' notes. The issues for 3 pence to 10 shillings were headed 'Civil Commissioner, Bulawayo' and printed on cardboard with a British South African stamp affixed to indicate their value. The notes derived their name from the Secretary to the Administrator, who signed the issues. These two emergency issues are as rare as the Mafeking siege notes and valued at about £35 ($70) each.

An Act of 1922 legalised all paper money issues of the commercial banks then circulating in the country. The very attractive notes of the Standard Bank of South Africa Limited and Barclays DCO, for example, were the last of the Commission's bank notes to circulate. They remained legal tender until 1942.

The Coinage and Currency Act of 1938 established the Southern Rhodesia Currency Board. Paper issues of the Board circulated in Southern Rhodesia, replacing the paper money of the commercial banks. The notes for denominations from 5 shillings to £10 had the portrait of King George VI on the face. These issues are extremely popular.

The Southern Rhodesian Currency Board changed its name to the Central African Currency Board in 1953, when the Federation of Rhodesia and Nyasaland was formed. In 1956 the Bank of Rhodesia and Nyasaland was established and the Central African Currency Board dissolved. The Bank began with its own issues in 1957. These notes circulated side by side with the early issues of the Currency Board. The £10 note is the most expensive and difficult to come by, valued at £30 ($60).

In 1964 the Federation of Rhodesia and Nyasaland was dissolved, and three central banks came into being: the Reserve Bank of Rhodesia, the Bank of Zambia and the Reserve Bank of Malawi; each was now responsible for issuing its own currencies. On 15 November 1964 the Reserve Bank of Rhodesia issued its own notes for 10 shillings, £1 and £5 and these are still legal tender.

As a result of Britain's sanctions on Rhodesia in 1965, Rhodesian bank notes were no longer printed in the UK, but in West Germany. These were embargoed by the British government, so the note issues had to be printed in Rhodesia, and the first of these came into circulation in 1967. Rhodesia adopted decimal currency in 1970.

Southern Rhodesia was united with Northern Rhodesia and Nyasaland in a Federation from 1953 to 1963. The Bank of Rhodesia and Nyasaland issued its own notes for denominations from 10 shillings to £10. (82 × 150 mm)

Bibliography:

Reserve Bank of Rhodesia *Bank Notes of Rhodesia*, IBNS, September 1973
Remick, J. *Complete Rhodesia Catalogue*, WC, October 1965
News Item *Rhodesia Prints Own Notes*, WC, September 1967

RUMANIA

Currency: 100 bani = 1 leu (lei)
(Austrian kronen dated 1919)

There are four distinct periods in the notaphilic history of Rumania. These begin with Rumania as an independent principality, when the first issues dating between 1866 and 1881 circulated. The 1877 state notes are headed 'Bilet Hypotheacar' and are the rarest Rumanian notes, of beautiful but simple design, valued at about £70 ($140) for denominations from 5 to 500 lei. These were replaced by the notes of the National Bank of Rumania in 1881.

The second period relates to Rumania as a kingdom. The notes under King Carol I were issued yearly from 1881 to 1914, invariably for values of 20, 100 and 1,000 lei, with one exception in 1914 when a 5 lei note was issued. The 1,000 lei is priced at approximately £20 ($40). The remainder, particularly those of this century, can be obtained at less than £5 ($10).

The first 1 leu and 2 leu notes were not issued until 1914, when Ferdinand I took over the throne. The National Bank of Rumania temporarily ceased its issues and was replaced by the Banca Generala Romana which, for the first time in 1917, began issuing fractional notes of 25 and 50 bani; the year 1917 also saw the issue of some fractional notes by the Ministry of Finance. These are all priced under £4 ($8).

In 1918 some of the notes of the former empire of Austria-Hungary circulated with an overstamp on them headed 'Romania Timbru Special'. This round overprint is not particularly common, and denominations of the 10,000 kronen dated 2 November 1918 fetch about £10 ($20). The smaller denomination and later issues can be obtained more cheaply. The National Bank resumed its note-issuing responsibilities in 1922 and the notes after 1941 are among the commonest, priced at below £2 ($4). Before the Kingdom of Rumania gave way to the People's Republic, there was a currency revaluation of 20,000 to 1 leu, and the new notes are still obtainable at prices below £8 ($16).

On 30 September 1947 the first of the issues of the National Bank of Rumania under the People's Republic were put into circulation. They were issued with red serial numbers printed in the USSR; after 1952 another series was printed in Czechoslovakia; all are extremely common. The Socialist Republic issues came into circulation in 1966 and they are the legal tender notes of today.

Bibliography:

Coman, V. *Catalogue of the Bank Notes of Rumania*, (in German and English), Frankfurt, 1967

RUSSIA – see USSR

S

SALVADOR – see EL SALVADOR

SARAWAK

Currency: 100 cents = 1 dollar

Early notes of Sarawak were printed by Perkins and Bacon and date back to 1880. They are extremely attractive issues and expensive, fetching over £100 ($200) each. These were issued under James Brooke in 1868 who hand-signed the early notes. He was followed by his son Charles, under whose auspices a series of notes was issued in 1917. They are cheaper, although still very hard to obtain, and priced at approximately £50 ($100) each.

The modern issues, after 1922, can be obtained at under £22 ($44). These, printed by Bradbury Wilkinson, portray Charles Brooke and continued to do so until the issues ceased in 1940.

Sarawak issued its first notes as early as 1880 and they usually depict a member of the Brooke family who were instrumental in promoting Sarawak's economic and political development. The issues from 1922 to 1940 depict Rajah Charles Vyner Brooke who ceded Sarawak to Britain as a Crown Colony in 1946 after ruling over the territory for some thirty years. (114 × 158 mm)

There are few cheap notes of Sarawak and there are no notes dated after 1940. Today, notes of the Board of Commissioners of Malaysia circulate.

Bibliography:
    Dodrill, T. *Sarawak 1918–1919 Notes*, WC, March 1967
    Weissbuch, T. N. *The White Rajahs of Sarawak*, WNJ, November 1965

SAUDI ARABIA
Currency: 100 halaah = 20 guershe (piastre) = 1 Saudi riyal

Saudi Arabia did not have its own official government currency until June 1961. The note issues before then were the unofficial "pilgrims' receipts" which circulated after 1953 throughout the country, after having been issued by the Saudi Arabian monetary agency, in order to alleviate the pilgrims' transport problems on the way to Mecca (see also Pakistan).

The pilgrims' receipts were issued in considerably large quantities. It is recorded that by 1954 over 150 million riyals were circulating in denominations of 1, 5 and 10. These issues are not particularly uncommon and can be obtained for about £10 ($20) each. The pilgrims' receipts were withdrawn in 1963.

Previous to 1953 the only paper currencies circulating, mainly in Saudi Arabia's coastal areas, were foreign notes including issues of India, the US dollar and other Middle Eastern currencies.

Bibliography
    Aramco Hand Book *Flags, Money and Stamps: A Résumé of Saudi Arabian Monetary History*, CC, Winter 1970
    Robin, P. *Pilgrims' Receipts*, CC, Autumn 1969
    Whitler, J. M. *Special Notes for Moslems to Mecca*, WC, September 1972

SCOTLAND
Currency: (as in England)
    12d. = 1 shilling
    20s. = 1 pound
    100 pence = 1 pound since February 1971

There are two kinds of rare Scottish notes: all those issued before the turn of the century, and those in very good condition from the Great War period.

Notes issued in the second decade of this century have been found in large quantities; many early commercial bank, national bank and Royal Bank notes have all appeared in well circulated condition. The notes are priced at about twenty-five per cent over face value, but in crisp condition they can fetch well over double face value.

David Keable, a leading British notaphilist, has established a pattern for Scottish bank notes in which one can recognise first and last issues by reference to the serial letters and numbers.

The 'Big Three' of Scotland still in existence today were formed from mergers of ten banks that were active in Scotland at the beginning of this century. Although records of eighteenth-century issues are in existence, the notes never come onto the market. Nineteenth-century private bank issues are also uncommon with a few exceptions, such as the East Lothian Bank and Leith Bank issues of the 1820s. They are easier to come by and now fetch prices between £15 ($30) and £35 ($70).

As Scottish notes are dated, there is far less emphasis on signatures, as opposed to the Bank of England and Treasury issues this century where the signature is the main factor.

Four standard sizes are recognised for Scottish £1 notes as follows:

Size A: Large square notes – the earliest to be issued up to 1926. Approximately $6'' \times 5''$.

Size B: Standard size notes – equal to Bank of England £1 of the 1928–60 issue $6'' \times 3\frac{1}{2}''$.

Size C: Elongated notes issued since 1962 simultaneously with the Bank of England notes $6'' \times 2\frac{3}{4}''$.

Size D: Small notes – current size of modern notes, approximately $5\frac{1}{2}'' \times 2\frac{1}{2}''$.

Bibliography:

Douglas, J. *Scottish Banknotes*, IBNS, Autumn 1962

Gibb, A. *Collecting Scottish Bank Notes*, IBNS, September 1970

  A New Pound Note for Scotland, IBNS, March 1970

Kerr, A. W. *History of Banking in Scotland*, Edinburgh, 1902

Morgan, L. *A Unique Scottish Bank Note*, IBNS, Summer 1967

Munro, N. *History of the Royal Bank of Scotland, 1727–1927*, Glasgow, 1928

Rait, R. S. *The History of the Union Bank of Scotland*, Glasgow, 1930

Wheeler, M. *Early Intrigues of Scottish Banking*, C&M, October 1972

Williamson, B. E. *History of the East Lothian Banking Company*, IBNS, June 1970

The notes issued in Scotland throughout this century have been particularly colourful compared with those of England. Because of the numerous issuing banks, there are a variety of notes to be collected, usually by date and signature. (152×84 mm, notes not in proportion to each other)

SOUTH AFRICA

Currency: rixdollar until 1831

   12 pence = 1 shilling until 1961

   20 shillings = 1 pound

   100 cents = 1 rand

Until the foundation of the Union of South Africa in 1910, paper money issues circulated on a parallel basis, but separately, in the four provinces: the colonies of the Cape of Good Hope and Natal and the republics of Transvaal and Orange Free State.

The earliest paper money of South Africa was issued in 1782, written out in hand in the rixdollar denomination for the Cape of Good Hope. These extremely rare issues are priced at over £250 ($500) each. The Cape of Good Hope was the earliest of the provinces to get organised from a monetary point of view. The first state bank, the Lombard Bank, was established in 1793. In 1837 the first of the private banks, the Cape of Good Hope Bank, was also established and survived until the end of the century. By 1880 thirty private banks had been issuing paper money of their own. These issues were now in sterling denomination. Banks in Transvaal and Natal had also begun issuing their own notes and the banks of one province often opened branches and issued bank notes in another.

The year 1857 saw the first issues of the London registered banks, known in the provinces as 'The Imperial Banks'.

The emergency issues of the Boer War at the turn of the century were circulated in Transvaal and these have been separately considered in Chapter IV.

All of the bank note issues by the private banks in the Cape from 1892 to 1920 were of a standard size and design known as 'uniform notes'. This was because a law had

Of all the Commonwealth countries, South Africa is one of the most popular among collectors because of the diverse banks, dates and signatures which make completing a collection a challenge. This National Bank of South Africa £1 note was issued in Johannesburg in 1920. (75 × 161 mm)

been enacted whereby funds had to be deposited in the Treasury by banks which issued notes. The appropriate name of the bank and dates were filled in on the standard bank-note form.

Further issues of the Reserve Bank of South Africa were circulated at regular intervals, and the last of these are the easiest to come by. Many of the notes can be obtained at just over the face value. South African notes can be collected by signatures and dates.

Bibliography:

> Aglfer, A. *The Transvaal; The South African Republics War Issue*, cc, Spring 1966
> *Bank Notes of the South African Union 1910–1922*, cc, Summer 1966
> Bergman, W. *A History of the Regular and Emergency Paper Money Issues of South Africa*, Cape Town, 1968
> Lawrence, J. *Paper Money of South African Republic*, IBNS, March 1967
> Levius, P. H. *Catalogue of South African Paper Money since 1900*, Johannesburg, 1972

SPAIN (España)

Currency: 100 centimos = 1 peseta
> (Also reales, escudos, reales de vellon and pesos fuertes)

It has been claimed that Spain was the first European country to issue what could be interpreted as paper money. These were issues circulated by King James I of Catalonia and Aragon in 1250, and furthermore, in 1401 the Exchange Bank of Barcelona also issued its own paper currencies. The same sources indicate that a number of siege notes were circulated in the latter half of the fifteenth century when the Iberian Peninsula was occupied by the Arabs. There is, however, no definite evidence to substantiate these claims.

The first notes of which examples exist were issued in 1799 during the reign of Charles III. The three denominations of 100, 200 and 1,000 reales that are known are extremely rare, valued at over £100 ($200) each. That part of Catalonia which had been incorporated into Napoleon's French Empire saw its first issues in 1813. The French were replacing the Spanish currency with assignats in the peseta denomination, but some of these efforts were frustrated. Spanish patriots attacked the military convoys transporting the assignats and destroyed the paper money issues in order to preserve the country's coinage. Not many examples of these Spanish assignats survive and they are valued in the £100 ($200) range.

During the course of the nineteenth century, beginning in 1829, a large number of private banks put their notes into circulation. Most of these issues were in the real denomination and circulated throughout the country. They are valued in the £75 ($150) to £100 ($200) range.

The Bank of Spain was established in March 1874 and was immediately granted a monopoly on paper money issuing. Many of the banks then in existence ceased their issues and their notes were withdrawn and destroyed. Surviving specimens are exceedingly rare. The early paper money issues after the establishment of the Bank of Spain were interest-bearing notes; from 1893 to 1913 quite a number of private banks issued bonds in denominations from 10 to 100 pesetas, some of which circulated as paper currency. The notes of the Bank of Spain under the first republic and the kingdom, that is up to 1931, are quite difficult to obtain. The higher denominations are very rare, and the 1,000 peseta note is priced at over £75 ($150). Some of the 25 and 50 peseta notes, however, dating between 1874 and 1889, can still be obtained at about £35 ($70) each.

The note issues from 1906 become fairly common; many of these, with the exception of the 500 and 1,000 peseta, are obtainable at £5 ($10). Until 1935 the lowest denomination note of the Bank of Spain was the 25 peseta. In that year, for the first time, 5 and 10 peseta notes were issued and these are priced at a few pence. The last of the issues of the Bank of Spain before the outbreak of the Civil War in 1936 are among the rarest low denomination Spanish notes of this century, valued at about £50 ($100).

Spain issued a number of bank notes for circulation in overseas territories; the issues for Cuba dating back to 1869, for Puerto Rico in 1813, for Santo Domingo in 1867 and the Philippines from 1852 are separately referred to.

Bibliography:

De La Riva, J. P. *History of the Paper Money in Spain*, IBNS, March 1965
De Loe, C. *The Basques and Their Money*, IBNS, June 1971

When notes leave the printer they have the counterfoil still attached and bear no signatures. The counterfoil is used to check the authenticity of the issue before redemption. This Spanish note of 1857 has the counterfoil extending to the top to allow for the matching of the four signatures. (195 × 320 mm)

Graeber, K. *Bono de la Libertad*, cc, Vol. 10, No. 3
Narbeth, C. C. *Early Paper Money*, ibns, Spring 1969
Pere, G. A. *Collecting Spanish Paper Money*, lansa, February 1974
Velez-Frias, S. R. *Bank Notes of Spain*, ibns, September 1972
Vicenti, J. A. *Catalogue of Spanish Paper Monies 1808–1971*, Madrid, 1971

SRI LANKA (formerly Ceylon)

Currency: 100 cents = 1 rupee

The earliest rupee notes for Ceylon were introduced by the British in 1917. (Before that, Indian notes circulated in the island based on the panam and paisa denominations.) Earlier still, the rixdollar had followed Dutch issues of the eighteenth century.

The early rupee notes are typical British colonial issue types of the Government of Ceylon. In crisp condition these fetch prices up to £80 ($160) for the higher denominations.

The Central Bank of Ceylon was the only other authority that ever issued notes, and until 1956 when independence was declared, the portrait of the British monarch appeared on the notes. Thereafter, the Singalese heraldic lion, alternating with a portrait of Prime Minister Bandaranaike or a statue of King Bahu, replaced it. Most of the issues are very reasonably priced, below £20 ($40).

A number of Singalese notes in book form have often appeared on the market, which has considerably lowered the potential value of a complete booklet. The last one to appear in an auction in 1972, containing one hundred 1 rupee notes dated 1943, sold for £38 ($94, at the time).

Bibliography:

Jayasekera, K. D. *Ceylon's Vanishing Rupee Notes*, wc, February 1968

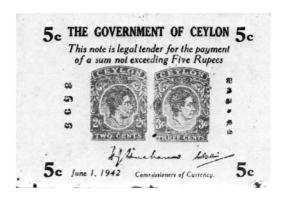

Reproductions of stamps on paper money are extremely rare. The Government of Ceylon 5 cent note is a great curiosity. (58 × 85 mm, actual size)

STRAITS SETTLEMENTS

Currency: 100 cents = 1 dollar

The legal tender currency of India circulated in the Straits Settlements until 1867. These early issues do not in any way indicate special validity for circulation in the territory and consequently they cannot be differentiated from ordinary Indian notes of the period. Thereafter the notes of several banks, such as the Hong Kong and Shanghai Banking Corporation and the Chartered Bank of India, Australia and China, circulated with the name 'Singapore' on them. The first notes with the heading 'The Government of the Straits Settlements' did not make an appearance until 1898, in the dollar denomination. These are among the rarer notes priced well in excess of £75 ($150). The notes which were issued at regular intervals from that date decrease in value and increase in availability in relation to their date. The modern notes can be obtained for about £10 ($20).

The most interesting notes of the Straits Settlements are the issues of 1930 which were intended for inter-bank circulation only. These notes for 1,000 and 10,000 dollars have fetched over £100 ($200).

The most available notes are the emergency issues of the First World War for 10, 25 and 50 cents which can be obtained at about £2 ($4) each.

Bibliography:

Law, C. C. *Malaya Bank Note Catalogue*, wc, September 1967
   Straits Settlements Currency Notes, ibns, Christmas 1967
Pridmy, F. *Coins and Coinage of the Straits Settlements and British Malaya, including Tokens Issued by the Merchants on Singapore*, London, 1968
Saw, W. *Paper Currency of Malaysia, Singapore and Brunei 1848–1970*, Kuala Lumpur, 1971

SUDAN

Currency: 100 piastres = 1 Sudanese pound

When Sudan is mentioned, the experienced collector immediately thinks of the Gordon issues of Khartoum which date back to 1884. (See 'Seige Notes', Chapter IV). In fact this is the only contribution to early paper currency that Sudan has made. Until 1956, when Sudan was declared a republic, the notes of Egypt and Britain were circulating in the area.

In 1956 the Sudan Currency Board was formed and issued paper money notes from 25 piastres to 10 Sudanese pounds. The central banking authority in Sudan, the Bank of Sudan, was established in 1961 and issued notes still in circulation today. The Sudan Currency Board notes also continue to be legal tender and consequently most of these will only fetch a premium over their face value.

SWEDEN (Sverige)

Currency: 100 öre = 1 krona

Sweden has the distinction of being the first European country to issue a bank note as such. Johan Palmstruch, of Swedish, Dutch and Latvian ancestry, had founded the Stockholm's Banco in 1652 and issued his first credit notes in 1661. Unfortunately the first note-issuing Bank of Europe, which relied entirely on public confidence, went bankrupt and closed its doors within six years. It was not until 1726 that the Bank of the States of the Realm, which was founded in 1668, issued its own paper notes. These were known as 'transfer notes' and although at first not officially declared legal tender, they were accepted in payment of all debts, including taxes to the Government.

Until 1759 the designs on Swedish notes were extremely dreary and colourless, and they were printed on white paper. However, some of the water-marks on notes printed after 1737 show very attractive vignettes. All these fetch prices in excess of £300 ($600) each.

The late notes of the eighteenth century are still available and are fetching prices in the £90 ($180) range; the notes after the currency reform of 1835 are more readily available. From this date full colours were used on the notes, but the quality of the paper was poor; consequently, the first issues, after 1835, are difficult to come by because the notes easily deteriorated.

In 1879 another currency reform led to new designs on the notes which, unlike those of any other country in the world, have persisted to date.

A huge number of signature varieties and dates exist for the issues since 1879. The rarest are the 100 krona denomination issues at the end of the last century; the high denomination issues of this century, too, will fetch prices in the £100 ($200) range. On the other hand the 5 krona issue of 1890, for example, is worth about £12 ($24) and the same denomination issue for, say, 1921, is worth no more than £5 ($10).

The earliest private bank recorded in Sweden issued its notes in 1830, having obtained government sanction to do so. By the end of that decade over thirty banks had issued their own private notes and although some of these are poorly printed

they are imaginatively designed. Well known among the private note-issuing banks is the Swedish Enskilde Bank which is still in existence today. This bank was issuing notes as early as 1894, in different parts of Sweden, through its branches. The notes of the Enskilde Bank are now valued at about £100 ($200).

In 1903 the paper money issues of all the private banks were called in and the Swedish Central Bank was given the sole right of bank note issue. During the first three years of this century some twenty private banks which had previously issued notes, closed their doors for ever. Denominations of 5, 10, 50, 100 and 500 krona have been recorded for most of these, and the high denominations are particularly rare.

Bibliography:

Heckscher, E. and Rasmusson, N. *The Monetary History of Sweden*, (3rd edition), Stockholm, 1964

Lindgren, T. *The First Swedish Paper Money*, IBNS, Summer 1967

Narbeth, C. C. *Early Paper Money*, IBNS, Spring 1969

   Sweden's Bank Notes, IBNS, September 1968

Nathorst-Boos, E. *Sweden's Paper Money*, IBNS, August 1969

Obojski, R. *Sweden's Royal Coin Cabinet*, WNJ, July 1966

Paltbarzdis, A. *Sveriges Sedlar Lund*, (in Swedish), 3 vols., Stockholm, 1963

Sveriges Riksbank *The Paper Money History of the Riksbank*, (in Swedish), Stockholm, 1968

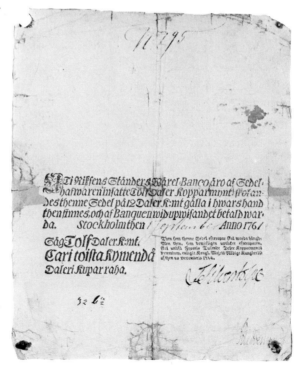

Opposite: **The portrait of George V on the Straits Settlements note appeared for the first time in the series dated 1914–1930; the earlier dated notes are the more highly priced ones. (128 × 205 mm)**

Right: **'Early' notes have become a collectors' theme in recent years, as they are a particularly good investment. The 12 daler Swedish issue of 1761 is valuable even in the poor condition illustrated. (158 × 194 mm)**

SWITZERLAND (Schweiz) (also Liechtenstein)

Currency: 100 rappen = 1 franc

100 centimes = 1 Swiss franc

Switzerland has had a very limited number of designs for its paper money issues. The notes that circulate today are, but for the date and signatures, identical to the type first put in circulation in 1954, and there was no change in design previous to that since 1910. All notes of the National Bank of Switzerland are still legal tender and these are the only low-priced Swiss notes one can expect to purchase. They are naturally priced over and above their face values.

The first notes of Switzerland date back to 1883. These were known as the Concordat Issues and placed into circulation by forty-two banks. All the notes are priced at over £100 ($200); the only denominations issued were for 50, 100, 500 and 1,000 franken. With the exception of the individual bank's name, the notes are again of a standard design with a portrait of Helvetia to the left and a child to the right. Most of the banks issued notes in 1883; the majority had ceased to issue by 1910.

The Central Bank of Switzerland was established in 1905 and its first issues which came into circulation in February 1907 were identical in design to the earlier Concordat notes. The denominations from 50 to 100 francs are again priced at over £125 ($250). In 1910 the new design was adopted. The 500 and 1,000 franc notes are very rare items. The lowest priced note is the 5 franc, first placed into circulation in 1913 and worth about £15 ($30).

Private bank notes from Switzerland are colourful and very rare when signed and issued. The 100 franc note is the highest denomination of this series of four. (110 × 177 mm)

The only other two government issues coincide with the outbreak of the First World War, when the State Treasury of the Confederation as well as the Loan Office issued their own bank notes in 1914. The latter issued only a 25 franc note, priced at about £60 ($120); the former issues of 5, 10 and 20 franc notes, are valued today between £10 ($20) and £25 ($50) each. There is a wide range and variety of signatures and dates on Swiss paper money.

Liechtenstein

The only note issues of Liechtenstein are the 1920 to 1924 notes in the heller (100 heller = 1 krone) denomination. The notes, in the German language, can be recognised because of the word 'Liechtenstein' appearing on them as a heading, but they are extremely common and can be easily obtained. Previous to these issues, Austro-Hungarian notes circulated in Liechtenstein, and since 1924 Swiss money has been legal tender in the country.

Bibliography:

Graf, U. *Swiss paper Money 1881–1968*, Bielefeld, 1970

    Competition for the Creation of the New Swiss Notes, IBNS, Summer 1971

    The Caisse d'Escompte de Genève, IBNS, June 1972

Muszynsky, M. *Notes of the Banque Nationale Suisse*, IBNS, Summer 1968

Schwarz, D. *Schweizerische Banknoten*, (in German), Zurich, 1946

SYRIA

Currency: 100 piastres = 1 Syrian pound (livre)

Since 1957 the Central Bank of Syria has been issuing the paper currency. Notes with texts both in English and in French and a separate series in French have been issued since the inauguration of the Bank.

No notes of any Arabic country have been fully catalogued and their background consists of a relatively short notaphilic history. Normally notes of occupying forces have circulated for specific periods of time. It is known that the bank notes of the Ottoman Empire up to 1918 circulated in Syria, as they did throughout the area.

From 1920 until 1925 Lebanon and Syria constituted one geographical area under the French Mandate and notes of the Syrian and Lebanese Bank were issued between 1935 and 1949. The earlier notes were those of the Banque de Syrie et du Grand Liban, overprinted with the word 'Syria', and showing the date 1939. The notes were very similar to the Bank of France issues and had been printed in England by Bradbury Wilkinson. (See also 'Lebanon'.)

With the 1939 issue the name of the Bank changed, leaving out the word 'Grand'; from that date until 1949 denominations from 1 to 100 Syrian pounds were issued, showing the values in French. All issues are priced at between £5 ($10) (for a very good condition 1 Syrian pound note) and £75 ($150) (for the 100 livre notes).

In 1942 a number of government notes were placed into circulation for 5 to 5,000 Syrian pounds and the issues for denominations in excess of 50 livre are very hard to come by.

T

THAILAND (formerly Siam)

Currency: 100 satang = 1 baht = 1 tical since 1912

The commonest and best known Thailand notes are the 1941 Treasury issues circulated by the Government of Colonel Luang Pibul Soggram. At this time, Thailand had annexed most of Laos, Cambodia (now Khmer Republic), four Malayan states and a number of the Shan States of Burma. These undated issues have a variety of signatures and water-marks but they all portray King Ananda Mahidol. The notes were issued in the satang and baht denominations and, of the several diverse issues, none are considered to be scarce; they are priced at under £25 ($50).

The earlier notes, of Siam, are those issued by foreign banks, such as the Hong Kong and Shanghai Banking Corporation, which issued notes in Bangkok between 1893 and 1902. The Chartered Bank of India, Australia and China did so, too, and these issues, indicating Bangkok as the city of issue, are highly priced at over £65 ($130) each. The Bangkok branch of the Banque de l'Indo-Chine also issued its notes during this period.

From 1902 a range of notes of the Siam Government were issued in the tical denominations; of these the 1 to 1,000 tical have been verified and are extremely rare. The 1,000 tical denomination is valued at well over £100 ($200).

From 1902 the Ministry of Finance of the Siam Government placed its own notes into circulation and some of them were dated and printed with an English text. These common issues continued until 1945. They were mostly printed by Thomas de la Rue.

The pre-1945 notes are a little scarcer, generally fetching prices up to £10 ($20). The issues from 1946 have the portrait of King Phumipol Adulvadej and they are now headed 'The Thai Government'.

Bibliography:
Anon. *The Paper Money of Thailand*, (in Thai).
Little, S. *Bank Notes of Thailand*, Virginia, 1973
Shafer, N. *Paper Money of Siam, Made under Contract with the United States Bureau of Engraving 1946–1947*, WNJ, July 1965

TIBET

Currency: Tangka (Tamka) 1915–45; 10 shokang = 1 srang since 1945

There are no common notes of Tibet and in view of the printing process involved in the earlier issues, it is not easy to find these in good condition. The srang notes, priced at £5 ($10) to £15 ($30) depending on the denomination and condition, are the 'edible' notes, so called because they have been block printed on rice paper. The earlier tangka notes are more expensive but even the most expensive should not exceed £45 ($90).

Other notes which can still be encountered are Chinese issues overprinted with Tibetan characters for use by Tibetan communities in China. These include the Manchu-Mongol Bank of Colonisation, the China and South Sea Bank, and the Central Bank of China. In each case the value of the overprinted note should not be fifty per cent more than the value of the Chinese note.

Bibliography:

Bowker, H. F. A. *Tibetan Paper Note*, Numismatist, February 1951

Hage, S. *Tibetan Paper Money*, IBNS, Autumn 1963

Helfer, A. L. *Tibetan Paper Money*, IBNS, Summer 1964

Muszynski, M. *Varities on the Tibetan 50 Tam Note*, IBNS, March 1970

Panish, C. K. *Tibetan Paper Money*, Parts I and II, WNJ, August and September 1968

TUNISIA

Currency: 1000 millimes = 1 Tunisian dinar from 1958
Piastre up to 1891; French franc from 1891

Dar-El-Mal is the name of the first Tunisian bank set up in 1847 in order to issue certain certificates used as paper money in the piastre denomination. This was done in order to alleviate a coin shortage. The notes, which were withdrawn by 1852, are the first paper monies of Tunisia and only a few known examples survive in museums.

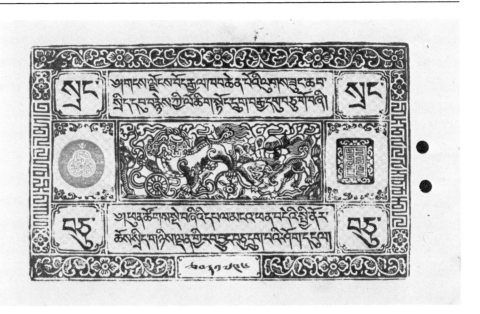

The bank notes of Tibet have a particular curiosity value because they were printed on rice paper from wood barks. The 10 srang note was issued between 1947 and 1950. (86 × 145 mm)

In 1891 a law was passed whereby the piastre denomination was changed to the franc, at par with the French franc.

Until 1904 Algerian paper money circulated in Tunisia under the French joint protectorate of the two countries. In that year, with French consent, the Banque de L'Algérie was established in Tunis and issued its own notes with the overprint 'Tunisie' and different serial numbers. These issues are extremely attractive and the high denomination notes should fetch no more than £50 ($100) in very good condition.

The majority of the issues of Tunisia are those of the Banque de L'Algérie. In later issues the word 'Tunisie' was no longer overprinted but actually drawn into the design of the notes.

In 1948 the Tunisian government established a law whereby the name of the Bank was changed to Banque de L'Algérie et de la Tunisie. This bank was the sole note-issuing authority in Algeria until 1958. The notes of the joint-bank can be obtained for a few pounds each in crisp condition. In 1958 the Central Bank of Tunisia was formed.

An interesting emergency series was put into circulation in Tunisia in 1915, printed on notebook paper and issued by the Company for Phosphates and Railways of Gafsa. These occasionally appear on the market and, as emergency issues, fetch up to £35 ($70). They were replaced in 1918 by issues of the Direction Général des Finances whose issues circulated throughout Tunisia as legal tender. They can be obtained at about £5 ($10) to £25 ($50) each in spite of their relative scarcity. They are dated between 1918 and 1922.

Bibliography:

Dudendre, N. P. *Catalogue des Billets de la Banque d'Algérie et de Tunisie*, BCD, 2nd semester, 1953

Farrudia, de Candia, J. *Billetes de la Tunisie*, BCD, 1st semester, 1951, English reprint in CC, Winter 1970 and Spring 1971.

Philipson, F. *Paper Currency of Algeria and Tunisia*, CM, February 1973

TURKEY (also Ottoman Empire)

Currency: 100 piastres = 1 livre Turque
100 gurush = 1 Turkish pound

Modern Turkish notes date back only as far as 1923 and consequently a complete collection of these issues can be obtained, although some of the high denomination notes are rare.

The far more challenging aspect of a Turkish collection, however, is the discovery of a wide range of Ottoman bank notes which circulated through a vast geographic area including the Middle East, northern Africa and part of south-eastern Europe. The best known of these issues are this century's colourful notes of the 'Debte Publique Ottomane'. These state notes of the Ministry of Finance were issued for denominations from 1 gurush to 50,000 livres (pounds). The notes appear in different colours, with different dates in the Muslim calendar, and a wide range of signatures. These issues first made their appearance on 30 March 1915 and circulated until 1920 when they were withdrawn. They remained redeemable for a long time and consequently the higher denominations are extremely rare.

There are no existing specimens in excess of the 100 pound note and one of these would fetch a price close to £300 ($600). The 1 pound note can sometimes be obtained at about £3 ($6) and the fractional notes do not fetch over £2 ($4) each.

Collectors of Turkish notes include in their collections the notes known in England as the 'Gallipoli overprints' for 10 shillings and £1. These are the English Treasury issues of the First World War signed by John Bradbury and overprinted in Turkish with '60 gurush' on the 10 shilling note, and '120 gurush' on the £1 note. The lower denomination can still be obtained at about £60 ($120) but in a recent auction in 1973, £450 ($900) was paid for the £1 Gallipoli overprint.

One important issue of Turkey this century is that of the Banque Impériale Ottoman. With the exception of the £1 denomination these are extremely rare,

The Ottoman Empire has only issued paper money since the 1850s although it has existed for five centuries. These early issues are very rare, but less colourful than those of the First World War. They circulated throughout much of the Empire, which included at that time a large part of the Middle East, Northern Africa and Southern Europe. (101 × 159 mm)

fetching prices of over £120 ($240). The £1 can be obtained for about £50 ($100). The notes, printed by Waterlow and Sons, were issued at the outbreak of the War but quickly lost their value. They were redeemable in gold until 1948, which is the cause of their rarity.

The commoner notes that one can come across are the Kaime issues, of 1876 and 1877, with denominations from 1 to 100 gurush. With the exception of the 1 gurush note, they are all vertical notes and can be purchased at about £10 ($20) each, in good condition. There was only one series of horizontal notes, for denominations of 50 and 100 gurush, issued in 1877. The 50 gurush of this issue is fetching a price of approximately £65 ($130).

The note issues of the Republic of Turkey mentioned earlier, are divided basically into two series: some bearing the portrait of Kemal Attaturk and others that of Ismet Inonu. The first of the Republican notes, issued by a decree dated 1925, were put into circulation by the Ministry of Finance and the lettering on them appeared in Arabic. These are fairly common in the 1 and 5 pound denominations but the extremely rare 500 and 1,000 pounds are valued at over £200 ($400) each.

The Central Bank of Turkey was established in 1923 and immediately issued the first series of the notes of the Republic.

Bibliography:

Erol, M. *Ottoman Empire Paper Money (Kaime)*, (in Turkish), Ankara, 1970

Haffner, S *The History of Modern Israel's Money*, (2nd edition), California, 1970

MacKenzie, K. *Ottoman Bank Notes*, AH 1293, IBNS, March 1972

Morrison, M *The Bank Notes of the Ottoman Empire 1875–1919*, MCBN, September/October 1971

Ölçer, C. *Paper Currency of the Republic of Turkey*, (English résumé), Istanbul, 1973

U

USA

Currency: 100 cents = 1 dollar

A collector who intends to begin collecting American paper money currencies is advised to concentrate at first on modern American bank notes, that is, the small-sized notes issued since 1929. The very wide range and vast quantity of paper monies that have circulated in the United States in her notaphilic history would be too wide a field to collect from the outset.

The earliest recorded collection of USA notes is that of Dr Joshua I. Cohen, dating back to 1828. Records also indicate that George Washington possessed sheets of Continental currency notes within the framework of a general 'Americana' collection. Some advertisements in numismatic publications last century offered American paper money for sale; there are also records of auctions of American notes before the outbreak of the Second World War. A large number of publications have appeared on American paper notes (see bibliographic listing at the end of this section) and the hobby has a long and continuous history of success in the United States.

Colonial

The first note issue of the United States was the paper money of the Massachusetts Bay in 1690. This was public money intended to finance the military expedition to Canada in King William's war, which lasted from 1689 to 1697. A second similar issue, intended for the military expedition against the Spaniards and Indians in Florida, was circulated by South Carolina in 1703. Many of these early colonial notes are extremely rare – they continued to be issued until 1775, after which date the geopolitical area was named the 'United Colonies'. A wide range of fractional denominations were issued. Notes for 2 pence, 4 pence, 12 and 14 shillings, and even 7-shillings-and-6-pence made an appearance. Maryland issued odd denomination colonial notes for one-ninth and two-thirds of a dollar.

The majority of the issues of the thirteen colonies, which are dated before 1775, are very difficult to obtain. There are some exceptions, however, which, because of the huge quantities issued, can be purchased at reasonable prices. The Pennsylvania issues printed by Benjamin Franklin between 1756 and 1764 are one example. They are priced at about £30 ($60) in good condition. After 1775 the notes become far more common. A representative colonial collection could be built up without having to spend more than, on average, £15 ($30) per note.

Celebrations for the 200th Anniversary of American Independence began early in the 1970s. Thus all the notes of the period have now become popular collectors' items. This 60 dollar issue was circulated in 1779. (73 × 90 mm, actual size)

Continental

The Continental Congress during the American Revolution issued notes headed 'The United Colonies' and for a number of years from 1775, these circulated quite successfully for their full value. Having no backing, however, the issues soon lost their value. The heading on the notes was changed from 'United Colonies' to 'United States' in 1777. These issues, like their predecessors, were soon almost valueless and the still current expression 'not worth a Continental' refers to them. By 1780 the 'Continental' currency had depreciated to one-fortieth of its original face value. At the end of hostilities the only consolation the government could find for the many citizens who lost everything by holding on to the currency was to request that they should look at their losses as being a tax on the entire citizenship of the United States to help finance the war!

The public's mistrust and its unwillingness to accept revalued Continental currency on any basis brought about a temporary moratorium on paper money issues. This government measure led to a number of private bank issues. These banks soon became known as 'wildcat banks'. They flourished, and collapsed, at regular intervals between 1780 and 1861. The many issues of some extremely colourful bank notes are today commonly referred to as 'broken bank notes'. They followed shortly after the American War of Independence and by the time the Civil War erupted in 1861, over 1,600 banks, in thirty-four states, had issued more than ten thousand different notes. Many are available to the collector at no more than £5 ($10) each but some of course are rare, and relatively expensive. The issues of Texas are an example. The 100 dollar note of 1841 is valued at over £25 ($50) and the 3 dollar at over £35 ($70).

Confederate

The outbreak of the Civil War in 1861 saw distinct issues which circulated simultaneously: notes of the Confederate States of America, which continued to circulate until 1864 and those of the Union in the north, which began with the 'demand notes' in 1861, and were the first official paper money issues of the United States of America.

The paper money issues of the Confederate States are extremely easy to come by. Many can be obtained at under £3 ($6). Several hundred notes and bonds were issued in denominations of up to 1,000 dollars. The higher denominations were naturally printed in lesser quantities and consequently have a higher market value – but not in excess of £25 ($50). A complete collection of Confederate notes could be aimed for, since all of the material has been well catalogued.

At the end of the Civil War Confederate currency was declared to be invalid. It has been recounted elsewhere how soldiers could find no better use for this money than to stuff the bills into their boots in order to keep their feet warm while marching during the winter months!

United States since 1861

The period relating to the paper money issues of the United States proper, that is from 1861 to date, is divided into two major sections. The first relates to large-sized notes dating between 1861 and 1929; the second period relates to small-sized notes, including the current issues in circulation, from 1929 to date. In spite of the standard uniformity of design in the small-sized issues there are a surprisingly large number of types and variations on the basis of which a collection can be formed. It is interesting to note that in the period of the large-sized notes – from 1861 to 1929 – 1,200 distinct note varieties were issued, whereas in the far shorter period since 1929, over 1,180 varieties of the small-sized note have already been recorded.

The first notes that were put into circulation by the United States in 1861 were the original 'greenbacks', known as 'demand notes' because the legend on the face of the note read 'The US promises to pay the bearer . . . dollars *on demand* . . .' There were only fifteen varieties of this issue and the 5, 10 and 20 dollar denominations are all very rare. The cheapest note in good condition is valued at over £150 ($300).

Also in 1861 five distinct varieties of US legal tender notes were put into circulation. Denominations of up to 10,000 dollars were issued on different dates, the last of these in 1913. In view of the fact that these notes – as is the case with all United States currency issues – are still redeemable, the higher denomination notes are the more expensive and always fetch prices above their face values. The 500 dollar note of 1862 or 1863, for example, is priced at £350 ($700). The 1 dollar note dated 1917, on the other hand, can be purchased at about £8 ($16), subject, as always, to the rarity of the seal and signature.

The issue of compound interest Treasury notes also began in 1863. These are extremely rare and valued at about £400 ($800) each. The 10, 20, 50 and 100 denomination notes which circulated for a limited period of time bore, respectively, the portraits of Solomon P. Chase, Abraham Lincoln, Alexander Hamilton and the standing figure of Washington. In the same year, one year interest-bearing notes were also issued and are again quite difficult to come by. The higher denominations are virtually unobtainable.

In 1879 refund certificates were circulated for the first time. Of these issues only the 10 dollar note, bearing a portrait of Benjamin Franklin, is valued at over £100 ($200). The remaining denominations rarely make an appearance on the notaphilic market.

The first silver certificates came into being in 1886. The 1 dollar denomination notes, at least, are surprisingly easy to obtain. The 1899 issue with the blue seal, for example, can be purchased in very good condition for about £20 ($40). As the denomination increases, so does the market value. The 20 dollar notes of 1878 are priced at about £400 ($800) each because of the uniqueness of the denomination. The 500 and 1,000 dollar silver certificates, issued between 1878 and 1891 are again outside a priceable range.

'Coin notes' for denominations from 1 to 100 dollars followed the silver certificates in 1890 and were in use for only one year. The name derives from the text on the notes which states 'The U.S.A. will pay the bearer . . . . . . dollars in coin'. The high denominations, again, are the rarest and most expensive, priced at several hundred pounds each. The least expensive of the coin notes is the 1 dollar denomination which, in reasonably good condition, would fetch around £110 ($220).

The National Bank notes are divided into different charter periods. These issues could form a collection on their own. They circulated from 1863 to 1929 and a vast series of issues were released during the three charter periods. In spite of the fact that these bank notes were issued by individual private banks, the National Banking Act of 1863 sanctioned them as official United States paper money. Many of them were printed by the Bureau of Engraving and Printing under terms identical to the printing of the regular Treasury issues. Furthermore, the notes were all fully negotiable.

The name of the State in which the issues were circulated appears on the face of the note. Because of the nick-name applied to the demand notes these issues with a green reverse also became known as 'greenbacks'. Some National Bank notes are obtainable from about £15 ($30). Most of them are much rarer.

The Federal Reserve Act of 1913 authorised the Federal Reserve issues. Two separate series were issued by authority of two specific laws; one dated 1915 and the other 1918. There is a distinction between the notes issued under the Federal Reserve system, known as Federal Reserve *Bank* Notes (which are issues of the individual banks in the system) and the Federal Reserve Notes (whose notes belong to the system proper). Only the former are headed 'national currency' and are somewhat similar to the National Bank notes. They are priced at about £15 ($30) for the small denomination 1918 issues but up to £350 ($700) or more for the 20 dollar 'Dallas' 1915 series, with the Teehee and Burke signatures. Again the higher denominations are always more expensive and the 50 dollar Federal Reserve Bank note, issued by the St Louis Bank only, is valued at £1,000 ($2,000). The 1920 New Orleans 100 dollar caused a sensation and established a new record in 1974 when it sold for no less than 25,000 dollars!

The Federal Reserve notes, on the other hand, were issued in the 1914 and 1918 series. They are similarly valued. Denominations upward of 5 dollars can be

With the establishment of the Federal Reserve system in 1913 a new series of notes was issued known as Federal Reserve Bank notes and headed 'National Currency'. These usually indicate the city of issue, in this case San Francisco. (66 × 157 mm)

purchased at about £20 ($40), while the 100 dollar notes fetch over £100 ($200). Unlike the Federal Reserve Bank notes, denominations for 100 and up to 10,000 dollars were issued in the 1918 series of the Federal Reserve notes. The higher denominations are not priced.

Among the most attractive notes in American notaphilic history are the reverses of the National Gold Bank notes of California. These have beautiful designs, a group of gold coins printed on yellowish paper, which associate the paper money issue with the Californian gold rush of the 1840s. The Bank issued denominations from 5 to 500 dollars, all of which are rare. The smallest denomination, the 5 dollar note, will fetch about £85 ($170) in fair condition. The issues are rarely encountered in mint condition. They circulated for only five years from 1870 to 1875, after which date they were withdrawn.

The Act of 1863 also authorised the issue of gold certificates which, to a certain extent, circulated simultaneously with the National Gold Bank notes of California. These are again considered to be very rare, except for the modern ones. The 10 dollar 1922 note can perhaps be obtained at about £35 ($70), but earlier issues are often three times as much.

Fractional currency of the United States was issued for a period of fourteen years from 1862 to 1876. The often tiny notes are for odd denominations from 3 and 5 cents to 10, 25 and 50 cent issues. Many are obtainable at very reasonable prices; some at just a few pounds each. Other issues are extremely rare, however. Several hundred pounds have been paid for some of the fractional notes such as the surcharged red reverse 50 cent notes with printed signatures. All fractional currencies became redeemable in silver coin by a Congressional enactment of April 1876. Fractional notes fall into the same periods as the large-sized notes issues.

Smaller-sized notes were issued entirely for economic reasons. As a result of the change, several million dollars were saved in paper costs. In this context it is interesting to note that as recently as September 1973, Crane and Company, the makers of all United States currency paper, were experiencing financial difficulties arising out of the increasing cost of basic materials. Although they were claiming that their huge production of seven million pounds of paper a year was not affected by wood pulp shortages, it was noted that raw cotton prices had also doubled. Thus they were in fact being hit, despite using a textile base as their raw material. (Could this be an indication of a further size decrease in United States currency notes in the foreseeable future?)

The first small-sized notes to appear in the American economy were placed into circulation on 10 July 1929. These were the legal tender notes for 1, 2, 5 and 100 dollars. The last of these is at present still in circulation. The smaller denominations have the series year on the face of the note, dating them between 1928 and 1963. The 1, 2 and 5 dollars can be obtained in mint condition for about £10 ($20) or so. Silver certificates of the small-sized notes were also placed into circulation from the outset. Many of these can be purchased for a few pounds. The modern ones of the 1957 series can probably be obtained at just over their face value, at less than £1 ($2). The higher denominations of the silver certificates are more expensive and some exceptionally rare issues, such as the 10 dollar notes of the 1933 series bearing a vignette of the head of Hamilton and signed by Julian and Woodwin, are valued at about £1,000 ($2,000).

The third sub-division of the small-sized notes are the National Bank notes. Again denominations from 5 to 100 dollars were issued and they are now all obsolete. They can, of course, still be redeemed by the Treasury and they are not too expensive, usually priced at under double their face value.

The Federal Reserve Bank note issues, also for denominations of 5 to 10 dollars, are now all obsolete (but redeemable) and fetch about double their face values. The small-sized Federal Reserve notes are in fact those presently in circulation in the United States. They form the largest quantitative issue of today's currency and are the backbone of the whole American economy.

Denominations of 1 to 10,000 dollars do exist, although the issues circulating among the general public do not exceed the 100 dollar bill. The 1928 series were redeemable in gold until the Gold Reserve Act of 1933. Only a particular combination of signatures gives a collectors' value in excess of, say, double the face to these issues. The 1928 series is probably the most expensive. There are as always a great number of rare issues due to misprints and errors, but these are exceptions.

The last sub-division of small-sized notes are the gold certificate issues; the 10, 20, 50, and 100 dollar notes with the gold seal are priced in mint condition at between £25 ($50) and £90 ($180) each, depending on the denomination. They are not particularly rare, and most of the issues can be obtained without too much difficulty.

All United States money from 1861 to date is redeemable at the Treasury; consequently there is no note which should be sold at below its face value.

Bibliography:
A vast number of books, catalogues and articles have been published covering American paper money and the following is only a select bibliography of the best known material and some of the more easily accessible articles:

Books and catalogues
General
> Criswell, G. C. *North American Currency*, (2nd edition), Florida, 1969
Early American Paper Money
> Chalmers, R. *A History of Currency in the British Colonies*, London, 1893
> McKay, C. L. *Early American Currency*, New York, 1944
> Newman, E. P. *The Early Paper Money of America*, Wisconsin, 1967
> Wismer, D. C. *Check List of Continental and Colonial Currency*, Hatfield, 1927
Confederate and Southern States Currency
> Criswell, G. C. *Confederate and South States Currency*, Vols. I and II, Florida, 1964
> Bradbeer, W. W. *Confederate and Southern State Currency*, New York, 1915, (reprint 1956)
United States Paper Money
> Friedberg, R. *Paper Money of the United States*, (7th edition), New York, 1972

Hessler, G. *The Comprehensive Catalogue of the United States Paper Money*, New York, 1974

Shafer, N. *A Guide Book of Modern US Currency*, (5th edition), Wisconsin, 1971

Articles

Ball, D. B. *Confederate Currency Derived from Banknote Plates*, Numismatist, March 1972

Banyai, R. *The U.S. Government Treasury Bill*, IBNS, September 1974
The Federal Reserve System of the USA, IBNS, Christmas 1965
Colonial Paper Money and Land Bank System 1710–1760, IBNS, September 1966
Old American Notes (Wells Fargo Bills), IBNS, June 1969

Bigsby, V. L. *Paper for Colonial Currency*, WNJ, Vol. I, No. 6.
The Story of a Ragged Note, WNJ, September 1964
U.S. Navy Money, WNJ, December 1967

Bresset, K. E. *A Patriotic Valentine*, WNJ, February 1968

Deloe, V. *Treasury Notes of the Confederate States of America*, IBNS, Christmas 1961
Free Banking 1837–1865, IBNS, Christmas 1966

Donn, A. I. *A Checklist of US Obsolete Notes by Denomination*, IBNS, Christmas 1961

Glaser, L. *Paul Revere and the Massachusetts Currency*, WNJ, February 1966

Hughes, B. H. *Sam Upham's Confederate Notes – The Saga of a Lawful Counterfeiter*, WNJ, Vol. 2 No. 5

Kagin, D. *First Attempts at Paper Currency in America*, Numismatist, April 1973

Kauth, J. *Early American Currency Notes*, IBNS, Christmas 1961

Lloyd, R. H. *A Treasury Miscalculation? (Writing Off Notes)*, Numismatist, January 1972

Obojski, R. *American Colonial and Revolutionary Paper Money*, IBNS, March 1973
Paper Money of the US, Southern States and Texas, IBNS, Autumn 1962
Collecting USA Currency, IBNS, Summer 1963
Collecting Uncut Sheets of U.S. Currency, IBNS, Spring, 1964

Pick, F. *Destruction of the U.S. Dollar*, WC, September 1973

Philipson, F. *The Crisis Note from America*, IBNS, Autumn, 1962
Collecting Colonial, CM, October 1970

Phillips, A. *Confederate Currency*, CM, December 1970

Shafer, N. *Interest Bearing Notes*, WNJ, January 1967

Smith, F. C. *A Tale of Three Banks (Broken)*, WNJ, January 1967
College Currency (Currency of Educational Institutes), WNJ, September 1967

NB The interested collector's attention is drawn to the Society of Paper Money Collectors – which specialises in USA currencies (see Appendix IV).

URUGUAY

Currency: 100 centesimos = 1 peso

The first notes issued by Uruguay are attributed to the 'Banco del Salto' which opened its doors in the 1840s, long before Uruguay was formally recognised as an independent state by neighbouring Argentina and Brazil. Very soon thereafter a large number of private banks issued their own paper monies, including some well known today, such as the Banco de Londres y Rio de la Plata. The notes of this Bank which were issued in the Montevideo Branch, were discovered in very large quantities in 1973 and the market value subsequently decreased to a mere £6 ($12) each, in spite of the notes being dated 1878.

The Exchange Control Society, known as the 'Sociedad de Cambios', issued notes in the reis denomination in 1856 and these are practically impossible to find.

Uruguay has a very profuse issue of last century notes and some of them are quite easily obtainable. The Banco de Credito Auxilar issued notes in 1887 for denominations up to 1,000 pesos and these are obtainable at about £10 ($20) or less, but the 1,000 pesos are worth £90 ($180). Similarly, the Banco Italiano del Uruguay, which portrays both Garibaldi and Christopher Columbus on its 100 peso issue, has notes obtainable in sheets at about £45 ($90) for the four notes. Some early issues, of course, are quite difficult to come by.

The Banco de la Republica Orientale del Uruguay began its issues in 1911. The dates appearing on these notes refer to the law authorising the issue. The first of these laws was enacted in August 1896.

The higher denomination notes are priced at £30 ($60). Later issues dated September 1914 are more readily available and those of January 1918 are quite common, at £5 ($10). Some specific issues, normally high denomination notes, make the exception. The 50 peso fetches prices in the £35 ($70) range.

In 1935 the special issuing department of the Bank was established and issued notes which are extremely colourful and attractive. They are dated 14 August 1935 and 2 January 1939. The latter, in particular, are extremely common, often obtainable at £5 ($10) each. The notes dated 2 January 1939 were issued again after 1970, by the Banco Central del Uruguay which took over the activities of the Banco de la Republica Orientale del Uruguay. These are current notes, presently in circulation for denominations of up to 10,000 pesos.

Bibliography:

Araujo, E. O. *Uruguay's Monetary System*, WNJ, October 1967
New Issues of the Central Bank of Uruguay, WNJ, October 1968
Matz, A. C. *Varieties of Uruguay Bank Notes*, CC, Fall 1969
Varieties on South American Bank Notes, LANSA, February 1973
Seppa, D. A. *Paper Money of Paraguay and Uruguay*, Texas, 1970
Willagran, E. O. A. *New Emergency Money of Uruguay*, Parts I and II, WNJ, February and March 1966.

USSR (Russia) (also Armenia, Azerbaijan, Estonia, Latvia and Lithuania)

Currency: 100 kopecks = 1 rouble
10 gold roubles = 1 tscherwonez
200 schamiw = 2 griwen = 1 karbowanez (rouble) – (Ukraine)
100 penni = 1 marka ⎫
100 senti = 1 kroon ⎬ (Estonia)
100 kapeikas = 1 rouble ⎫
100 santimu = 1 lats ⎬ (Latvia)
100 centu = 1 litas (Lithuania)

In the novice collector's mind, Russian notaphily is directly associated with the revolutionary period in the first quarter of this century. There is considerable justification for such thoughts. Huge quantities of paper money were issued by three authorities during 1917 and immediately afterwards. Many of these entered the collectors' market in the West as a result of individual Russian families fleeing their country with, literally, suitcases full of paper roubles which, of course, were soon found to be worthless.

Russian paper money issues can be divided into several distinct periods. The Tzars were the first to issue their own currency and it is on record that as early as 1725 serious consideration was given to the possibility of issuing paper money, but as a result of loans obtained from foreign banks, the idea was rejected. It was only under Catherine the Great, in 1769, that the first official paper money issues of Russia came into circulation in the form of assignats (i.e. backed by land). These notes remained redeemable in copper coin until 1777. In that year the right of redemption was withdrawn but the issues continued to circulate well into 1785. These issues of the Imperial Assignat Bank from St Petersburg and Moscow are now rarely encountered, but they do occasionally appear on the market and fetch prices in the £80 ($160) to £150 ($300) range.

The nineteenth-century Russian notes, however, are commoner and assignat issues dated 1811, and up to and including 1829, were recently offered for sale in an auction at prices ranging from £17 ($34) to £35 ($70). These are relatively inexpensive because between 1801 and 1825, for example, a different series of notes was issued every year.

In 1841 Russia issued the first credit notes and in 1847 the Credit Chancellory took over the note-issuing activities of the Imperial Bank. Because of the inflation following the first notes of 1841, these credit notes are among the rarest of Russian paper currencies, as are those of a similar issue in 1876.

The State Bank, known as 'Gosudartvennye Bank', was established in 1860 but it did not issue its own paper money until several years later. In the meantime, between 1855 and 1881, paper money had been issued every year. Some of these notes therefore can be obtained with relative ease and comparatively cheaply – within the £20 ($40) range.

Notes from 1898 onwards really become more common. Most dealers' price lists offer notes issued between 1898 and up to 1917 at under £4 ($8). The notes cannot normally be differentiated by their dates, as they may well be all dated in the same year. This is because the provisional government, when it came into power, used the paper money plates of the Tzars without change except for the signatures and serial numbers. The different signatures of the directors of the Bank, therefore, allow for chronological identification.

With the outbreak of the Great War in 1914 over 9,500 million roubles were circulating. Most of the notes of this period are obtainable at a few pence each. In spite of the economic crisis the government had no choice but to continue issuing more money. Orders were placed with foreign printing firms, such as the American Bank Note Company. whose notes were hurriedly placed into circulation without dates or signatures. Uncut sheets, known as 'beer stamps', were also placed directly into circulation in a bid to save time and expense. At this time, furthermore, twelve-month bonds bearing an annual interest of five per cent were circulated for total values of up to one million roubles. These too, by force of circumstance, changed hands as legal tender currency. All these issues are common and, in mint condition, can be obtained at under £5 ($10).

The First World War had bled the Russian economy and brought with it the downfall of the provisional government. In November 1917 the socialists took over. This was followed by a period of Civil War and foreign intervention which lasted until 1922. The only set of rare notes during these years was issued by the Union of Russian Stock Commerce Banks, for denominations between 25 and 500 roubles. They are valued at over £60 ($120).

Paper money issues following the Bolshevik take-over are of a wide and diverse range. The White Guards and other opponents of the socialist-communist government in Moscow issued paper monies which were intended for use within

Issues of Russia and the USSR include a wide range of notes from states with diverse backgrounds. The Emirate of Bukhara issued its notes in tingas and kopecks and in Arabic script. The 1919 note for 10,000 tingas was circulated in the period of inflation and has the highest denomination ever circulated by the Emirate. (128 × 270 mm)

limited geographical areas. This assured the Bolsheviks that their currency would not circulate in White Russia. Individual White Guard generals warring against the Moscow government also financed their military efforts with paper money. Simultaneously, the new provincial governments and the republics mushrooming throughout Russia printed their own paper currency. Well known among these are the notes of the republic of Georgia and Armenia; and to add to the widening diversity of issues, foreign countries who sent military forces to help the White Guards (Japan, Germany, the United States of America and Great Britain) also circulated military currencies in the territories they occupied.

The shortage of official government bank notes induced municipalities to issue city notes and a wide range of commercial entities, including shop owners, traded with their own improvised paper notes. The range of these notes is so widespread that it would be difficult, without highly specialised knowledge, to specify any one issue that is particularly rare.

The central Government based in Moscow, the Russian Socialist Federated Soviet Republic, issued in 1918 a series of notes from 25 to 500 roubles. These originally circulated under the Empire and are dated between 1908 and 1916; they do not fetch more than £5 ($10) each. The same period saw the re-issue of 1,000 to 1 million rouble short term Treasury notes, bearing five per cent interest. These are obtainable within the £10 ($20) range; the 100,000, 500,000 and million roubles, however, issued between 1916 and 1917 are priced around £25 ($50) each.

Between 1921 and 1924 three currency reforms took place: the first saw the establishment of the new rouble equivalent to 10,000 old roubles; the second, in 1923, converted 1 million old roubles to 1 new rouble; and the last, in 1924, led to the establishment of the gold rouble, equivalent to 5,000 old roubles. In 1923, a year after the defeat of the White Guards, the Union of Soviet Socialist Republics was formed and initially issued high denomination credit notes of 10,000, 15,000 and 25,000 roubles. During this inflationary period a limited number of fractional notes in the kopeck denomination were printed and are keenly sought after by collectors of Russian bank notes. They are valued at about £35 ($70) each. 1915 issues for different denominations were perforated in 1923 but are priced only a little higher than the original notes. A total of over 10,000 distinct notes had been issued by 1924 and many have still to be catalogued.

The well known 'stamp money' also appeared at this time. Printed on thin cardboard, they specifically stated on the reverse that they were intended to circulate as currency. A wide variety of stamps were used as money in this manner and most can be purchased at a few pence each. When they appear in complete sheets of thirty-six or more they have been known to fetch £15 ($30) per sheet.

On the whole, Russian notes and particularly those of the USSR are considered to be inexpensive. Error notes are the exception but a surprisingly large number of these found their way into circulation. Furthermore, the difficulty in identification has kept the prices comparatively low, at no more than a few times the value of the original notes. Many of the republics in the USSR have known a period of independence during which distinct paper money issues were circulated:

Armenia

A collector of Russian paper money will find that Armenian issues of 1919 will form an integral part of his collection. Notes of up to 10,000 roubles were issued during that year. On proclaiming itself the Soviet Socialist Republic of Armenia in 1920, the country went into an inflationary period when notes of a denomination as high as 5 million roubles were circulated. Most Armenian issues are common and are priced at a maximum of £15 ($30).

Azerbaijan

Another independent republic which issued its own currency in 1919 was Azerbaijan. In 1920 it became a Soviet Republic and, like Armenia, was hit by inflation. The high denomination notes circulated until 1923 including, again, a 5 million rouble note, issued in that year. The majority of these are valued at 50p ($1) to £2 ($4) each.

Estonia

Until the 1917 Revolution, leading to independence in 1920, Russian issues circulated in Estonia. The first recorded notes of the Republic are some private issues dating to 1918 by cement and cloth factories, which issued small denomination notes, the former for 50 kopecks and 1 rouble. These issues are considered to be paper coupons rather than notes. The earliest issues are the State Treasury notes of 1919 and up to 1940 when Estonia became a Constituent Republic of the USSR. Denominations from 5 penni to 1,000 markaa were circulated. All these issues are within the £15 ($30) range.

The Bank of Estonia also issued notes during the same period, changing the name of the monetary unit from marka to kroon in 1928. Again, these can be obtained at very reasonable prices. The rarest Estonian notes are those issued by 'Sons of the North', notes of Finland and Estonia overstamped with the seal of the cashier and placed in circulation in southern Estonia and northern Latvia in 1919. They sell for as much as £35 ($70) each.

Latvia

Latvia became independent in 1918. The notes first circulated there were the army issues applicable to other parts of Russia. In 1919, however, the Latvian State Bank was established and issued notes from 1 to 500 lats. The highest denomination of these is worth about £45 ($90); the remainder are valued at under £10 ($20).

The Bank of Latvia came into being in 1920 and issued its own notes until 1939. The notes of this Bank dated prior to 1924 are difficult to come by. Issues after that date do not fetch more than £10 ($20) each.

The different State Treasury notes of Latvia, issued prior to her joining the USSR in 1940, are obtainable for less than £5 ($10) each.

Lithuania

Lithuania, together with Latvia and Estonia, was a Russian province from 1795 until 1918, when she declared herself an independent republic. The issues circulating in Latvia also circulated in Lithuania, including German state loan notes, up to 1922.

In that year the Bank of Lithuania was established and issued notes in litas (100 centu = 1 litas) until 1938. All Lithuanian issues higher than the 100 litas denomination are priced at over £25 ($50). The smaller notes are a little easier to come by, fetching prices in the £10 ($20) range. The rarest Lithuanian note is the 1938 10 litas issue which is priced at about £50 ($100).

In 1940 Lithuania was incorporated into the USSR.

A great number of additional Russian and USSR issues have not been mentioned. There are notes of Georgia, both as an independent republic and a Soviet Socialist Republic. Note issues of the Crimean district government are attractive items, as are those of the Kuban district government. Furthermore, there have been several army issues, as well as those of some emirates. All of these, with only a few exceptions, are fairly easy to come by. The collector should not expect to pay more than £5 ($10) for any of these issues. One special exception is the North Caucasian Emirate: The 50 and 100 rouble issue of 1919 are priced at over £20 ($40) each.

Bibliography: *books*

Arnold, A. I. *Banks, Credit and Money in Soviet Russia*, New York, 1973

Baykov, A. *The Development of the Soviet Economic System*, London, 1947

Chizhov, S. *Russian Paper Currency – First Par-Value Credit Notes*, Moscow, 1914

Hollerbach, E. *History of Engraving and Lithography in Russia*, Moscow Petrograd, 1923

Hubbard, L. E. *Soviet Money and Finance*, London, 1936

Johanson, E. *Type Register of Cheques, Money Orders, Bonds, Tokens and Coupons in the Soviet Union, 1917–1924*, Helsinki, 1971

Kardakoff, N. *Catalogue of Russia and the Baltic States 1769–1950*, (in Russian and German), 1952

Katzenellenbaum, S. S. *Russian Currency and Banking 1914–1924*, London, 1925

Platbarzdis, A. *Coins and Notes of Estonia, Latvia and Lithuania*, Stockholm, 1968

Tiitus, M. *Paper Currencies of Estonia*, California, 1971

Articles

Atsmony, D. *Notes of the Jewish Communities in Russia*, IBNS, Autumn 1961

Bark, P. *The Imperial Bank of Russia*, Bankers' Magazine, 1921

Bramwell, D. *Forgotten Notes of a Turbulent Past*, CCW, 25 December 1971

Gribanov, E. D. *Money of the Tanu-Tuva Republic*, IBNS, Autumn 1963

US Printed Paper Money of Provincial Government of Russia, IBNS, September 1965

Private Cheques of Russia and the USSR, IBNS, Summer 1967

Money of the North Caucasian Emirate, WNJ, April 1967

Howlett, P. *From Tsar to Soviets*, C & M, December 1971

Philipson, F. *Adm. Kolchak – His Story and Currency*, IBNS, September 1970

Bank Notes of the Czars, CM, October 1970

Pick, A. *313 Varieties of 5 Rouble of 1917 Provincial Government*, IBNS, March 1970

Raffalovich, A. *The Note Issue under Bolsheviks' Rule*, Bankers Magazine, 1921

Rice, D. S. *The North Russian Currency*, Economic Journal, 1919

Seibert, V. C. *History of Russian Paper Money*, I, II, III and IV, IBNS, September and Christmas 1964, March and Summer 1965
Introduction to Paper Money of North Russia, IBNS, Autumn 1969
The Iman Usun-Chadschi's Bank Notes, IBNS, December 1971

V

VENEZUELA

Currency: 100 centimos = 1 bolivar(es)

Of the Latin American countries Venezuela is among the latest to issue its own paper currency. Although notes of Colombia circulated in the area it was no longer part of Gran Colombia after 1830, and yet its own first issues, with one exception, did not appear until the end of 1870. The Banco Venezolano de Credito circulated notes for denominations of 10 to 500 bolivares, all of which are rare issues, valued at over £100 ($200).

Many private banks came into being towards the end of the last century. Famous among them is the Banco Caracas. The easiest notes to encounter are those of the Banco Mercantil y Agricola which issued only three denominations: 10, 20 and 100 bolivares. All three unissued notes can be obtained as a set for under £25 ($50). The remaining private issues are far more difficult to come by, often fetching prices over £50 ($100).

The Central Bank of Venezuela issued its first notes in 1947. The 5 bolivar

Many Latin American banks closed down before they had issued all the notes they had printed. The Banco Mercantil y Agricola of Venezuela was closed in the 1930s. Its unissued notes are more common than the signed and issued ones. (60 × 154 mm)

commemorative issue celebrates the four hundredth anniversary of the founding of the capital city of Caracas on 10 May 1566. The note is valued at about £5 ($10). Venezuelan money is among the most highly valued currency in South America (in 1973 it was 4.35 bolivares to the dollar) and consequently the face value of the notes is very high. All of the issues of the Central Bank of Venezuela remain redeemable.

Some of the early notes at the beginning of this century are interesting because of their strange denominations: 200, 400 and 800 bolivar notes were issued by several private banks. The highest denomination ever issued is the 1,000 bolivar, also by two private banks, in the first decade of this century. All of these notes are valued at about £115 ($230) each, almost irrespective of condition.

Bibliography:
Loeb, W. *Commemorative Bank Notes*, WNJ, February 1967

VIETNAM

Currency: North Vietnam: 100 xu (ian) = 10 hao = 1 dong
South Vietnam: 100 cents = 1 dong

The collector who has decided to concentrate on the paper money of Vietnam must be able to define for himself the geographical area as well as the historical one. Paper money issues of Vietnam proper did not make an appearance until 1954 when Vietnam was divided into the Democratic Republic of Vietnam in the north and the Republic of Vietnam in the south. Considered historically, however, the first notes in the area were issued under French Indo-China in 1875.

The place of issue of these was Saigon and a similar issue was placed into circulation in Hanoi in 1900. If the collector's interests stretch this far back, it is suggested that he studies the paper money issues of the French overseas territories. A complete collection of South Vietnamese notes can be compiled quite easily and cheaply. Not more than twenty notes have been issued; the 200 and 500 dong denominations of the first series were withdrawn in 1958 and 1964 respectively and these two issues can be purchased for approximately £6 ($12) each. They are, in fact, the most expensive of the South Vietnamese notes. The remaining denominations are still in circulation.

North Vietnam has no record of its own issues of paper money until 1958 when notes with the Soviet star appeared. Most of the denominations can be obtained for about £4 ($8). However, as late as 1972, the acting director of the Control Division of the United States Treasury announced that notes of both the Democratic Republic of Vietnam and of the Vietcong insurgent forces were forbidden to be imported into the United States. This led to a higher notaphilic value for these issues, which do not frequently appear on dealers' price lists.

The designs on the North Vietnamese notes appear to have been directed at the United States, with subject matter almost provocative in nature. The 10 dong note, for example, shows camouflaged troops on the reverse attacking an infiltrating (American?) patrol! Similarly, the 50 dong note depicts several soldiers firing at a helicopter.

The notes of the Democratic Republic of Vietnam which have gained great popularity are those of the 1958 issue portraying Ho Chi Minh. They are comparatively common and can be bought at about £10 ($20) in mint condition.

The notes of Vietnam go hand in hand with the issues of Laos and Cambodia which formed the French Federation until 1954.

Bibliography:
News Items *National Bank 1972 Issues*, WC, November 1972
  Vietcong Insurgent Notes, WC, September 1972
Philipson, F. *Paper Money of Laos, Cambodia and Vietnam*, CM, May 1971
Shafer, N. *Vietnam Propaganda Notes*, WNJ, December 1965

W

WALES

Currency: (as in England)

In May 1969 the world's attention was turned on a Mr Richard Williams of Llandudno, Wales, who had succeeded in persuading the British banking authorities to allow him to print his own Welsh money for denominations of 10 shillings, £1, £5 and £10. In spite of almost simultaneous talk of the formation of a Bank of Wales during this period, these notes came to nothing. Many collectors have succeeded in obtaining them, although at times paying prices well beyond their notaphilic value.

The well designed notes (the reverse shows an attractive water colour of Caernarvon Castle) made quite a stir in view of the fact that the Board of the Inland Revenue placed its official stamp on the notes. As a result, the notes were considered to be legal tender for a limited period of time. Today a set of them can be obtained for just a few pounds at the most.

The notes described above were intended to recall two early Welsh independent banks that were formed in the late 1700s. In 1810 the Aberystwyth and Tregaron Bank, known as 'the Bank of the Black Sheep', had issued notes for denominations from 10 shillings to £2. In order to aid the illiterate populace of the area, the 10 shilling note portrayed a lamb, the £1 a sheep and the £2 two ewes. The second note-issuing bank was known as the Chester and North Wales Bank which was launched in 1792. The circumstances leading to the establishment of the Welsh note-issuing banks were almost identical to those surrounding such issues in England. London, the centre of financial activity, was too distant and travel too inconvenient for the ordinary trader; so debts were often settled by bills of exchange. This led to the establishment of a number of local banks, with London agents. Although there are earlier records stating that Welsh banks were formed as early as 1762 it would seem that the first proper bank was established in Wrexham in 1785.

Between 1793 and 1825 many of the county banks failed, as they did throughout the country. The situation was aggravated by large numbers of forgeries of Welsh notes. The 1826 Act established the principle of joint-stock banking and the vesting of Bank of England notes as legal tender issues in 1833 led many Welsh banks, and other provincial banks, to use Bank of England notes as backing for their own issues.

The last of the Welsh bank notes were issued by the North and South Wales Bank, founded in Liverpool in 1836. The Bank withdrew all its notes in 1908 when it merged with the Midland Bank. Notes of the North and South Wales Bank are therefore still redeemable at face value. All these issues described up to 1908 are very rare and fetch prices well over £50 ($100) in good condition. There are no common Welsh notes except for Mr Williams' efforts, which are treated with some scepticism by many collectors.

Bibliography:

> Anon *Welsh Pound Notes*, IBNS, June 1969
> Jones, I. W. *Money for All, the Story of the Welsh Pound*, Llandudno, 1969
> Morgan, L. *The Bank Charter Act of 1844*, Parts I, II and III, IBNS, Summer, Autumn and Christmas 1963

Y

YUGOSLAVIA (also Serbia, Montenegro, Croatia and Fiume)

Currency: 100 para = 1 dinar
100 banica = 1 kuna – Croatia
(also lit – partisan notes of Second World War, perper(a) – Montenegro; lira – military issues of 1944.)

In the comparatively short period during which Yugoslavia issued paper money, a very wide range of distinct notes were put into circulation. The first notes of the Kingdom of Yugoslavia were provisional issues of 1919 circulated by the Ministry of Finance. These Austro-Hungarian bank notes were overprinted in 1919 with 'Kingdom of Serbia, Croatia and Slovenia'. The notes for 10, 20, 50, 100 and 1,000 kronen are valued at about £35 ($70) each. The original notes date between 1902 and 1915. A large number of provincial and private overprints exist on similar Austro-Hungarian bank notes. These are normally priced within the £5 ($10) range.

The first official issues of the Kingdom of Serbia were printed in 1919 for denominations of $\frac{1}{4}$, $\frac{1}{2}$ and 1 dinar. They again are extremely common notes valued at not more than £2 ($4) each. These same issues were later overprinted with the denomination in three different languages and the higher notes of the 100 and 1,000 dinar are quite scarce.

In 1920 the National Bank of the Kingdom of the Serbs, Croats and Slovenes was established and immediately put into circulation notes of which the 1,000 dinar remains the most expensive, at £20 ($40). The 10 and 100 dinar, the only other denominations, are priced at £5 ($10) each.

The National Bank of the Kingdom of Yugoslavia was established in 1929 and again notes from 10 to 10,000 dinars were placed into circulation by 1936. The highest denomination of these is rare, priced at £40 ($80).

With the German occupation of Yugoslavia, when a great deal of pressure was placed by the German occupying forces to separate Serbia and Croatia, the Serbian National Bank issued its own notes and here, too, up to 1943, most of the notes in the Serbian dinar denomination are very easy to come by.

A year later, in 1944, the Democratic Federation of Yugoslavia was formed and some state notes were issued under that same heading.

With the end of the War the National Bank of the Federated People's Republic of Yugoslavia was established and issued its notes until 1955 after which the National Bank of Yugoslavia came into being. In 1963 appeared a 5,000 dinar issue of the National Bank, now valued at approximately £25 ($50). All other previous and lower denomination issues are within the £7 ($14) range. The currency reform in 1965 gave rise to the new dinar, equivalent to 100 old dinars. These notes are at present in circulation.

There were a large number of other issues of Yugoslavia. During the German occupation notes printed in Slovenian and German were circulated in the lira denomination.

The well-known Yugoslavian notes overprinted with the 'Verificato' stamp were issued in 1941 during Italy's occupation of Montenegro. All these notes are fairly common, valued at about £20 ($40). The only exception is the 100 dinar note of 30 November 1922. This was a note already out of circulation, mistakenly overstamped by the Italians. There is only one known issue at present in the hands of a Yugoslav collector. Many partisan notes were also issued during the War and a wide range of other local issues is recorded.

Serbia
Collectors of Yugoslav paper money will also be interested in the currencies of the separate independent provinces which issued their own notes prior to the formation of Yugoslavia in 1918. Although the First Bank of Serbia was established as early as 1869, state Treasury notes did not enter circulation until July 1876. These were followed by issues in 1884 of the National Bank of the Kingdom of Serbia.

The denominations of 1 to 100 dinars are still obtainable and are priced at about £100 ($200) each. The gold-backed issues of the National Bank, which circulated in conjunction with silver issues, are rarer. Those from the last century are priced at about £150 ($300).

Serbia also saw, in 1917, the notes of the Austrian imperial and military government which circulated through the provincial command; these were issued in different cities of Serbia, including Belgrade, Cacak and Sabak. The notes perforated with the word 'Cancellation' are worth only half their face value.

Montenegro
Montenegro has a very short notaphilic history. This lasted from 1912, when the first Treasury notes came into being, until 1918, when it united with Serbia, Croatia and Slovenia to become part of the Kingdom. The first Montenegrin notes were the Treasury issues in the perper denomination. With the exception of the 50 and 100 perper, valued at about £125 ($250), these are quite easy to buy at less than £20

($40). The 1914 issues are more common, but it is the Austrian military government issues between 1916 and 1918 that give a wide range of Montenegrin notes; these are all overstamped, and indicate the city in which they were issued. Out of some 150 different issues only those relating to the cities of Belgrade and Cetinje are uncommon and valued at about £22 ($44); the remainder are valued at under £5 ($10) although the 100 perper denomination may be worth £10 ($20).

Croatia

Croatia was declared an independent state during the German occupation of Yugoslavia, between 1941 and 1944. Before that time it had been part of Yugoslavia and up to 1918 part of the Kingdom of Hungary. Consequently, the only paper money issues of Croatia are those dated from 1941. The earliest issues in the kuna denomination were the first ones to circulate. They continued to do so until 1944, headed 'The Autonomous State of Croatia'. In 1943 the Croatian State Bank was established and issued high denomination notes of 100 to 5,000 kuna. There are no rare issues of Croatia.

Fiume

The port city of Fiume, now known as Rijeka, was occupied by the Italians in 1919 before becoming temporarily independent in 1920. In that year paper money issues of the Austro-Hungarian Bank were overstamped and circulated as legal tender in the area. The original notes dating between 1902 and 1916 had two kinds of overprints. The 'City of Fiume' hand- and machine-overstamped notes are relatively common, priced at about £10 ($20); the rare notes, valued at over £50 ($100) each, are the issues overprinted 'Instituto di Credito Consiglio Nazionale, Città di Fiume'. No other notes were issued by Fiume.

Bibliography:

Atsmony, D. *Verificato Notes*, IBNS, Christmas 1961

Jerkovic, B. *Partisan Money for South Slovenia*, CC, Autumn 1964

Keller, A. *Verificato and the Story it Tells*, IBNS, Summer 1964

Loeb, W. *New Issues of Yugoslavia*, WNJ, March 1967

News Item *Ljubljana Dinar Note*, WC, September 1972

Pokrjcic, B. *The Paper Money of the Military Administration of the Yugoslav Army in Zone B and Allied Military Currency in Italy*, Sarajevo, 1971

Paper Money of the Yugoslav Democratic Federation Issued in Moscow in 1944, Sarajevo, 1971

Paper Money in Yugoslav Liberation Movements, Sarajevo, 1971

Forged and Printed Money in Sarajevo, Sarajevo, 1971

Spajic, D. *Paper Money of the Yugoslavian States*, Belgrade, 1969

Emergency and Supplementary Issues

Emergency paper money, almost without exception, has proven to have been issued as a direct or indirect result of military conflict – be it the First World War, plunging Germany into an inflationary spiral without precedence, or be it internal civil war, such as the one at the turn of this century which crippled the Colombian economy. The following table may illustrate the diverse range of emergency issues:

Emergency Paper Money Issues

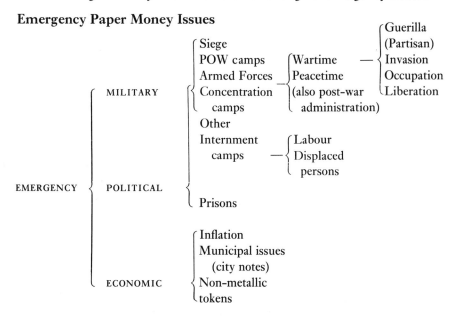

SIEGE ISSUES

Khartoum, Sudan

The mismanagement of the Sudan by the Egyptian Government in the 1870s brought about the famous revolt headed by the fanatic 'Mahdi'; thirteen thousand Egyptian soldiers had been sent to suppress the revolt and were massacred. In January 1884 General Charles ('Chinese') Gordon was despatched to the Sudan. His heroism in Khartoum and death at the hands of the 'Mahdi' are historically recorded. The notaphilic aspects of these events are not so widely known. The Treasury at Khartoum, in February 1884, was empty. Within a few weeks Gordon issued emergency notes. These were personally signed by him and accompanied by a statement that he was:

'personally responsible for the liquidation, and anyone can bring action against me, in my individual capacity, to recover the money.'

Mr Martin Parr, who has researched very thoroughly into these issues, has reached the conclusion that the total value issued by Gordon was equivalent to approximately £168,500. A large number of the notes were destroyed during the siege and immediately thereafter. But the issues were still redeemable in Cairo and some were smuggled from the Sudan to Egypt. After considering the quantities redeemed and the likely number destroyed, it is estimated that only two thousand Gordon notes are still unaccounted for.

Over fifty thousand of the notes were hand-signed by Gordon himself. The remainder bore Gordon's printed signature. Piastres and Egyptian pounds had been issued; the denominations are not immediately decipherable, so there has not been a great deal of price differentiation. The Gordon notes with the manuscript signature are always more expensive and have been sold at over £140 ($280). The hectographed signatures are not much cheaper, priced in the £100 ($200) range.

Mafeking, South Africa

Robert Stephenson Smyth Baden-Powell, the founder of the Scout movement, made his mark on history during the 217-day Siege of Mafeking by the Boer forces at the turn of the century. The publicised invention of the locally manufactured 'Wolf', an improvised weapon made of an old iron steam pipe and a water tank lined with fire bricks, was used during the siege. It symbolised the ingenuity of Lord Baden-Powell's forces under constant attack. The initiative taken by the British forces equally extended to a temporary currency system consisting of notes which were to be 'exchanged for coin on the resumption of civil law'. Baden-Powell's sketch of 'the Wolf' was used as the design on the £1 note; the drawings were mounted,

General Gordon of Khartoum issued hand-signed notes, during the siege in which he was killed. Of the two thousand surviving specimens, most are now in collectors' hands. They are very expensive. (63 × 107 mm)

photographed and locally printed by the Ferro Prussiate process on ordinary writing paper.

The first Mafeking issues were the 1, 2 and 3 shilling notes. A little later 10 shilling and £1 notes were also circulated. J. P. Ineson, a Scout District Commissioner, has researched the subject in depth, and his findings reveal that the 3 shilling and £1 notes are the scarcest of the series. These can be purchased for about £125 ($250) in good condition. The next rarest issue is the 10 shilling note, worth about £30 ($60). One of the series of the 10 shilling notes, however, was printed with a mistake: the letter 'd' was left out of the word 'Commanding' which appears at the foot of the notes. The 'Commaning notes', are valued at £45 ($90). The 1 and 2 shillings are only comparatively cheaper. In good condition they fetch about £20 ($40).

There is another Boer War issue recently discovered. O'Okiep was the scene of the last battles of the War. The garrison was under siege from April to May 1902. Finally, a peace treaty was signed between the belligerent parties. Notes roughly typewritten on plain paper and hand signed were found to have been circulated. The only available note for inspection was the £1 sterling denomination headed 'On His Majesty's Service, O'Okiep Siege Note' and dated May 1902. It is unique at present.

Lyons, France

The earliest known siege notes came from the French Revolution period. The Revolution had, by 1793, spread to war with Europe. Internal strife was also rampant and it led to many civilian as well as military sieges. The split in 1793 between the Parisian Republicans and the Royalists in the provinces brought about one of the most famous sieges in French history. With the execution of Louis XVI in January of that year, the supporters of the King based in Lyons revolted. The city was almost immediately besieged and for sixty days found itself under daily attack by Republican forces. While under siege the Royalists issued notes crudely printed on cardboard with the design based on the assignat issues. Some of the notes were headed 'Resistance to Oppression' and others 'The Siege of Lyon'. The former are extremely rare. The latter, some of which have appeared in uncut sheets, are dated 1793, and referred to as 'The Reign of Terror' notes after the name given to the siege.

Mainz, Germany

The German town of Mainz (Mayence in French) was under French control in the spring of 1793 when it was besieged by Austrian forces. The Austrians took the city in July of that year, and found that the French forces had issued internal currency. Under the emergency conditions the coinage of the town had been hoarded and fractional currency notes had been hastily printed as a substitute. The best known of these issues are the 5 and 10 sol notes bearing three signatures and a round red overstamp. The text on the small square notes reads: 'Siege de Mayence Mai 1793 2e de la Rep. France'. This note can still be purchased, in excellent condition, at about £75 ($150). Other issues, in the sou denomination, were also printed and circulated and can be bought for about the same price.

The more expensive notes are the Mayence livre issues. Many were hand written on the reverse of the contemporary assignats. A note for 50 livres dated May 1793, printed on the reverse of a 25 livre assignat dated October 1792, fetched £135 ($270) in a sale recently.

Coburg, Prussia

Coburg, the city which is now within Germany's boundaries, formed part of the East Prussian Dominion when it was first besieged in 1756. It did not finally surrender until 1761. It was during the Napoleonic campaigns in 1807 that the most famous of the sieges took place; an unsuccessful concentrated effort that lasted nearly six months. Napoleon's forces were being harassed by the Prussians and his military efforts were being disrupted by the Coburg garrison. The French Navy's blockade and siege in February 1807 was successfully resisted by the garrison. It is not clear under what circumstances the notes were put into circulation, but one presumes that the coinage began disappearing from circulation and notes were printed to alleviate the shortage. The cardboard issues were written out by hand in Prussian on one side and stamped; the date of issue appeared on the other. The groschen denomination was used for all the issues. These notes are extremely rare.

Venice, Italy

The Venetian Moneta Patriotica of 1848 is not siege money as such. It was issued in between sieges of the city by Austrian forces.

Daniel Manin was responsible for issuing notes in Venice between the Austrian sieges of the 1840s. (80 × 115 mm)

Hatred for Austria, in both Lombardy and Venetia, was gradually increasing in the mid-nineteenth century. Toward the end of 1847 Daniel Manin, jurist and statesman, took the leadership on behalf of Venetian patriotism; his arrest by the Austrians the following year, on charges of high treason, led to further agitation until Manin was released. He formed a provisional government and forced the Austrians out of the city. This was followed by the issue of 'patriotic notes' in large quantities; a morale-boosting measure no less than an economic one. By March, the Austrians had completely evacuated the city, and the Independent Republic of San Marco was proclaimed, with Daniel Manin as President. But this was a short-lived freedom. King Charles Albert abandoned the city to the Austrians, who, strongly reinforced, reoccupied Venetia. The true siege was to begin only now, during the early months of 1849, when Manin heroically led the defence of the city. Venice was bombarded from all sides for eight months. Cholera broke out in the city and food and provisions were exhausted. Finally, in September, an honourable capitulation was negotiated by Manin. He left Venice for ever and died in exile nearly ten years later.

The 1 to 100 lira issues are among the easiest siege notes to come across, both because of the large quantity printed and because most never circulated. They are priced at no more than £3 ($6) to £35 ($70) for the two high denominations.

Palmanova, Italy

During this same 'Year of Revolutions', 1848, the Austrian armies successfully reoccupied Padua, Mestre and Palmanova. The last of these, a fortification, came under cruel siege and heroically held out against a much stronger Austrian force. In the last few days a series of siege notes was issued, backed by immovable properties in the city – not unlike the earlier French assignat notes. The notes for denominations from 1 to 6 lira, and two fractional notes of 25 and 50 centisimi were also circulated. Hand signed and numbered, the notes were headed 'Finance Commission of Palmanova – in a state of siege.' They are dated 1848 and very few specimens have survived. The city capitulated shortly thereafter.

Mantova, Italy

At the beginning of the eighteenth century Mantova (Mantua) in Lombardy, northern Italy, had been ceded to Austria and in 1796 it came under siege by Napoleon. He took the city at the beginning of the following year. 'Cedole' notes had, in the meantime, been issued headed 'Moneta di Mantova'. The only recorded denomination is the 3 lira note, extremely scarce; it is the only surviving evidence of the siege currency of the city.

There are a few more siege issues known to the collector. In the 1850s Austria was encountering mounting dissension: the Hungarians' fight for independence under Lajos Kossuth was directed as much at the Austrians as at the Russians. The besieged Austrian garrison in the town of Arad issued ten different denominations of siege money, from 1 to 20 kreuzer as well as 1, 5 and 10 gulden notes. Additional siege issues are reported to have circulated in the town of Temesvar in Rumania and in the town of Esset in Slovenia. Komarom in Czechoslovakia, too, issued its own siege notes. These are only a few more examples of many issues that undoubtedly remain to be discovered.

INFLATION

Since the end of the Second World War in particular a continuous world-wide inflationary trend has become part of our lives. Inflation has been defined as 'too much money chasing too few goods'. That is, an abnormal increase in the quantity of paper money issues, relative to the availability of goods. Economic chaos sets in when the fall in the value of the currency (because of the rise in prices) is more rapid than the quantity of paper money put into circulation. The first half of this century alone saw catastrophic inflation take a grip of some European countries, while in the present decade, inflation has brought havoc to some Latin American countries from which the governments continue the struggle to recover.

China

Inflation is not a modern phenomenon. The Chinese, who used paper money in lieu of metallic currency a millennium before the first issues circulated in Europe, had, by the fifteenth century, experienced several unhappy incidents resulting from unrestricted note issues. Lessons learnt from early inflationary experiences, however, were to no avail. China's hostilities with Japan, and involvement in the Second World War brought on an inflationary spiral which was not successfully subdued until the 1950s. All currency issues in post-war China suffered from inflation. The customs gold units, brain-child of Dr E. W. Kemmerer, were introduced in 1931. By 1948 denominations of 250,000 customs gold units were in circulation (although still dated 1930). The new currency law of 1948 brought into being the gold yuan, this time pre-dated 1945. In spite of government promises that these issues would not exceed the 100 yuan denomination, within one year denominations of 1 million gold yuan were found in circulation!

The vertical customs units notes are attractive and easy to come by at just a few pence each. The early Chinese inflation notes of the fifteenth century, issued by the Mongols in order to finance the expansion of their Empire, are impossible to obtain. But the vast quantity, the colourful designs and easy availability of the modern issues make Chinese notes easy to collect.

France

The French Revolution and the subsequent issues of assignats was the first crisis to bring chaos through inflation to Europe. Based on John Law's thesis of 'money backed by land', the notes were initially issued by the French Revolutionary National Assembly in April 1790.

John Law had already left his mark on the French economic scene some seventy years earlier. He had failed to have his theories accepted in his native Scotland, so he turned his attentions to France. A financial genius in his own right, he theorised that only a land currency would not be subject to a fall in value. It would remain unaffected by the fluctuation of precious metals backing the nation's wealth. Greatly assisted by fortune and friendship with members of the French government, Law was granted a charter and established the Banque Générale in 1716. Notes redeemable in coin of fixed weight were immediately issued. Within two years the Banque Générale became a state Bank. His success was beyond his own expectations. He was able to sell his company shares in ever-increasing quantities to the public; and he used these funds to make loans to the state, which became increasingly reliant on his enterprising activities.

In the spring of 1719, John Law's bank had 100 million livres outstanding in bank notes; in the late autumn the figure had increased to 800 million! On 5 January 1720 John Law became the Comptroller General of France, but the inevitable was to happen. As large stocks were put onto the market, the public began to panic. Even John Law's arrogant brilliance could not prevent a run on his Bank. In October of the same year, his note issues were formally declared to be invalid. The failure and desperation that followed needs no elaboration; and yet France, not unlike China this century, was to learn nothing from this harsh lesson. Within just a few decades the Republicans of the French Revolution again issued assignats, based on John Law's original theory of land-backed paper currency.

John Law's notes of the early 1700s bore his signature and are rare pieces, fetching over £150 ($300) each. The modern assignats of the French Revolution, however, were issued in huge quantities. Many types are easy to come by. The diverse range of assignats included notes known as 'Billets de Confiance' and 'Territorial Mandats', among which a few rare issues are to be encountered. Complete sheets of assignats have also made their appearance on the market, but these do not fetch much over the total value of the individual notes on the sheet.

While the Revolution was in progress, the devaluation of the assignats led to their redemption at only three per cent of their face value. By 1796 over 50,000 million assignats had been circulated. Government efforts to curb the inflation were to no avail. The plates were destroyed after limited quantities of assignats were issued, but

Playing card money was issued in 1792 during the French Revolution. The French 'Billets de Confiance' were a repeat of the earlier Canadian playing card money. (55 × 83 mm)

new notes were still printed.

The Territorial Mandats were intended, originally, to replace assignats, but these, too, suffered from uncontrollable depreciation. Some were no longer worth their own face value by the time they had been printed. It is understandable, therefore, that they are today obtainable at reasonable prices. With a few exceptions, the expensive French assignats should not exceed £25 ($50).

Canada

Jacques de Meulles was the Intendant of the French garrison in Quebec at the end of the seventeenth century. Having exhausted his own funds and borrowed all he could from friends, he still needed money to pay his garrison in the new French colony. The French Government had failed to supply coinage and this led to near mutiny by the troops. Force of circumstances, and De Meulles' enterprising ingenuity, led to the first paper money of Canada in 1665, written out on the back of playing cards which had been requisitioned from the soldiers and civilians. The blank reverses of the cards were filled in by hand for denominations of 15 and 40 sous and 4 francs. They were dated 1665 and signed by the Intendant. The issue had been in breach of the royal prerogative and Louis XIV ordered the destruction, after redemption, of all the notes in existence; thus there appear to be no surviving specimens of the first issue.

Further issues, however, circulated right through the last two decades of the seventeenth century. The money made an appearance each time shipments of coins from France were delayed. Furthermore, whereas gold and silver were being quickly exported from Canada, the card money, in its durable state, continued to circulate for over forty years. New issues on playing-card paper were printed and circulated at almost regular intervals up to 1719. Throughout this period they were gradually being replaced by ordinary cardboard notes. By 1720 a total of 1.6 million livres in cardboard notes was in circulation in the province.

In an attempt to control the obvious inflationary trend, the colony was authorised to redeem all the card notes – but for only half their face value!

The practicability of these notes led to additional issues being placed in circulation as late as 1741. When Britain ousted the French authorities in 1759 several contemporary issues were still in circulation. Some of these are still obtainable but highly priced. The ordinary cardboard notes, which succeeded the first playing card issues in the first decades of the eighteenth century, are priced above £150 ($300). The 12 and 24 livre cardboard issues of 1729 are very rare indeed, valued in EF condition at over £400 ($800). The 1728 playing cards themselves never come onto the market.

At the end of the century, the French, who had improvised with revolutionary assignats by printing them on all kinds of material from parchment to the back of book plates, also used playing cards for the 'Billets de Confiance'. The 1790 issues are not of course as rare as the Quebec counterparts but still highly sought after, priced at about £150 ($300). The majority were destroyed in accordance with a government law passed in November 1792.

Mexico

President Pastrana of Mexico claims that the spirit of the 1910 revolution is the inspiration for modern Mexico's dynamism – to the exclusion of the inflationary spiral of the period no doubt! Economically, the revolution was the worst ever experienced in the Latin American sub-continent. Francisco 'Pancho' Villa in the north and Emiliano Zapata in the south were the revolutionary patriots who have been compared to Robin Hood in English legend. They were both greater revolutionaries than they were economists and created havoc with the country's finances.

The paper money of the Mexican Revolution proper was first placed into circulation in August 1913. At this time Zapata and Villa, as heads of their respective movements, were authorised to issue notes and strike coins; they both decreed that the circulation of their currencies was to be compulsory in their respective territories.

It appears that the Mexican inflation was a deliberate step taken by General Victoriano Huerta. When he came into power in August 1913 he ordered the immediate closure of all banks. As a result, the paper money already in circulation remained without backing. The country's coinage in the meantime began to disappear from circulation and cardboard issues made an appearance.

Pancho Villa had been greatly frustrated by the ever-increasing cost of essential goods in the area under his control. He was, apparently, unable to appreciate that his own inflationary measures had brought about the situation. To finance further his revolutionary troops millions of pesos in notes were placed into circulation by him. These issues are known as the 'Sábanas de Villa' – 'Villa's Bed Sheets' – so called because they bore his name and increased in size in relationship to the denomination. Complete sets of eleven Villa notes sell for approximately £75 ($150).

The cardboard issues of the period are very easy to come by and although quite unattractive in appearance, they are historically fascinating material.

USSR

While Mexico was in the very midst of its Revolution, the First World War was raging in Europe. Russia's war effort in 1914 was costly; and what better way of 'creating' money than by resorting to the printing presses? Over 1,600 million roubles were in circulation in the country during the early months of 1914. By September the figure had reached nearly 3,000 million roubles! The intrinsic value of coins, mostly silver and gold, had led to the speedy hoarding of the metallic currency. In order to gain public confidence the government decided to try to deceive the people into believing that the currency was sound and long-lasting. They did this by printing notes from plates originally used by the Tzars, and therefore still dated 1898! This only led to further confusion.

The provisional government which came into power on 2 March 1917, was able to do nothing with the chaotic state of affairs. Strikes and demonstrations by the working class and Lenin's demands imposed on the provisional government caused additional trouble and finally led to the Government's downfall.

In April 1917 the War debt alone amounted to a fantastic figure of 30,000 million roubles, and yet the Russians needed more money. When the coalition government was formed on 5 May 1917, 1 rouble was worth no more than 27 kopecks!

The rapidity with which the paper issues ordered from the United States and England had to be put into circulation forced the government to issue the notes in uncut sheets, without serial numbers or signatures. David Atsmony of Tel Aviv, Israel, told me of his recollections as a youngster in Russia. Under the provisional government his father used to bring wages home in bundles of uncut kopeck sheets; it was David's duty at home to cut the sheets into individual notes as the need arose!

Russian notes of the period are available to the collector everywhere because many Russians escaping from the country filled suitcases with notes to take away with them. They had hoped to be able to redeem them somehow, in the West; instead they found themselves with thousands of pieces of worthless paper. Ironically, these notes are now beginning to have some collector's value. They have a wide variety of signature combinations which can form the basis for a large collection. The inflation also led to the issue of interest-bearing bonds. These were later overprinted for circulation as legal tender currency. They form an integral part of a Russian collection of the period.

Germany

The German inflation followed closely on the First World War and was the most obvious evidence of the after-effects of world conflict. Unlike the Greek and Hungarian war-time inflations in the 1940s, Germany's economic depression and final catastrophe after the First World War have been publicised a great deal in recent times in relation to our own economic situation.

Inflation had been soaring in Germany from 1919. When it reached its climax in 1924 paper money had been issued for denominations of 100 billion marks, equivalent at the time to £7. Year by year and month by month the mark was dropping in the foreign exchange markets to incredible lows. By December 1922 the dollar was worth 18,000 marks!

The whole output of Germany's industry, her only source for foreign currency income, had been expropriated by the French, whose troops physically occupied the industrial sector. In retaliation the German Government ordered a general strike, prohibiting labourers to work for the French. The workers agreed but they still had to be paid. The solution was again found in the printing presses. The established pattern followed and chaos, gradually but firmly, took a grip.

The late Dr Arnold Keller of Berlin, a leading world authority on paper money, has written of his experiences during this period in many numismatic articles; he has recounted for example a stay of several days in a hotel in Berlin in 1921 which, all inclusive, cost him 15 dollars. Another Berliner, speaking of the period when the inflation was at its peak, recounted:

'One fine day I went to a café by our house and ordered a cup of coffee which was costing 5,000 marks. I sat down and read my paper and had my cup of coffee and an hour or so later asked for the bill. The waiter came and gave me a bill for 8,000 marks. I said "Why 8,000 marks?" "Oh well, the mark has gone down!" he replied. The mark had changed, in the meantime, so much that the price of one cup of coffee had gone up by 60 per cent within the hour.'

Factory workers were being paid regularly during the day. Paymasters collected the wages from the banks, in trucks and by cartloads! When pay time came, the employees were assembled in the forecourt and, standing on the lorry, the cashier called out the names of individual employees, throwing at them bundles of bank notes. As soon as the workers were able to lay hands on the money, they rushed to the nearest shop and purchased *anything* still available for sale. A policeman's widow was awarded three months of her late husband's salary. The woman carefully checked all the papers and forwarded them to the required destination. There they were rechecked, rubber-stamped and returned to her with the sum due. By the time the money was paid, all she could purchase with the total amount were three boxes of matches!

Such stories of depression are innumerable. The banks notes handled by the German people were literally not worth the paper they were printed on. They are also among the cheapest notes available to the collector today.

Hungary and Greece
Hungary and Greece suffered equally and almost simultaneously the inflation inflicted upon them by the advent of the Second World War. Greece, beset first by the Italians and then the Germans, was to find herself in 1945 in a very weak position and at the mercy of the world. The incredible hyper-inflation in Hungary led to the issue of bank notes with grotesque denominations and cyphers that were beyond pronunciation by the ordinary man in the street.

Four years after the outbreak of the War, the total currency in circulation in Hungary had quadrupled. The first series of notes, put into circulation in July 1944,

By 1944 the Greek inflation notes had reached the 100 billion drachma denomination. The 5 million drachma was considered small change at the time! (62 × 140 mm)

were comparatively modest, but by the time the inflation had reached its climax in July 1946, the new monetary unit called the forint, was equivalent to 200 million adopengo (a 2 followed by twenty-nine noughts!). The Central Bank of Greece, too, on 20 October 1944 had issued its highest denomination inflationary note equivalent to 10 billion drachmai.

The inflation notes of both countries are extremely easy to come by. Some very high denomination Greek Treasury bills, issued in October 1944 at government branch offices, are rarity exceptions priced at over £20 ($40).

The Greeks also produced an interesting thirty-two page bilingual booklet entitled *Financial Facts of World War II*. A brief description of the devastation brought about by the Nazis and the Fascist occupation is included, with a complete collection of the twenty-five Greek inflation notes dating from 1941 to 1944. A collectors' item in itself, the book occasionally comes up in auctions priced at about £12 ($24).

CITY EMERGENCY NOTES (NOTGELD)

Soon after he begins the collector will encounter the word 'notgeld'. He will be told that notgeld are inflation issues of German cities which circulated after the First World War. This is not quite correct. The literal translation of the German word is 'money of necessity'. The exact definition of the word, however, when used as a notaphilic term, has not been finally determined. Dr M. R. Talisman has suggested that notgeld be defined as 'that medium of exchange issued in times of emergency by sponsors *other than* the duly recognised government or its representatives in that country'. Such a definition would include all the issues which are about to be considered in the following few paragraphs.

Although the word 'notgeld' is of German origin, its use need not be limited to the city issues of Germany. Austrian, Spanish and French notgeld, among others, have also circulated in times of emergency.

Germany

The best known and most popular notgeld are those of the 1920s issued in German cities. At the outbreak of war, in 1914, the German populace began hoarding coins, losing all faith in paper money issues. The coins which were fast disappearing from circulation were mostly small denomination fractional values and before long individual citizens were facing the problem of being unable to obtain change when paying for anything in paper currency. A number of cities took the initiative and began to print their own small denomination notes for local circulation. The 50 pfennig, 1, 2 and 5 mark notes were the first to appear. These original notgeld issues were crudely printed and only intended as a provisional measure; they were not legal tender as they had been issued without government consent. They were, however, withdrawn by the municipal authorities within a few months. 'Loan notes' had been put into circulation by the Government in order to alleviate the shortage of currency, and the necessity for notgeld had disappeared.

A number of German citizens had by then begun to collect the different issues of some of the cities in spite of the distinctly unattractive designs. The early withdrawal of the notes whetted the appetite of the contemporary collector, who began to create a market for the issues. Within a short time some 462 cities placed into circulation a total of well over five thousand distinctive notes; and this was *before* inflation had taken a grip on the economy! As a result of the rise in the value of silver in the first months of 1916, a new shortage of coin occurred. The cities, guided by experience, immediately began to print their own issues.

A far more attractive range of designs was chosen; many of the city authorities began to receive enquiries from collectors regarding their issues. It became clear to the municipalities that the sale of notgeld to the collector could be valuable revenue; thus an even wider range of subject matter was used on the notes; they were now being colourfully printed on silk and linen and leather, depicting every aspect of life and myth.

These issues are seen in retrospect to have been justified after the War; but at the time, many cities did not need to issue notes and only did so to satisfy the collector. On 17 September 1922, a general prohibition was imposed on the issue of notgeld. Some, which had been printed prior to the prohibition but had not yet circulated, made their way into the hands of collectors. It has been reported that no less than eighty thousand different types of notgeld were issued in Germany alone. The late Dr Arnold Keller had a collection nearing fifty thousand notes; many collectors still regularly report on previously unrecorded issues. Literature on the subject matter is abundant.

Large notgeld collections, comprising several thousand notes, come onto the market regularly and sell, comparatively cheaply, at £50 ($100) to £200 ($400). Individual notgeld issues can be purchased at a few pence, but one must not be misled into thinking that there are no rare issues whatsoever. The first true notgeld, of 1914, are comparatively difficult to find. Some are priced at over £40 ($80); the issue of Alsace is an example. The city was partially occupied by French troops and government notes could not reach the city so the authorities issued their own notes. Most of the issues were redeemed at the end of the War.

There are many curiosity items among the German notgeld issues, famous among them the 1916 Cologne notes on which the signature of Adenauer as mayor appears.

Germany had a notgeld issue much earlier in its history. During the Franco-Prussian War, emergency notes dated 31 July 1870 were issued in the city of Kaiserslautern. They were signed by the city cashier and the mayor. These issues, authorised by the Government of the Palatinate, were intended to aid industry. Notes for 1, 2 and 5 gulden were issued for a total value of 100,000 gulden, with repayment guaranteed in South German currency three months after cessation of hostilities. These are true notgeld issues and very expensive. Their values range between £75 ($150) and £100 ($200) each.

Notgeld were also circulated in some German cities in the last days of the Third Reich. As the Second World War was reaching an end an order was sent to the authorities of districts that were cut off, permitting them to issue notgeld. Hamburg and Stuttgart issued emergency notes of 1 and 50 marks in April and May 1945. Some have survived; others were intentionally destroyed by Allied military authorities after Germany surrendered and were replaced with occupation money.

The Hamburg issues of the same period are all said to have been destroyed by the British occupying forces. Second World War notgeld are priced at over £5 ($10) each.

Austria

The emergency city issues of Austria were very similar to the German notgeld. In general, fewer Austrian notgeld were issued and they have been less well documented; consequently, their popularity is not as high as their German counterparts. Because of the geographical and political proximity of Germany and Austria, their notgeld issues were similar. The basic difference – which aids the collector in identification – is the denomination on the note. The Germans used the pfennig and mark monetary unit, whereas Austrian notgeld were issued in hellers.

Inflation had the same devastating effect on the notgeld of Austria and Germany, as it had on the normal government issues. The words '100 milliard mark' feature on some notgeld!

Dealers' price-lists, which are always a good guide, have appeared listing emergency notes by subject matter: propaganda, history, poetry, humour, Martin Luther, for example; each could form a theme for a collection of notgeld. (City issues should be collected in crisp condition only.)

Left: At the end of the German hyper-inflation, the gold mark was established at 4.20 to the dollar, in November 1923. The first notes were circulated in the same year. (62 × 115 mm)

Below: Austrian notgeld issues of the 1920s were almost identical to the German issues except for the heller denomination. They are extremely common, and come in a variety of sizes, shapes and colours. (46 × 72 mm)

France

Coin hoarding by the people and the allocation of available metal to the War effort, led to the issue of substitute emergency currency on cardboard and paper. The chambers of commerce of many French cities issued fractional notes within weeks of the outbreak of the Great War. Departmental authorities soon followed suit; some were forced to do so under German occupation.

Most notes are dated between 1914 and 1918. Mr Fred Philipson, 'grand old man' of the IBNS, stated for example, that the 1915 Bouvignies issues of 100 and 1,000 francs 'may well be listed among the world's most rare issues.' Because French city notes were emergency currencies intended for circulation, and not directed at the collector, they are much rarer than the German counterparts. Some municipal issues were printed on circular pieces of cardboard to resemble the coins which they were replacing. This is in strong contrast to the oddly-shaped German notgeld issues, printed in order to give wider variety to the collector. Although the different city notes in France continued to circulate after hostilities terminated, they were not affected by inflation. Thus, both the variety and quantity of French notgeld are more limited than German and Austrian issues.

Early French emergency notes circulated simultaneously with the German issues in the Franco-Prussian war in the 1870s. Many cities that were to issue notes in the First World War had done so before Napoleon III's surrender to the Prussian forces in Sedan. The issues are rare items, valued at over £45 ($90). Some collectors have been aiming at matching the notes of the same city during the two different periods of issue.

Spain

Several thousand different emergency issues were put into circulation during the Spanish Civil War, from 1936 to 1939, by both sides in the conflict. The territories still controlled by the Republicans, which had been cut off from their source of supply, issued notes under the government's auspices.

The Bank of Spain formally issued notes for the cities of Madrid, Bilbao, Gijon and Santander. Simultaneously other Republican authorities circulated notes with government backing. They were printed on low-quality paper and saw heavy use; consequently, they are not easily found in good condition. They are still quite common, and most are not priced much over £4 ($8). A collection of the Bank of Spain issues of the Civil War can be completed at reasonable cost.

The vast majority of the profuse issues of the Spanish Civil War are those of the Anarchist Popular Front movement: the centre and left parties which brought General Franco into power. Before the outbreak of the Civil War secret organisations had been formed and under an organised administration notes were issued in the different areas which fell into the hands of the anarchists.

Municipal and Political Corporations issued internal currencies to alleviate the shortage of money, this time brought about by the government's withdrawal of coins from circulation. Soon these issues began to be accepted as currency for general circulation within limited areas. Most can easily be recognised by the initials 'FAI' or 'CNT' which appear on the notes, standing for the Iberian Anarchist Federation and the National Confederation of Labour, the two bodies to which the anarchists were affiliated. A catalogue on all these issues is in preparation by Kenneth Graeber of the USA.

The issues so far considered are all examples of the more popular emergency currencies. They were put into circulation by authorities directly associated with specific cities or towns. Similar issues, less well known and more difficult to collect, belonging to Hungary, the USSR, Rumania, Tunisia and many more nations were also circulated. The collector should enjoy the research involved in order to cater for his own interests.

CAMP MONIES

Internment camps, be they prisoner-of-war or concentration camps, frequently derived their names from the towns at or near which they were located. Internment camp monies which have been defined as 'currency issues for special-purpose circulation within limited boundaries imposed on the interned', comprise a surprisingly wide range of paper notes. The most obvious to come to mind are the military prisoner-of-war camps; paper money has also been issued for use in concentration and labour camps and in displaced persons' camps, all of which would more correctly be classified as 'political' rather than 'military' issues. Furthermore, some such issues were intended for use by the officials running the camp and not by the inmates. In any case the definition given easily accommodates these qualifications.

The main reason for issuing internment camp money is the isolation of the prisoners. In the case of POW camps, one of the overriding considerations has been the prevention of escape. Japanese POW camps, which were formed in the first two years of the Second World War, were run on ration cards and never issued any internal currency at all.

Many internment camp monies cannot be considered 'money' in the accepted economic sense of the word. These may often be referred to as 'scrip' or 'tokens', and other terms have been put forward to describe them. Many notaphilists do not collect these items, so their values are entirely dependent on the individual collector's interests.

Prisoner-of-War Camps

Considering the fact that paper money in Europe came into being only some 250 years ago, it is quite amazing that POW monies printed on cardboard and paper were in evidence as far back as the Seven Years War in 1756. These first recorded issues, circulated by the Austrians and Prussians, are impossible to encounter; they were followed, however, by camp issues during the Napoleonic Wars. The American Civil War in the 1860s also brought about similar notes. Most are extremely rare, unlikely to be offered for sale. On the other hand, POW camp notes of the Franco-Prussian War of 1870 come onto the market priced at about £175 ($350). Boer War issues (1899–1902) for POW camps (established by the British in South Africa), however, are fairly readily available to the collector. They circulated concurrently with the siege notes mentioned earlier. Recently, a small hoard was discovered, all in crisp condition. A set of three South African POW notes of 1900 was being offered for sale at about £10 ($24) in 1973.

During the course of the First World War over one million prisoners had been interned in camps; it is only natural, therefore, that huge quantities of internment currencies should be available to the collector. Belgium, France, Italy, the Netherlands, and many more European countries issued such notes; Britain had camps both in Egypt and France and the issues are among the rarest of the First World War period. The USA, too, had POW camps established in France; the notes are still encountered at about £55 ($110), half the value of their British counterparts. Irish Free State POW camps were established in 1921, intended for prisoners during the Irish struggle for independence from Britain. Their note issues are undoubtedly the scarcest of POW notes. They all fetch well over £100 ($200) each.

POW camp money in the Second World War was different in one important aspect from the issues of the previous war. Notes of a standard size and colour were printed and used both by the Germans and British authorities for all camps. The issues were merely overprinted with the name of the camp within which they were intended to circulate. The first camps of the Second World War were set up by the Germans for Polish prisoners in September 1939 just a few weeks after the Wehrmacht overran Poland. The British and German POW camp notes can be obtained in sets. The latter are easier to come by and many have the swastika on them. They fetch £25 ($50) per set of four. The British notes cost about the same.

The British also issued notes for the POW camps in her colonies. The ones that appeared in India and Egypt, for example, are exceedingly rare, in the £100 ($200) range.

The United States POW camp issues are the cheapest of the internment camp notes. This is probably because they do not look like real paper money. The ticket-shaped notes, whether issued in Japan, Italy or Germany, do not exceed £15 ($30).

Concentration Camps

The Second World War brought with it a type of internment camp hitherto unknown: the concentration camp, where millions of innocent prisoners of both sexes and all ages were systematically and indiscriminately put to death. Some hundred camps were established by the Germans, mainly for political prisoners. The majority of the concentration camp issues were intended for use by the SS guards in the camp canteen. The best known issues, however, are the exception to this general rule.

Theresienstadt and Litzmanstadt were two ghetto camps for internment of Jews. The notes were receipts exchanged for the German currency that had been confiscated from the interned. They comprised part of a monetary system intended to give a pretence of normality to the ghetto. Observers from official bodies of other governments enquiring as to the whereabouts of prominent Jews could thus be satisfied of the supposed well-being of the people involved. Complete sets of notes were issued in both camps; a facsimile of the signatures of the respective Jewish elders of each of the ghettos appeared on the notes. The Theresienstadt issues are the better known because large quantities, from a Swiss source, appeared on the market in recent years. The set of seven notes depicting Moses holding the Ten Commandments can be obtained for £15 ($30). A complete set of the notes of the Litzmanstadt concentration camp is far more difficult to encounter and has been sold for £90 ($180). The 50 pfennig, 1 and 20 mark notes often appear as a set, but these three are worth no more than £1 ($2) each, at the most.

Famous camps like Buchenwald, Dachau, Ravensbruck, among others have issued internal currency for use by the SS guards in the camp canteens. The prices of the individual notes have increased in recent years, which now fetch over £40 ($80) each.

Some concentration camps were set up in France by the Vichy government, and the notes reported are all very rare issues.

Other internment camps

In spite of strenuous efforts by the Russians to deny the existence of 'correctional labour camps' in the Soviet Union, information about these has infiltrated to the West. One must presume internal camp money was issued. Similar unconfirmed reports have come of camps established in places as remote from each other as Cuba and China. It is the duty of a collector to keep an eye open and report any information that may be made available regarding such issues.

Internal money issues in criminal penitentiaries and other civilian prisons have

Concentration camp money was used by the German forces in local canteens. The 10 pfennig note from Oranienburg is of interest because this was the camp later known as Sachsenhausen, where pounds and dollars were forged during the Second World War. (70 × 95 mm, actual size)

also been excluded from this section. Many types of issue have been reported from such places as the Sing-Sing Prison in New York and Holloway Prison for women in London. These are all coupons and paper chits not intended for circulation. We must await an increase in interest in this field of collecting before greater emphasis is placed on these items.

Displaced persons' camps, established after the Second World War, were less restricted establishments; nonetheless they were confinement camps and the issues placed into circulation within their boundaries are collectable items, some of them keenly sought after. The most valuable of these are the American Jewish Joint Distribution Committee notes from camps that were set up in Cyprus. These were established for Jewish refugees fleeing from war-torn Europe on their way to Palestine. The camps, in existence from August 1946 until March 1949, were closed down after the establishment of the State of Israel. The text on the notes is significant. It states: 'Good for purchase in the canteen in Cyprus or for exchange for cash in Jerusalem.' The set of three notes in the shilling denomination is now considered to be very rare, priced at over £150 ($300).

Similarly, Hungarian refugees that fled from Budapest across the Austrian border in 1956 were placed in temporary camps, where internal money was in circulation until it became unnecessary as a result of American generosity. All essential goods were given free of charge to the refugees. The notes were mostly destroyed.

Austria, Germany and the Netherlands are among other known nations to have set up camps to accommodate displaced persons after the Second World War. Many more undoubtedly remain to be recorded and so the alert notaphilist has an opportunity to make new discoveries.

MILITARY ISSUES
Military notes are taken as those issues intended for use by regular or irregular military forces, for specific military efforts whether peaceful or belligerent. Although such notes have been issued in limited quantities throughout modern military history, emphasis is placed on the Second World War.

Pre-Second World War
The earliest issue on record which could be termed a military paper item is a pay warrant made out in favour of Sir Francis Godolphin, commander of the garrison on the Scilly Isles off Cornwall. It is dated 14 March 1635. It was signed by the then Treasurer of England, Francis Baron of Cottingham, and addressed to Richard Kinsman, the Royal Auditor for the area. The note requires the payment to Sir Francis of a sum of £400 'for and towards the payment of himself and his soldiers as Captains of the Garrison of the Isle of Schilley'. This, of course, is an extremely interesting and unique item and any hope to possess such a note in one's collection will turn out to be futile.

Some early military notes, however, are available. Unconfirmed reports indicate Japanese military issues distributed to soldiers fighting in the Sino–Japanese War of 1894–95. More accurate information is available on the Russo–Japanese War at the beginning of this century. These notes in the sen (100 sen = 1 yen) denomination were issued by the Ministry of Finance of the Imperial Japanese Government. They are comparatively easy to come by at prices not exceeding £25 ($50). These issues have four Japanese characters on the obverse indicating that they are intended for military use. Later issues, dated 1918, had the text and value of the note overprinted in Russian vertically along the side of the note.

All the major belligerent parties in the Great War issued military currencies for use by their respective armed forces. The collector will not have difficulty in identifying these issues, by checking the date on the note. The French issues show portraits of French soldiers and the word 'Armée' appears on many of them; similarly, the Germans issued a series of notes from 50 centimes to 100 francs, dated 1915, for occupied France. It is the Second World War military issues, however, that have caught the imagination of the majority of 'militaria' collectors today.

Allied Military Currency

The commonest and best known of the Second World War military notes must be the small square French Invasion notes. These caused a great deal of indignation on the part of De Gaulle who was in exile in London in 1942, because the French Tricolore appeared on the reverse of the notes, and the obverse stated that they had been issued in France. The new issues were soon amended. These, and many more of the Allied military currencies, were printed in enormous quantities to satisfy the requirements of the Allied armies numbering several million soldiers. It is not surprising, therefore, that the issues are very common; most can be obtained in perfect condition at just a few pence each. These same series were issued for use in Austria, Germany, Italy and Japan and the respective issues appear with the corresponding languages on the notes.

When Germany, at the end of the war in Europe, was divided into the British, American, French and Russian sectors, each had its own notes circulating. These had been printed in the USA and the USSR. The identical looking notes can be identified for each sector by the prefixes to the serial numbers.

The initial effect of the outbreak of war on British currency was a repetition of the steps taken at the beginning of the First World War, when a shortage of coins and paper led to the conversion of postal orders to legal tender issues. But these were hardly military notes. Military issues were circulated by the British military authorities in Tripolitania (Province of Libya) and a separate series in other areas of North Africa and Greece. It is by force of circumstance that British issues are scarcer to find. Greater demand for these notes has been created by the numerous collectors throughout Britain and the Commonwealth. A set of British military authority invasion notes will fetch up to £20 ($40) and some of the individual high denomination notes, such as the £5, £15 ($30) each. British military issues also include the NAAFI (Navy, Army and Air Force Institutes) 'Special Vouchers' from 3d to £5, used in canteens by Army personnel.

Towards the end of and immediately after the War, the Allied countries issued a series of notes for the liberated territories; these may be termed 'administration issues'. The United States issued such notes in Greenland, in 1941, in the skilling denominations; these are comparatively difficult to come by. Similar issues by the United States Army for circulation in Japan are the longest lasting military notes ever issued. They were in use from 1945 to 1958. Several distinct series exist and bear different dates. Most are priced at under £10 ($20). The cheapest issues of the administrative notes are those of the British military administration of Hong Kong, issued in September 1945 and portraying King George V. The two high denomination notes, the 5 and 10 dollar issues, were printed under emergency conditions and in one case electrical power was obtained from a British submarine stationed in Hong Kong harbour. (These should not be confused with the Hong Kong emergency notes of May–October 1941 signed by R. R. Todd, the Financial Secretary, who at the time acted for the Government of Hong Kong.)

The French also had profuse issues of administrative notes for Africa and Oceania. These include issues for French Equatorial Africa, French Guiana, West Africa, Indo-China and Algeria, Tunisia and Morocco (the last being known as the 'De Gaulle' notes). The rarest are the 1,000 franc issues exceeding £50 ($100), but most are obtainable at under £4 ($8).

Within the category of Allied military currency must be included the several notes issued by European countries after liberation. Belgium, the Netherlands and Luxembourg are just three examples. The 1944 issues of Luxembourg, which had a German text on the reverse, were circulated with the text obliterated from the note. In the Netherland Indies, Allied troops were paid with the 1943 liberation currency in the rupiah and gulden denominations. In the Netherlands itself troops were paid with liberation notes which, however, were withdrawn by order of the Government after circulating for only three months. Because all those issues were of a transitional nature they were often used for several consecutive years. The higher denominations are the more expensive.

In mid-1942 a special series of US notes was issued as military currency for the

The liberation notes of 1943 were withdrawn by the Netherlands Government a few months later. The high denominations were printed in small quantities and are therefore rare. (72 × 150 mm)

Pacific area. Ordinary dollar issues of the 1934 and 1935 series were overprinted on both sides with the word 'Hawaii' and replaced the issues throughout the Pacific. This was done as a precaution against any of the currency falling into Japanese hands. The interesting notes of this series are the 'freak' issues. The 20 dollar note appeared with the word 'Hawaii' reversed, and on some notes the word had been completely omitted from the reverse. These fetch prices in excess of £45 ($90) but most of the normal overprints are valued at between £10 ($20) and £30 ($60) depending on the denomination.

Among the more exciting liberation issues are the Philippine 'Victory' series. General MacArthur personally led the US marines who formed the beachhead at Leyte Island on 20 October 1944. The soldiers who landed were carrying in their pockets their monthly wages with a special morale booster: the notes had all been overprinted with the word 'VICTORY' on the reverses, in bold black capitals. These were the No. 66 series about to go into circulation for the first time. The notes happened to be the last of the Treasury Certificates that the US were to supply to the Philippines and they were identical to the previous notes but for the overprint. The notes were also overprinted 'Central Bank of the Philippines', a formal indication, for the first time, of the recognition given to the Central Bank by the United States. These notes, originally issued in 1944, continued in use until 1949. Very many were kept back as souvenirs by both the local populace and the American GIs. Thus, the notes are comparatively common.

Axis Military Currency

On the Axis front, the Germans, in almost sinister preoccupation with exactitude, devised a system for the distribution of their military currency which greatly contributed to the welfare of the German soldier but not of the civilian. These were the 'Behelfszahlungsmittel', the undated Auxiliary Payment Certificates, issued as of August 1942. The notes were distributed to soldiers in Germany and were worth ten

The Hawaii overprint is probably among the best known of the American wartime issues. All the 1934 series circulating in the Pacific were withdrawn and overprinted as a precaution against their falling into Japanese hands. (67 × 157 mm)

times their face value. With 1 Reichsmark 10 Reichsmarks' worth of goods could be purchased, but the shopkeeper had to subsidise the difference!

The other German Army issues, those of the Wehrmacht, are the 'Reckoning Notes' from 1 to 50 Reichsmarks given to troops prior to their departure for occupied territories, where the notes were exchangeable for the local circulating currency. The notes, although dated 15 September 1944, were not issued until January 1945; they are therefore easily available.

The auxiliary notes, too, with one exception, are within the same price range, approximately £5 ($10). The exception are the 1940 auxiliary certificates used in France. The 1 Reichspfennig note, the only denomination known and printed on white paper, is worth about £25 ($50).

Japanese notes for use by the military forces in the Second World War were limited to the 'Bird and Dragon' notes, so called because of the main design on the issues. The vertical notes were issued in 1937 and were overprinted in Japanese characters, indicating their purpose as military notes. They circulated primarily in China and are all extremely common.

All remaining Axis military issues can be classified under 'Occupation Notes'.

The Japanese dream of complete rule in the Far East within the framework of a 'Great East Asia Co-Prosperity Sphere' must have appeared very near realisation to many of the Axis powers in mid-1943. Japan was at the peak of her military campaigns. She had successfully invaded British, American and Dutch possessions in the area. She had also destroyed, almost completely, the American Asiatic Fleet in Pearl Harbour and the Philippines, as well as the British fleet in Singapore. Some of

The Verrechnungsschein were used by German Wehrmacht troops for conversion into local currency in occupied territories during the Second World War. The swastika on the emblem to the left is not as common as one would expect on German notes of this period. (84 × 155 mm)

the colonies welcomed the Japanese, a fact which considerably hindered the Allied forces, and the US in particular, when they gradually began to recapture the territories.

Japan had issued occupation notes in local denominations in every invaded country. Many remained in circulation until August 1945 when the Japanese finally lost their possessions. Naturally, the exceptions make the rule. Some rare or even never-seen notes exist, but none likely to be sold for over £5 ($10) on the open market.

Collectors popularly refer to these issues as 'JIM' – the initials for Japanese invasion money. Abundant literature and easy availability have made JIM a popular subject matter. Japanese invading forces placed their money into circulation almost as soon as they arrived in the occupied territories. All issues had been previously printed and can be identified by studying the serial letters. The first letter is invariably the initial of the name of the country involved. The occupied countries were Burma, Malaya, Sumatra, the Philippines and Oceania. Almost all the occupation notes have a 'B', 'M'. 'S', 'P', or 'O' at the bottom left-hand side of the face of the note. The occupation notes were intended to circulate at par with local currency; but within months, as inflation began to soar and the Allies increased their successes in their counter-offensive, the Japanese occupation money began to lose its value. Attempts to force the circulation of the notes were to no avail, and forgeries, executed by the USA and locally distributed, merely added to the existing chaos.

A definite interest in 'JIM' notes has developed in the last few years, and an increasing number of specialised collectors are publishing material regarding the diverse range of series and serial letters in which the JIM were issued. This has

Jersey and Guernsey were the only part of the British Isles to be under German occupation during the Second World War. In 1942 a set of Jersey notes was issued by Edmund Blumpied. The 10 shillings is now valued at over £100 ($200) in crisp condition. (85 × 130 mm)

naturally increased the value of these notes. If the collector's interest is aroused by the prospect of collecting the JIM by serial numbers and letters now is the time to start. What may pass from one specialised collector's hands to another's at, say, £5 ($10) can probably still be obtained for just a few pence from a dealer.

At the other end of the scale price-wise are the Channel Islands (Jersey and Guernsey) issues during the German occupation. The 10 shillings and £1 in particular are expensive and very desirable items. The individual notes of the Guernsey set are more expensive. The cheapest note of either island will still exceed £4 ($8). The Channel Islands was specially commended by the British Government through the Bank of England in 1945. They were the first in the United Kingdom to resume normal banking functions after the War. They were the only part of the British Isles to be occupied – and yet, under the very noses of the German forces the Channel Islanders printed their own £1 notes dated 1945, putting them into circulation within hours of the end of the War.

Italian occupation notes are among the most expensive in the field. The Casa Mediterranea per L'Egito issues had been printed in 1941 but Italy was never able to circulate the notes as intended. A complete specimen set of nine was offered at £2,000 ($4,000) each in 1975, and three notes sold in the United States in 1972 for $850. A similar set of five notes, from 1 to 100 Egyptian piastres, was printed for the occupation of Sudan. Mussolini had originally intended to divide Egypt into two distinct occupied territories, Egypt proper and the Sudan, but this never came to be. Most of the notes were destroyed by Allied forces without ever having been circulated. They are very similar in design to the notes issued by the Italians for the occupation of Greece and the Ionian Islands. The Casa Mediterranea per la Grecia set go up to 20,000 drachmai, and a set of nine notes in crisp condition has been offered for sale at £400 ($800). The eight notes of the Ionian Islands are less

It is reported that the Italian occupation of Greece saw an extremely rare issue of 50,000 drachmai; this 20,000 drachma note, portraying Michelangelo's David, is the second rarest of the series and very expensive. (93 × 195 mm)

expensive, and change hands at £70 ($140). The Greek and Ionian issues were in circulation for about a year and were withdrawn by the time Italy capitulated in September 1943.

Poland was the first victim of Nazi aggression in the Second World War and she was the first to experience occupation money. The Bank of Poland notes of 1932 and 1934 were withdrawn by the occupying forces in 1939, overprinted in German 'For the Occupied Polish Territories' and then put back into circulation. The notes are comparatively common because the Polish populace refused to use them.

The German currency used for all the occupied territories was of standard dimensions. The Reichskreditkassen circulated in eleven countries. In Poland they were issued on 2 October 1939 by order of the Supreme Army Command and as the German armies advanced through Europe additional issues for the capitulating territories were circulated.

There is still some controversy regarding the German paper money issues for the occupation of Russia. In spite of confirmation of the existence of such notes, their printing by the State Printing Works of Berlin is being vehemently denied. The notes in question are outside the collectors' reach. It appears that the only existing set of three notes is in the hands of the USSR Treasury.

The swastika does not appear as often as one might have expected on German war-time notes. The notes issued in the Ukraine under occupation from 1941 to 1944, have the Nazi symbol appearing on the obverse, but they are the exception rather than the rule. In more recent times collectors of military and occupation notes were anxiously awaiting official announcement, in 1967, that Israel had issued special notes for the occupied territories after the 'Six Day War'. They were disappointed. Many, however, may not realise how close Israel came to putting into circulation such issues. Immediately after the occupation of the West Bank of the River Jordan, Israel decided to exchange the local currency for Israeli pounds (at a rate, incidentally, favourable to the Arabs) in order to enable life to continue as normally as possible. However, almost simultaneously, King Hussein of Jordan announced that all current Jordanian bank notes were to be declared obsolete. This, naturally, put Israeli monetary authorities in an uncomfortable position. Israel had by now accumulated a large quantity of Jordanian dinars. A decision was reached therefore to issue occupation notes for circulation in the West Bank and the printing of such notes was ordered. Notes for Gaza and the Golan Heights were also prepared.

By the time these occupation issues were ready for distribution international bodies, including the IMF, had pressured Hussein to revoke his earlier announcement. This was followed by Israel's decision not to issue its own occupation notes, but continue with the original plan of exchanging dinars for Israeli pounds. Not all the issues have been destroyed, and there is evidence of existing specimens in the Israeli authorities' hands.

There are many more issues within easy reach of the collector, who should keep in mind that the dividing line between the distinct types of military issue should not be too strictly adhered to.

Military Payment Certificates

As the world found its way back to stability after the War, the governments of countries with military forces still overseas made special arrangements for payments to their troops. Most were paid in the currency of the country in which they were based; alternatively, they were paid in the currency of their home country and given facilities to exchange their dollars, pounds and francs for the local legal tender issues.

Many military bases in Europe used internal vouchers and chits, but the US Government soon found it necessary to issue a series of special notes to all its military and authorised civilian personnel away from home. These issues were, and still are, referred to as MPCs – Military Payment Certificates. They can form an extremely interesting and varied collection. They have been issued since August 1946 in those areas of the world where there has been a special need for currency control, necessitated by black marketeering in most cases. Immediately after the War US forces were able to purchase dollar currency credits, with Russian-printed Allied military marks (in Germany, for instance) and a lucrative black market began to operate. Soldiers were effectively draining the Army budget by amassing huge profits well in excess of their normal pay. Neil Shafer in his *Guide Book of US Modern Currency* has reported that at its very height the 'overdraft' to the Army budget was a staggering 530 million dollars!

MPCs were used in the territories of twenty-one countries in which USA forces were stationed. That number has gradually diminished and Vietnam and Korea were the last two countries where the notes continued in use. The MPCs are collected by series, and as of October 1973 a total of thirteen series have been issued.

The smaller denominations of the later issues are, of course, more easily available. Most fractional notes (5, 10, 25 and 50 cents) can be obtained in good condition for under £2 ($4); with the exception of recent issues, they are rarely encountered in perfect condition. They were well used and very few were kept by discharged soldiers as souvenirs. The higher denominations increase in price in direct ratio to their face values. The 20 dollar, the highest denomination issued, can fetch as much as £45 ($90).

There is another kind of certificate, peculiar to the circumstances under which it was issued. In the period between Japan's capitulation, at the end of the Second World War, and the departure of the US forces from the area seven years later, the supreme command of the Allied Powers authorised the issue of Foreign Trade Payment Certificates. The purpose of these issues, which were discontinued in June 1950, was to accommodate visiting foreigners to Japan. The certificates were a temporary measure for use by visitors to pay their hotel and other bills after exchanging their own currencies into these certificates. They were issued in dollars and pounds sterling and were readily convertible into current notes of the respective countries when the visitor was ready to leave Japan.

The Foreign Trade Certificates are a type of MPC. They were issued in fractional denominations, as well as whole units up to 10 dollars, and were abolished with the introduction of the convertible yen system. Today collectors will pay up to £50 ($100) or more for a complete set of seven notes.

Guerilla and Partisan Issues

In contrast to the officially issued notes distributed by military authorities under government approval, many types of unofficial notes have been used. The five year period of the Second World War was fertile ground for such issues. Most of these were partisan and guerilla monies issued without backing and, usually, hurriedly printed in circumstances which did not allow for any standardisation. Almost without exception, the issuers relied on the support of the populace to circulate the notes. Many were coarsely printed bordering on propaganda issues; but they served their purpose, both financial and patriotic.

The discovery of large quantities of forgeries, and the recent publication of a comprehensive catalogue of Philippine currencies, have led to a great deal of interest in the issues of the guerilla forces of that country. The notes were first circulated during the Japanese Occupation, in 1942 and 1943, and were widely used at great risk to the possessor. The Japanese had decreed that possession of guerilla notes was punishable by death! The clandestine notes of the local provinces and municipalities were sanctioned by the Philippine Government in exile, and many were redeemed after liberation. The notes were intended for everyday use, and many of the smaller denominations were circulated in large quantities. They are, therefore, easily available at about £3 ($6). The higher values are much scarcer. There is only one known 500 peso note, hand-signed by members of the authorising board. The Philippine issues are the classic patriotic notes, successfully used in spite of strong opposition.

Yugoslavia is probably the most profuse of the issuers of partisan paper money with at least fifteen districts issuing several hundred notes in different denominations. Most were issued in 1944 by specially set up liberation committees, but were not considered as 'diminishing notes'.

The partisan issues of Montenegro, however, were. These, in order to force their circulation, had a certain percentage of the value deducted every month. They were crudely printed but portrayed some vivid scenes. The 1,000 dinar, for example, shows a wounded freedom fighter on horseback. Croat and Serb guerillas also printed their own notes, some with propaganda slogans on them such as 'Death to Fascism'. These are always a little more expensive.

The Civil Committee of National Liberation issued notes in northern Greece, in 1944, and colourful guerilla notes were issued in Italy by the Garibaldi Brigade, whose very name must have inspired many eager partisans to join the force. These and other similar notes are all evidence of a fascinating history of resistance to oppression. The expensive partisan notes of the Second World War are not usually priced at over £35 ($70).

CARDBOARD TOKENS, CHITS AND SCRIP

Cardboard tokens, chits and scrip have all been mentioned in passing. Many collectors are strongly against their inclusion in a paper money collection. The decision will finally remain with the individual. Some mention must be made, however, of the many issues on cardboard in improvised form.

Israel's kibbutzim – the collective farms – have internal scrip which is widely collected. They form an integral part of a non-metallic collection of tokens. Their use was widespread in the country between 1940 and 1954. They are still in use, to some extent, today. The earliest tokens of Israel were issued by the transportation companies under emergency. The World War and subsequent inflation, had led to a shortage of small change. When the bus fare was raised from 5 to 7 mils there was a critical shortage of 1 and 2 agurot coins. To alleviate the situation, bus companies such as Dan and Egged issued cardboard notes which were effectively 'change tokens' (as opposed to transportation ones). There was a further improvisation on similar issues after the devaluation of the Israeli pound in 1952, when Israel Red Cross tokens appeared for use as small change. Here the benefit from destroyed and lost tokens went to the institution and considering the short life span of the tokens, this must have led to considerable contribution to a worthy cause. The early issues of 1942 are expensive.

In war time, tradesmen – from grocers and bankers to barbers and taxi drivers – all issued personal tokens which circulated within small communities.

The Soviet Union, too, issued a great number of similar coupons for transportation, restaurants, delivery of goods and other purposes. These are cheaper than the Israeli issues but in line with most similar tokens; many can be obtained for just a few pence. Bank notes will have to be firmly established as a collectors' hobby before paper tokens can begin to be seriously considered; until then advantage should be taken of the easy availability of many of these pieces at very low prices, because of the limited demand. Within the framework of a collection of cardboard tokens would come all internal coupons issued by shops and stores, farms, mines and factories. A good collection can be easily formed at very small cost.

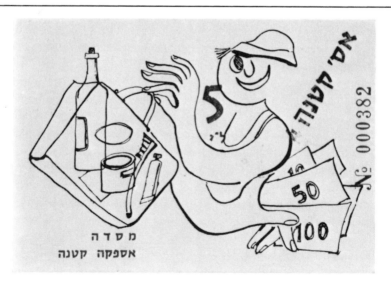

The 1,300 foot high fortress in Masada, Israel, was besieged by the Romans in AD 72. The card illustrated is a canteen chit, used as currency by visitors to the site. (69 × 99 mm, actual size)

Oddities

Bisected Notes

In 1944 an acute shortage of small fractional coins in Colombia (caused by illegal hoarding of coins for conversion into buttons and sale to tourists at inflated prices) led to the withdrawal of 1 peso notes dated 1942 and 1943. The Banco de la Republica de Colombia then overprinted each half of the notes, headed them 'Provisional Notes – half peso', cut them in two and recirculated them. These issues being provisional are more difficult to come by than similar bisected notes without government approval. The Colombian 'half peso overprint' recently fetched £25 ($50) in an auction.

The bank notes of the Ottoman Empire, issued near the end of the First World War, were cut and overprinted not only in half but in quarters of a livre! The denomination and signatures were overprinted on the notes. The one-quarter livre is among the rarest issues of the Ottoman Empire.

The bisection of a note to generate small change at times of necessity seems a logical action. However, the Greek Government in between the wars divided notes in a three-quarter/quarter ratio. The 1915 100 drachma note was cut into 75 drachmai and 25 drachmai. A Greek law of 25 March 1922 authorised the Government to float an obligatory loan which was based on the circulating paper currency. This led to the bisection of notes; half were recirculated at fifty per cent of their face value and the other half was held by the issuing authority, as a loan certificate. These notes were not overprinted, so their value is low.

Similar action was taken by Finland just after the Second World War. In 1946 Finnish fiscal authorities bisected note issues of 500, 1,000 and 5,000 markkaa, valuing the left halves, temporarily, at 250, 500 and 2,500 markkaa respectively. This applied to the 1939 issue of the 5,000 markkaa and to 1922 issues of the 1,000 and 500 markkaa. Again the notes were not overstamped; the right half of the note became a forced loan issue, the Government considerably benefiting by the temporary measure.

Notes have sometimes been cut merely to ensure their cancellation. These are usually mutilated and not worth collecting unless the original is a rare note.

Generally, the overstamped bisected note is scarcer than the original whole. The Australian currency of Fanning Island Plantation Limited is an exception; in a London auction in 1973 it sold for £110 ($220). At the outbreak of the Second World War almost the total population of the Fanning Islands (about 740) was working for Fanning Island Plantation Limited, a company handling the export of coconuts from the island. As the Japanese expansion in the Pacific increased, problems arose with the supply of essential commodities, including currency, to the isles, and employers were finding difficulty paying labourers. The enterprising local manager found a solution to the problem. With the help of the US military forces he arranged for £1 notes to be printed in Hawaii. A few weeks later one thousand notes were delivered and placed into circulation. With the ending of the hostilities in 1945, normal Australian currency began to circulate and the 'plantation notes' were withdrawn, except for a few left behind after being cancelled by bisection. The separate halves of these notes were used by the inhabitants as admission tickets to the local cinema! All the left halves are marked '1 shilling' and the right halves 'two shillings'. These figures indicated the entrance price to the theatre. The halved notes are being sold as a pair at about £15 ($30).

A note may also be bisected for safe transportation. Last century, issues being delivered to the Bank of England from its branch offices throughout the country were often halved and sent by separate wagons in order to ensure that they would not be attacked by highwaymen. A more recent example is the Palestine Currency Board

Left: A severe shortage of small change led the Colombian Central Bank in 1953 to withdraw the 1 peso notes of 1942 and 1943 and overprint ½ peso on each half; after bisection the halves were circulated until new coinage was minted. (55 × 56 mm, actual size; notes not in proportion to each other)

Below: The Fanning Islands were forced to issue their own currency during the war. After redemption these were cancelled by bisection, but the halves were used as cinema tickets. There are very few surviving specimens of the whole note. (80 × 150 mm)

bank notes. These were used by Israel as a Sovereign State after 1948, to back its own currency. The halved sterling notes were returned to England for redemption, in separate shipments on different dates. This had been previously arranged with the Crown Agents to whom the bank notes were being returned; it also partially accounts for the relative scarcity of the issues.

There are more examples in notaphilic history of such issues; they include Lajos Kossuth's Hungarian notes of 1848 as well as Polish and Austro-Hungarian issues earlier this century.

Stamp Money

Thirty-four countries throughout the world have used postage stamps as currency, but limited literature has been published on the subject. The comparative unattractiveness of the issues has undoubtedly been a major reason for their lack of popularity.

Stamps have been used in lieu of currency in times of emergency when there was hoarding of coinage or during a shortage of paper. Stamps were a good substitute because of the ease with which they could be introduced as a temporary monetary measure. The populace recognised in stamps their specified value and accepted authorisation to treat them as legal tender currency.

The overprinting of ordinary stamps as currency units was dismissed by the authorities because of the frailty of the paper on which stamps were printed. The protection given to stamps, therefore, consisted of printed cards, on one side of which instructions were given; on the other side, an actual postage stamp was affixed to the card.

Although these were the main types of stamp currency there have been a number of additional improvisations. The USA used cardboard pockets with denominations printed on the outside; the backs were often used for advertisements. In Germany and Austria stamps were slipped between two parallel slits on a piece of cardboard. Other countries protected the stamp in special transparent envelopes. With the exception of Ceylon and the Boer War issues in Rhodesia, the British Commonwealth has not issued any postage currency. The two examples mentioned, however, are extremely rare specimens, priced at over £35 ($70) each. Ceylon was not the only Asian country to use stamps as money. A Korean issue for 5 sen circulated when Korea was a Japanese province. The stamp is Japanese, but mounted on cardboard headed in Korean 'Special Postage Stamp'. It is priced at £5 ($10).

'Ayer's Sarsaparilla to Purify the Blood' was one of the slogans used on the back of the American encased postage stamps in the 1860s. The stamp was usually placed in a metallic container with a transparent front. The monetary value of the item was equivalent to the value of the stamp inside. Mr Ken Roberts of the United States, an expert researcher on the subject matter, has stated that a complete collection of the American encased stamps issued only in 1861 and 1862 would total exactly 222 items. Thus, it is not inconceivable that there may well be over one thousand varieties of the US encased postage stamps. Many of these are extremely rare, but the demand is limited and they can be obtained at around £10 ($20) which is well below their potential value.

Spain issued stamp currency during the Spanish Civil War. In 1938 a series of cardboard issues in fractional denominations circulated with government consent. First the stamp was drawn on one side but gradually, as several enterprises began using the currency, stamps were attached. Germany, too, in conjunction with many notgeld issues, placed stamp money into general circulation as did Turkey, when still under Ottoman rule, Uruguay and Greece, among others.

The commonest stamp monies on the market are those issued by the Russians. They reprinted postal stamps on cardboard with an explanation on the back of each stamp authorising its use as circulating money. Some of the stamps were further overprinted with the value in kopecks on the obverse. The same stamps were also used as normal postal stamps and are far more valuable as such, mainly because many of these issues form an established branch of philately, whereas they are only now beginning to interest paper money collectors. The Russian issues often appear in sheets, sometimes consisting of two hundred stamps. The complete sheets have been sold for about £15 ($30), while individual USSR stamps can be obtained for a few pence each.

Proofs and Specimens

A great deal of material associated with paper money is not in itself currency. Some of this material can be a desirable collectors' item. Essay-proofs and specimen notes are specific examples. Most paper money auctions held in London from 1972 to 1974, devoted the first part to a wide range of printers' proofs. They are almost identical to those used for philatelic purposes.

The hand-drawings of vignettes appearing on a bank note are undoubtedly the rarest of the collectable items under discussion. Very few appear on the market;

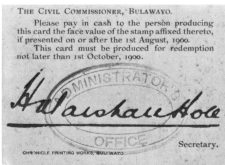

Rhodesian stamp money was issued in 1900 on plain cardboard and stated 'The Civil Commissioner Bulawayo. Please pay in cash to the person producing this card the face value of the stamp affixed thereto, if presented on or after the 1st August 1900.' It is hand-signed by the secretary and known as 'Marshall Hole Currency'. The illustration on the left shows the reverse with the stamp affixed to it. (55 × 75 mm)

when they do their value varies, depending on whether they can be matched to a specific bank note issue. The drawings alone may fetch between £150 and £500 ($300 and $1,000). The engraver works on his plates from these drawings. He will test the plates on which specific sections of the design have been engraved by making 'pulls' in order to examine his work in print. These are the dye-proofs. They may be of the coat of arms, an allegorical figure or a vignette for example. All have appeared in dye-proof form on the market, priced between £15 and £50 ($30 and $100). The value doubles if the dye-proof can be related to a printed bank note.

The engraver's complete work is then sent off to the printers who make final proofs for presentation to the client. The difference between an essay and a printers' proof is merely commercial. If the latter is accepted by the client and bank notes are produced from it the proof becomes a printers' proof. If they are rejected and a new design is engraved, the proof is termed an essay. Essays are designs that have not been used for bank notes. Unlike dye-proofs and essays, printers' proofs can be matched with bank notes on which identical designs appear. This, however, does not affect the prices. All three are in great demand by collectors.

Normally there might be a total of only two or three essays, up to about six dye-proofs and some twelve printers' proofs. Hand drawings are often unique and even two drawings of the same design are unlikely to be identical. Matching proofs to bank notes is a worthwhile challenge, but not an easy one. All proofs appear in black and white (blue is occasionally used), but they are often printed on paper of a different texture. A number of printers' proofs are hand-coloured before being presented to the customer-bank. Subject to these being genuine, their prices will usually be above the cost of a black and white proof of the same design. (See details of the 'Essay-Proof Society' in Appendix IV).

The preparation of a 'specimen' note is the step that follows the printers' proof. After final approval by the government or bank concerned, the bank notes are printed and the finished product is delivered to the customer. Some six hundred of these notes are overprinted with the word 'specimen' and usually neither signed nor given a serial number, before being distributed by the note-issuing authority to all banks in the country, to inform them of the new issues about to be circulated. As a courtesy, specimen notes are also sent to other central banks throughout the world. These notes often fall into collectors' hands and can occasionally be obtained from Central Banks by request. The notaphilic world is small and rumour spreads rapidly. Names of government banks who unofficially make their issues available to collectors is soon public knowledge. These must not be confused with commercial ventures undertaken by some Central Banks, whose demonetized notes, overprinted or perforated with the word 'specimen', are officially offered for sale to the public. East European countries, Czechoslovakia in particular, and other central banks like those of Cuba and the Dominican Republic, have offered sets for sale at prices ranging from £1 ($2) per note to £100 ($200) per set.

There is considerable controversy over whether a specimen note is a true notaphilic item. Some claim that a paper money collection should consist only of notes which had (or have) a monetary value and which were intended for circulation;

specimen notes are only samples, always in perfect condition and marked in a manner indicating that they are not bank notes in the economic sense of the word. Conversely the main argument put forward by those who advocate the inclusion of specimens in a notaphilic collection, is that the note, whether or not overprinted with the word 'specimen', has all characteristics identical to the bank note itself. Furthermore, it is known that only very limited quantities of specimen notes are printed and this therefore increases their value. These two opposing views influence the pricing of specimen sets and individual notes.

There are three established sources for specimens – the banks that place their notes on the market at a fixed price, those that will contribute a set free of charge to a bona-fide collector and banks which refuse to release specimens but whose sets still occasionally appear on the market. There is one additional source; the printing company itself, which invariably keeps some specimens and overprints them with its own name. Specimen notes with 'Bradbury Wilkinson' or 'Waterlow and Sons' overprinted on them are popular items. Prices will finally be determined by the interest the collector has in the issues, but clearly the specimen notes of a bank known to refuse the release of such notes will fetch higher prices than the commercially inclined banks whose specimens can be purchased on the open market. In 1973 a set of Bank of England specimens came onto the market. Nine notes all dated 1934, from 10 shillings to £1,000, changed hands at over £1,000 ($2,000). In the course of publicity given to the sale, the Bank of England expressed the utmost surprise at the appearance of these notes on the market. It transpired that these

Printer's proofs are not bank notes in the strict sense of the word, and yet they are very popular with collectors. This proof of 1864 was prepared by Perkins Bacon and Co. for the Peruvian London and South American Bank. The company was famous for its notes and printed for most countries of the world. (205 × 115 mm)

specimens were distributed to Central Banks on the clear understanding that the recipient would return them on request.

Forgeries and Counterfeits

Conscious possession of forged notes is a punishable offence in the UK. (The law emphasises the words '*knowingly* possessing'.) However, early last century the ignorance that a note was forged was no defence. Being caught in possession of one was a criminal offence punishable by public hanging. During the early 1820s hanging of innocent men and women took place at regular intervals. This led to the 1819 publication of the *Report on the Mode of Preventing the Forgery of Bank Notes* by the Society of Arts. It was supported by the increasing public outcry against the severe punishment of people who, by chance, had received forged notes. The law courts, too, gave considerable support to this report and many judges and juries were already beginning to bring in verdicts of 'not guilty' even in obvious cases. The book pointed an accusing finger at the Bank of England, in that insufficient care was being taken to ensure that bank notes could not easily be forged. The Society proposed a distinct set of copper plates and the employment by the Bank of highly qualified artists to design the notes and engrave the plates. But the Bank of England took no action, and innocent people continued to hang.

In 1826 the famous artist George Cruikshank produced an 'anti-hanging note'. He had been induced to do so by what he saw when passing Newgate Gaol one morning. Several persons, two of them women, had been hung by the authorities for passing forged £1 notes. In his own words Cruikshank 'determined, if possible, to put a stop to such terrible punishment for such a crime' and made a sketch of the note which was then printed. It was headed 'Bank Restriction Note' and depicted Britannia with a skull instead of a head, and with several human beings in postures of

Specimen notes were issued by different printers in different ways. This Ottoman note of the First World War was printed by the German firm Giesecks and Devrient, and perforated with the word 'SPECIMEN' in the lower centre of the note. (97 × 175 mm)

despair in the background. The note was signed by J. Ketch, who was the well known hangman at the time. It also depicted eleven individuals hanging from scaffolds. Many of these 'skit' notes were distributed to the public, becoming a far greater contribution to the abolishment of hanging than had ever been expected. Directors of the Bank of England were harassed by crowds, which had to be dispersed by the police. Since 1826, when the notes appeared, hanging for passing forged notes and similar minor offences has been abolished. The 'Cruikshank' note is not easy to come by and has fetched prices well over £75 ($150). It is not regarded as a forgery.

Much has been written about 'Operation Bernhard', the German forging of British bank notes in the Second World War. The notes were falsified by the Germans at the Sachsenhausen concentration camp. Three hundred prisoners, many of them experts in forgery, managed to falsify a total of £130 million in Bank of England notes. The forgers, after preparing engraved plates for United States dollar and French franc notes, were sent for extermination to Ebensec. A large number of these notes have been discovered in Toplitz, Austria, and the perfection of the notes is such that detection is impossible. The only way to be sure is to allow the Bank of England to inspect them – forged notes will be confiscated and a receipt issued to the owner in return.

The Bank of England too has been involved in forgery. During the French Revolution the Bank forged large numbers of French assignat notes, an activity which, although it was government approved, is not frequently publicised.

The forging of notes is as old as paper money itself. The famous Ming notes of China were forged in the nineteenth century because of their obvious historical

Francisco 'Pancho' Villa issued his famous 'Bed Sheets' in uncontrolled quantities. They were freely accepted because of his popularity with the ordinary folk, but this led to their frequent falsification. The forged notes were withdrawn and stamped 'FALSO'. (91 × 172 mm)

interest. But since it was the collector that the forger aimed to deceive, the notes were of a very low quality, and yet forgeries of Ming notes have been selling at over £50 ($100) in London auctions!

Contemporary forgeries are not uncommon. The revolutions and counter-revolutions in the Latin American sub-continent gave rise to many such issues, which complement original notes in many collections. The 1,000 Day War in Colombia, saw a great number of falsifications, particularly of the locally printed notes, which were much easier to counterfeit. The bank notes of Pancho Villa, the Mexican revolutionary, were falsified and overprinted by the authorities who had originally issued them, because they were unable to obtain government approval to increase the issue!

Bank notes of the Ottoman Empire were printed in the USA in the 1850s. They were simultaneously forged in the States and circulated with the originals! The latter are valued at approximately the same price as the forgery. Some collectors of Turkish material may well pay more for the contemporary forgery than for the original note, since both circulated, albeit unintentionally, as legal tender issues. There is a strong case for collecting such forgeries. They are as relevant to the historical study of the period as the genuine issues.

In the late 1860s forgeries of New Zealand £5 notes showed the use of fine straight lines instead of the words 'five pounds' which appears repeatedly on the text of the original note! The forged issues of one bank sometimes bore the signature of the manager of a different bank! And yet they successfully circulated for a short period of time. The paper was invariably of a different texture to the original notes; and so the detection of an early forgery becomes extremely simple when the suspected note can be compared to an original.

In a study on forged bank notes published in the IBNS journal in June 1972, Mrs Beate Rauch comments that forgeries were often officially recorded and documented by the relevant authorities. Mrs Rauch cites the Austrian 1 gulder note of January 1858. Records of the forgeries of this issue show that by May 1863 over 2,344 forged notes had been discovered. Clearly the document itself is a collectors' item.

In most countries of the world it is an offence to deface, obliterate or otherwise interfere with the currency in circulation. These laws, the contravention of which are not as serious as those applicable to forgeries, are enacted in order to prevent counterfeits. One aspect of falsification by alteration relates to the counterfeiting of overprints, on revalidated notes. The notes are genuine but the forger can easily counterfeit a relatively simple overstamp. This happened in Czechoslovakia immediately after the First World War. The currency was revalued by being overstamped. The kronen notes of the former Austro-Hungarian bank were still obtainable and a counterfeit of the overprint, showing the year 1919, was easily achieved.

Even easier to falsify, however, were the Allied military currency notes used in Italy. The 1943 series carried the denomination in figures only, so that it was relatively simple to alter the figure 50 to read 500. The procedure involved a 'transplant'. The zero was cut out from lower denominations and skilfully placed in position on the 50 lira issue. The 'donor' notes, which can still be encountered, often have the missing zero replaced by a white piece of paper stuck on the back and the figure 'o' filled in in pencil. This was a common practice, which led to the new issues showing the denomination in both numerals and letters.

Three factors are important in the modern preparation of a bank note: appearance, durability and security. Every attempt is made by the printing company to avoid possible counterfeiting. One of the results of such efforts is the beautiful designs of bank notes printed on very high quality paper. The paper is specially treated in a manner which will disclose any attempts to alter the writing or printing on it. These methods of prevention apply to all security paper, whether bank notes, cheques, or other documents. The watermark and metallic thread which appear on many notes are other measures taken to prevent forgeries. By force of circumstance the notaphilist is efficiently protected from being deceived in his own hobby.

Hidden Meanings

What may be referred to as 'notaphilic oddities' form an integral part of a paper money collection. Many of them will be bank note issues which have within their designs a 'hidden' meaning, whether intentional or otherwise.

The Canadian 'Devil's Face' note has now become famous. The Portrait of the Queen on the 1954 issue of the Canadian Central Bank had the hidden face of Satan appearing in the folds of her hair. When these notes were put into circulation, the public refused to handle them and they were withdrawn. The hair was delicately retouched on the plates before the notes were issued once more. People still speculate as to whether the hidden face was added intentionally by a politically motivated engraver.

Another example is the 'Stranglehold' on several German inflation issues of the early 1920s, which can be interpreted either as a hand or a vampire clutching at the throat of the young man on the notes; there is also the liberty head on the 50 centavo of the Republic of Argentine issues of 1948, which, when appropriately folded shows a rotund backside! The 1 peso Mexican note of July 1970 has De Gaulle's profile outlined in one of the signatures on the notes.

Last year, the government of the Seychelles considered the withdrawal of some of her notes because the leaves on the palm trees on the right side of its 50 rupee note spelt the word 'sex'!

One of the many fascinating stories relating to the German occupation of the Channel Islands refers to the German decree which forbade the use in any form or manner of the 'V' sign, representing 'Victory'. When Blumpied, the Jersey painter, was requested to prepare drawings for a set of bank notes under German occupation, he chose to merely paint the word 'SIX' across the back of the 6 penny note. This was quite intentional, with the 'X' widely drawn. The Jersey populace folded three quarters of the note before passing it on so that only half of the 'X' showed and a clear 'V' was handed over.

All these have a value in excess of the original note if the hidden meaning does not appear in all the issues in question. For example, both the Canadian 'Devil's face' and the Mexican 'De Gaulle Signature' have notes of the same series which are identical in design but for the oddities mentioned. In such cases, the note depicting the oddity may fetch up to twenty per cent over the collecting value of the original note.

Propaganda Notes

There are two kinds of paper money issues which served as propaganda. The first are normal notes in circulation which have been retouched. Most people realise that bank notes can be an effective medium for dissemination of propaganda. A good example are USA 'peace bills'. A large number of US students began the practice of drawing symbols and slogans on circulating dollar bills urging the withdrawal of American forces from Vietnam. 'Peace' was their key word and thousands of dollars began appearing with the symbol. They were withdrawn by the Federal Reserve Bank at the peak of the American dissension against the Vietnam War.

It is not an uncommon practice for individuals to scribble their name or a message on a bank note to see whether they will encounter it again. From some notes in my

Canadian 'Devil's Face' issues were the first notes of the new 1954 series. The populace's refusal to handle them led to the withdrawal and replacement of the notes. They did, however, remain in use for several years and are therefore fairly common. (70 × 155 mm)

own collection, dating back to the last century, it is clear that this is not an innovation of our generation.

Slogans used on notes by the Chinese Communist Party during the Hong Kong riots in 1967 are another example of the circulating media being converted into an instrument of propaganda. The main purpose of the issues was to discredit the Government, but the notes were soon withdrawn and are now highly desirable collectors' items. The communist riots sparked off by local strikes coincided with the devaluation of sterling in 1967, when the Hong Kong government followed suit. The Chinese Communist Party capitalised on the delicate state of economic affairs, and overprinted the 10 and 100 dollar issues of the Hong Kong and Shanghai Banking Corporation circulating in the colony. The 10 dollar denomination was overprinted, at the top left-hand corner, with the famous figure of John Bull. He had an outstretched hand and a wide open mouth. Six Chinese characters behind his head read 'He is so greedy that he swallows money'! Two additional large Chinese characters overprinted on the bottom of the notes stated 'Devaluation'. The reverse of the note was overprinted with a text explaining the influence the British had in Hong Kong, which culminated in the devaluation of the Hong Kong dollar. The words 'imperialism', 'banditry', and 'fascism' were abundantly used.

The 100 dollar note, also overprinted with a caricature, showed a pirate carrying a large sack on his back with 100 dollar bills protruding. The four Chinese characters on the sack spell out the words 'Open Banditry' and the central frame, which gives the denomination in English, is overprinted with Chinese characters reading 'Worth only 94.30 dollars after devaluation'. The reverse is overprinted with a Chinese text, ending with the words 'fellow brothers: in order to survive we must unite together

The designs of the notes issued by Chinese puppet banks under Japanese rule were one method of expressing dissension. The sage in the portrait is making a clearly understood sign with his hands. The designer, a Chinese patriot, was caught and decapitated by the Japanese authorities. (175 × 90 mm)

and fight to the end against the British in Hong Kong.' These issues are hard to come by because they were confiscated by government authorities and destroyed. The few that infrequently come on the market fetch up to £45 ($90).

England was involved in a similar propaganda issue when German military notes were defaced by British government authorities and the reverse overprinted with derogatory comments about Adolf Hitler and his supporters. Those issues are priced at over £25 ($50).

Other kinds of propaganda notes are imitiations of real paper money issues. Exceptionally interesting items of this kind are the notes that were issued by Anthony Hall, who claimed to be legal heir to the British throne. He issued specimens of paper money entitled 'The Royal Mint of England' signed 'Anthony'. The notes were serial numbered and similar to early Bank of England issues of this century. They were distributed by Hall in the 1930s. He canvassed dockers on Thameside, harangued throngs of people leaving West End theatres and held regular meetings by the Irving Statue in Charing Cross, endeavouring to persuade the people of his rights. He promised that the notes would be redeemable when he became King of England, King of Ireland and Prince of Wales.

Frequently propaganda issues are related to elections. At other times there are political matters where public support is sought and opinion expressed through the printing of pseudo-monies. Israel is a case in point. In 1972 the Bar-Ilan University distributed a huge quantity of imitation 5 and 10 dollar notes, with the slogan 'Let My People Go' and a portrait of Brezhnev. The notes were directed at the Americans, requesting support for World Jewry in their endeavours to release Russian Jews for immigration to Israel.

Odd Currency Units and Denominations

The advanced collector who may have exhausted all conventional subjects could find the idea of identifying currency units and odd denomination issues a challenge.

In Chapter III, mention was made of the Fiji 1 penny note, dated 1 July 1942,

Israel's Bar-Ilan University students circulated a '100 dollar Federal Reserve Note' portraying Brezhnev, as a propaganda issue. The reverse of the note states 'To the 20th century land of slavery Ransom for the people of our nation. Let my people go!' (65 × 151 mm)

which circulated in the island during the Second World War. This note is valued at £5 ($10) in perfect condition. It is considerably rarer than its dollar counterpart – the government of Hong Kong 1 cent note, also issued during 1941. The latter was printed in huge quantities and valued so low that the World Paper Currency Collector Society in 1973 had all the membership cards printed on the blank back of the notes!

The high denominations of the inflation money of Germany, Greece and Hungary would also be included in such a collection. English provincial bank issues have circulated notes for values of 1 guinea, for 10 shillings and 5 pence and some with even stranger denominations. A wide variety of values is to be found in the American colonial issues, in the 1770s. American notes for 4 shillings and 6 pence, equivalent to 1 dollar, and 1 shilling, equivalent to two-ninths of a dollar, were common. The text on a Maryland March 1770 issue states:

> 'Six dollars, equal to 27s. sterling. This indented Bill of Six dollars, shall entitle the bearer hereof to receive bills of exchange payable in London, or gold and silver *at the rate of four shillings and six pence sterling* per dollar per said bill . . .'

Many of these are expensive, but some may be obtained within the £12 ($24) range.

Examples of odd denomination notes are infinite. The Bank of England issued notes for £2 and £30, £40, £200 and £300 sterling. Many of these issues are now unobtainable but some occasionally come into the market and are sold at several hundred pounds. The ½ denomination is comparatively common and the 2½ is probably among the popular odd denomination issues. In South America notes indicating two distinct currencies, where two denominations appear on the face of the same note, have been issued by several countries. The Banco Central de Chile indicates both the peso value and the converted condores on its issues. Early Colombian notes give the value in reales and pesos and in the 1880s when the peso was at par with the dollar both denominations appeared on the notes. No less fascinating are the Burmese kyats and pyas, Chinese renminbis, Guatamalan

The Hong Kong 1 cent *(above)* and Fiji 1 penny *(opposite)* are respectively the smallest-sized notes and lowest denominations in notaphilic history. They were both issued in the first years of the Second World War as an emergency measure to alleviate the shortage of small change. (Hong Kong 41 × 74 mm, Fiji 51 × 79 mm, both actual size)

quetzales, or the pa'angas and senitis of Tonga, a few examples of some currencies in daily use. Many currency units are no longer in existence, but they can be found on early paper money issues.

One issue of exceptional interest in notaphilic history is the Robert Owen Labour Note, which circulated in 1833, for values calculated in hours! Historically Robert Owen was known as one of England's great reformers. His efforts to establish a National Equitable Labour Exchange are recorded by the existence of his two series of notes issued in Birmingham in denominations of 1, 2, 5, 10, 50 and 80 hours. The last of these are now expensive and priced at over £50 ($100), but the lower denominations can be obtained for about £20 ($40). A socialist at heart, Owen felt that there must be a way for the workmen to cut the 'middle man' out of everyday trading activities. He devised a scheme whereby 'Exchange Bazaars' were to be established. Here, producers and manufacturers could exchange products with each other at a value based on the amount of labour that went into producing an item. Owen publicised his views through his journal *Crisis* and gained sufficient support to issue 'Labour Notes' to producers who placed their commodities at his 'London Exchange Bazaar'. Valuers determined the cost of the raw material and labour that went into production and issued the paper money in 'labour hours', for an equivalent amount of time. The utopian idea was short lived. Within a year, the system failed and Owen had to make good £2000. The deficit had accrued when the Bazaar closed down in March 1834.

Overprints and Overstamps

Change of denomination in the national currency of a country has often been implemented by overprinting the new monetary unit and value on the notes in circulation. This practice is basically for educational purposes, allowing the populace to become accustomed to the change in value of the money; it is also more economical.

Overprints are often executed to combat inflation. Under such circumstances, it is far cheaper to use existing currencies than to incur the expense of new designs, plates, paper, ink and printing processes. In the case of the German hyper-inflation of the 1920s when the value of the money was decreasing hourly, there was not time

to print new notes, and this is why on several the inflated value is overprinted.

The difference between overprinting and overstamping a note is self-explanatory; the latter is used under greater emergencies. The effect in both cases is identical. Any kind of alteration to a note, officially executed in order to alter the original intent of the issue, can be considered an overprint. Normally an overprint will only occur after a note has circulated but cases are known where obsolete plates have been used to print a new issue, which is simultaneously overprinted with an additional legend. Such notes would appear as new, but the overprint remains the legal authorisation of the issue.

There are many varied and unconnected reasons for overprinting money. Overprints can indicate that a note is a 'specimen' or that it is 'cancelled'; or they are used in order to validate notes, assuring the public that the issues are in order. The Yugoslav 'Verificato' notes already mentioned in Chapter III are a good example of the latter case. Yugoslav paper money deposited in the Belgrade National Bank was temporarily hidden in April 1941. Local officials fleeing from German troops entering the city, stored them for safekeeping in a cave in Montenegro. A group of peasants discovered the loot by chance and within a few hours crowds were at the site, sharing out the treasure and stuffing large sums into their pockets! When the Italians occupied Montenegro a few weeks later, they were amazed at the peasants' wealth. Having discovered its cause, they invalidated all notes other than the current circulating issues, by overprinting existing notes with the 'Verificato' stamp, to indicate they were the only legal tender currency. These notes can be obtained at about £5 ($10) in good condition.

Similar examples of much rarer issues are the Peruvian 'Legitimo' notes of the 1880s in the inca denomination. At this time the new series of issues was being

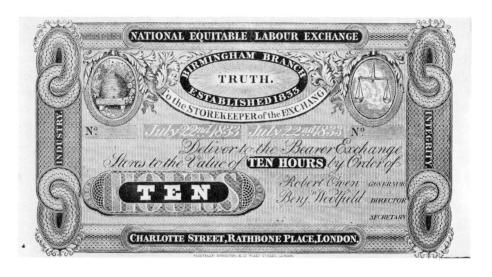

Robert Owen's Labour Notes of the 1830s were issued to fulfil a utopian ideal that was doomed to failure. This is a 'Ten Hour' note. (112 × 209 mm)

falsified at a tremendous rate. The original notes were withdrawn by the authorities and prior to re-issue overstamped with a comparatively intricate design with the word 'Legitimo' in the centre. The notes in question are valued at over £40 ($80) in crisp condition.

The exact opposite, the cancellation of a note with an overprint, is also a popular practice. Many War issues were forged and when discovered by the authorities overprinted 'False'. Notes overprinted in different languages could form a collection on their own.

The Portuguese 'War of the Two Brothers' in the eighteenth century (also known as the 'Miguelite Wars') is an exception to the common rule that overprints are due to inflation. A shortage of paper and printing facilities due to the wars forced the Portuguese government to keep the same notes in circulation for a period of almost thirty years. The issue had first been circulated in 1798 and initially printed on flimsy paper. The first alteration to the notes was the handwritten change of the date. The figure eight on the '1798' was crossed out and replaced by a nine.

Then the first 'true' overstamps began to appear. As of 1800 the blank reverses of the notes were overstamped with a dated seal of the crown. This was repeated yearly with each seal showing the initials of the authority under which the notes continued to circulate. The issues were often officially repaired. The notes, as evidenced on the obverse, were valid for payment after a period of one year; the reverse stamp extended this validity period in order to allow the continued circulation of the issues and thus some of the notes have up to thirteen stamps on the reverse, all dated differently! Yet the truly interesting overprints appear on the obverse of the note, in the form of a red stamp with a radiating pattern below which the name of either Dom Pedro IV or Miguel I appears, followed by the date: 1826 or 1828 respectively. The former must have been overprinted at the death of John VI when it was expected that his son, then in Brazil, would return to reign. When this did not materialise and Pedro's younger brother Miguel, who had been appointed Regent and declared King

A shortage of currency in the Fiji Islands during the course of the Second World War necessitated overprinting legal tender notes of New Zealand. These circulated as currency in Fiji during the war. (88 × 178 mm)

in 1828, took to the throne, the notes were overprinted with the same seal but with Miguel's name and the date of his reign. The ensuing bloody war, and Miguel's final defeat, are part of Portugal's turbulent colonial history. There are few very fine specimens of these issues and they are valued at about £14 ($28) each.

The economy of many countries has at times of internal strife or revolution fallen into a state of chaos because of indiscriminate paper money issues, both by authorised and unauthorised bodies. The rebuilding of a country's economy has frequently been achieved by the government nationalisation of the country's private banking institutes. The paper money issues of the banks concerned were withdrawn, overprinted with the decree by the authority of which they were now to circulate, and re-issued. The reason for such a procedure, rather than the printing of new issues, was twofold: economy and education. In countries where there were many illiterate people they could only recognise the values of paper money issues by the design of the notes.

Colombian notes during the 1,000 Day War are possibly the best examples of such overprints. The National Bank withdrew and took over the note-issuing activities of many private banks in the 1890s, overprinting the reverses with a range of laws and dates.

Replacement Codings

Even with today's ultra-modern systems, a note can be damaged or mutilated during the printing process. Such a note is nearly always discovered by an official checker before it reaches circulation. The note is removed and replaced by another, the serial number of which is no longer in sequence. It would be uneconomical and unnecessary to replace a note with one bearing an identical number. Thus a special set of replacement notes is used, usually identifiable as such. In the case of the USA they are the 'star notes', so called because of the star replacing the prefix letter of the serial number. (On Federal Reserve notes it is the suffix letter which is replaced.)

England is a different case. Although it is known that special cyphers are set aside for replacement notes and that such cyphers are placed in prefix positions on the serial letters not normally occupied by letters, it is also clear that the same cyphers are subsequently used in the normal run of later issue. Thus, a Bank of England replacement note is supposedly unidentifiable except by the Bank of England which is aware of the special cypher-number combination on the notes. Painstaking research by Vincent Duggleby, the BBC commentator and now a noted paper money collector, was published in 1975. It identifies in detail all the replacement notes of the Bank of England £1 and 10 shilling notes issued by serial letter combinations. (See 'England', Chapter III.)

Other countries use similar systems; the Canadian 'asterisk notes' and those of Australia, South Africa and New Zealand are all based on the pattern of a star preceding (or following) the serial number, established by the USA in 1910. Latin countries use the letter 'R' which appears completely separate from the serial number, usually between the signatures on the face of the note. Modern Danish notes have a set pattern whereby the last number and letter in the serial numbering system are repeated in very small print at the bottom right-hand corner of the note. In replacement notes the serial number endings at top and bottom do not correspond. Higher prices for replacement notes will only be paid by the specialised collector.

The observant notaphilist will sometimes discover symbols and signs on his notes that are a mystery to him. But they all have an explanation which may be obtained by writing to the central banking authorities or the bank note printer involved. The English 'G' and 'R' notes are now quite well known; the tiny letters appearing just above the words 'Bank of England' on the reverse of some English £1 notes since 1963 were quite a mystery initially. It was then disclosed that the 'G' appearing on notes with the Hollom (1963) and Fforde (1966) signatures was the initial of the name of the machine used for printing the notes: a Goebbels. These fetch about £2 ($4) each in crisp condition, where the note without the letter 'G' is worth only a fraction over its face value. The 'R', however, was a much rarer instance, on £1 notes of O'Brien's time only (1961) and proved to have indicated an experimental printing to test the resistance of the notes in circulation. These issues are valued at about £40 ($80) in reasonable condition.

Some issues have tiny semi-hidden numbers on the very edges of the design. These relate to plate numbers and the more scrupulous collector may want to amass notes identical but for this number.

There are many discoveries of this kind still to be made. Alberto Lozano of Bogotá, Colombia, has listed over two hundred distinct tiny marks appearing on the reverses of the Banco de la Republica's 1 peso 1953 issue. In spite of his personal contacts with the banking authorities and frequent communication with the printers, he has been unable to trace the reason for the symbols. His theory is that they are for detecting counterfeits. An obvious possibility.

Error Notes
In spite of all precautions taken by printers, an occasional misprinted or mutilated note will escape notice by the checker and hopefully end up with a collector. As

'Misprint' notes which were printed on surplus paper are among the most 'collectable' error notes because they do not appear too frequently. (72 × 145 mm)

curiosity items such notes are of interest to all collectors but some specialise in error notes only.

There are a number of errors with a specified rarity value directly related to their frequency. Notes with missing serial numbers, for example, are rare because these numbers are carefully checked and omissions are obvious. Some modern notes have fetched £50 ($100) above their catalogued value because of such missing serial numbers. Partially missing serial numbers are more common and cases of different serial numbers appearing on the same note are even more common, because serial numbers are printed by a secondary process in which errors can sometimes be made. Such notes will fetch about £10 ($20) above their catalogued value.

A note on which the whole of one side has been left blank is among the error rarities, valued at £50 ($100) above its catalogued value. Partial omissions, caused by the paper folding on the machine, are not so rare. Even more common still are double printings or misplaced print. The cutting processes can also lead to errors, some examples of which are highly sought after by collectors. Surplus paper on a note, either as a fold or as part of another note, are two examples. These would fetch about £10 ($40) over the normal value, although the specialised collector may pay much more. Among the fascinating error notes reported in the United States are issues where the obverse of a 5 dollar note has the reverse of a 10 dollar note! Error varieties are infinite and many can be found among the notes in circulation. The numismatic press often reports on new discoveries by alert collectors.

Although errors are usually 'one of a kind', the exception proves the rule. The 1957 dollar silver certificate issues of the United States, when placed into circulation in 1963, had 10,000 notes with mismatching serial numbers showing the same mistake on all of the notes. The second digit of the number was a '5' in the lower left-hand number and a '4' in the upper right-hand one. The publicity given to this incident caused the almost immediate absorption of all the error notes, not just by collectors, but by mere curio seekers.

It is hoped that the reader, whether a beginner or an advanced collector, will not only have derived some useful information from the preceding pages, but that reading them will have added to his enjoyment and appreciation of the hobby.

Grading Systems

STANDARD GRADING SYSTEM

The conventional coin grading system has been applied to paper money issues by many collectors and dealers. It is based on the following terms:

UNC **Uncirculated** (Crisp) A perfect note, new and clean, never having seen circulation. Without any creases or blemishes whatsoever.

EF **Extremely Fine** Crisp and clean as when new, but with minor creases possibly from folding. Colours have original brightness. A note that has the slightest indication of wear is no longer 'Uncirculated'.

VF **Very Fine** Fairly crisp and clean, with some creases and other signs of having been in circulation for a short period of time. There must be no serious soiling or any fading of the original colours.

F **Fine** A well-circulated note showing considerable wear but still firm and with little soiling or fading or colours. No tears or damage to the paper.

VG **Very Good** A note, with some signs of frayed edges, damaged corners, perhaps soiling and fading colours. Wear evident at the creases but no part of the note is missing or torn.

G **Good–Poor** Unless very scarce, notes in these conditions are not collectable items. Worn, dirty, faded and generally unattractive, often with tears and pieces missing.

BRAMWELL SYSTEM

This is a numerical system and has been accepted by many collectors and dealers in recent years. The main advantages are that it provides a continuous and convenient scale of grading, from the best to the worst notes, without the gaps in the standard method. In this system a perfect uncirculated note is graded as 100. For imperfect notes the degree of damage is assessed numerically as described below, and the result subtracted from 100 to give the grade of the note.

To facilitate assessment, note damage is considered under five headings, namely:
(1) cleanliness;
(2) folding;
(3) surface;
(4) edges;
(5) body.
The table defines three levels of damage under each of these headings and gives, for each level, a 'damage number'. Intermediate levels and their damage numbers can be estimated. To grade a note, the damage number under each of the five headings is estimated and the five damage numbers added together. This gives the 'total damage number' which is the number to be subtracted from 100 to give the note grade.

Table of Damage Numbers

Cleanliness:

Just detectable soiling	5
Considerable soiling and/or bankers' marks	10
Very dirty notes with legibility considerably reduced	20

Folding:

One or two folds that leave only a just detectable crease when flattened	5
Several folds which are clearly visible	10
Many and repeated folds which have caused damage to note surface	20

Surface:

Just detectable damage to surface, probably by folds or crumples	5
Damage to surface at several places or over considerable areas	10
Considerable damage to surface over extended area	20

Edges:

Just detectable roughness or indentation of edges	5
Considerable damage to edges and/or tears not extending beyond margin of notes	10
Badly damaged edges, or tears extending into edges of design	20

Body:

One or two pin holes	5
Several pin holes, or one or two larger holes	10
Several larger holes	20

APPENDIX II

Signatories to Bank of England Notes

From 1752 until 1782, the name of the payee, usually hand-written on the note, was that of the Chief Cashier; from 1782 until 1855 it was only the name of the Chief Cashier which appeared as the Payee. In that year 'pay the bearer' was substituted. From 1798 the Chief Cashier's name was printed on the issues.

1782–1807	Abraham Newland	1918–1925	Ernest Musgrave Harvey
1807–1829	Henry Hase	1925–1929	Cyril Patrick Mahon
1829–1835	Thomas Rippon	1929–1934	Basil Cage Catterns
1835–1864	Mathew Marshall	1934–1949	Kenneth Oswald Peppiatt
1864–1866	William Miller	1949–1955	Percival Spencer Beale
1866–1873	George Forbes	1955–1962	Leslie Kenneth O'Brien
1873–1893	Frank May	1962–1966	Hasper Quintus Hollom
1893–1902	Horace George Bowen	1966–1970	John Standish Fforde
1902–1918	John Gordon Nairne	1970–	John Brangwyn Page

Regular publications

The paper money societies listed separately issue their own journals to members. The following is a list of journals and other publications which are on sale to the public.

BANK NOTE REPORTER, Florida, USA, monthly
COINAGE, CALIFORNIA, USA, monthly
COINS AND MEDALS, London, monthly
COIN COLLECTING WEEKLY, London, defunct since April 1972
COIN MONTHLY, Essex, England, monthly
COIN WORLD, Ohio, USA, weekly
COINS, MEDALS AND CURRENCY DIGEST, London, monthly, defunct since 1969
COINS, STAMPS AND COLLECTING, London, defunct since January 1972
INTERNATIONAL COIN PRESS, New Jersey, USA, monthly
L'ECHANGISTE UNIVERSEL, Paris, monthly
MODERN COINS AND BANK NOTES (Spinks and Sons Ltd) London, bimonthly, defunct since 1973
NUMISMATIC NEWS, Wisconsin, USA, weekly
NUMISMATIC SCRAPBOOK, Ohio, USA, monthly
PAPER MONEY NEWS, Ontario, Canada, irregular
WHITMAN NUMISMATIC JOURNAL, Wisconsin, USA, monthly, defunct since 1968
WORLD COIN NEWS, Wisconsin, USA, monthly
WORLD COINS, Ohio, USA, monthly
WORLD PAPER MONEY JOURNAL, Florida, USA, monthly, defunct since 1960

Societies

The American Numismatic Association

Address:
Executive Director, Edward C. Rochette
Box 2366
Colorado Springs
Co. 80901
USA

Membership: $8 per annum plus $5 registration.

Publication: Monthly, *The Numismatist* (for sale to non-members for $10 per annum or $1 per copy).

Coverage: Worldwide coins, medals, tokens and paper money.

American Numismatic Society

Address:
Secretary, Leslie A. Elam
Broadway
Between 155th and 156th Streets
New York, NY 10032
USA

Membership: $15 per annum.

Publications: *Numismatic Notes and Monographs* – separate publication on single topics several times a year.
The American Numismatic Society Museum Notes – notes and papers on the Society's collections, published irregularly.
Numismatic Literature – listing numismatic and notaphilic publications with abstracts, twice yearly.
Numismatic Studies – series accommodating works in a larger format.
Coverage: Preservation and investigation of international coins, medals, decorations, and paper money.

The Canadian Paper Money Society

Address:
Secretary, Earl P. Briba
P.O.B. 35110
Station E,
Vancouver 13
British Columbia
Canada

Membership: $10 per annum or $150 life membership.

Publication: Quarterly, *The Canadian Paper Money Journal*.

Coverage: Canadian paper money.

The Cheque Collectors Round Table

Address:
Secretary, James F. Stone
Box 125
Milford
New Hampshire 03055
USA

Representative for Europe: Vincent Pearson
39 Skipton Avenue
Southport
Lancashire PR9 8JP
England

Membership: $7 per annum.

Publication: Quarterly, *The Cheque List*.

Coverage: Cheques, bonds and fiscal documents excluding paper money.

The Emergency Money Society

Address:
Richard Upton
P.O. Box 1541
Dallas
Texas 75221
USA

Membership: $5 per annum.

Publication: Quarterly *JEMS*.

Coverage: Emergency coins, paper money and tokens of the world.

The Essay-Proof Society

Address:
Secretary, Kenneth Minuse
1236 Grand Concourse
Bronx
New York 10456
USA

Membership: $10 per annum.

Publication: Quarterly *The Essay-Proof Journal*.

Coverage: Historical and artistic background of stamps and paper money of the world.

The International Bank Note Society

Address:
Secretary, Phillip B. D. Parks
15821 Quartz Street
Westminster
California 92683
USA
Fred Philipson
5 Windermere Road
Beeston
Nottinghamshire NG9 3AS
England
Membership: $7 per annum plus $1.25 registration fee.
Publication: Quarterly, *The IBNS Journal*.
Coverage: International Paper Money.

The Latin American Paper Money Society

Address:
President, Arthur C. Matz
3304 Milford Mill Road
Baltimore
Maryland 21207
USA
Membership: $10 per annum plus $2 registration fee or $200 life membership.
Publication: Three times yearly, *LANSA* (bilingual, English and Spanish).
Coverage: Latin American, Caribbean and Iberian paper money.

The Psywar Society

Address:
Honorary General Secretary, P. H. Robbs
8 Ridgway Road
Barton Seagrave
Kettering
Northants
England
Membership: £3 per annum.
Publications: Quarterly, *The Falling Leaf Magazine*.
Coverage: Psychological warfare documentation, Ariel propaganda leaflets (including propaganda paper money).

The Society of Paper Money Collectors

Address:
Secretary, Vernon L. Brown
P.O. Box 8984
Fort Lauderdale
Florida 33310
USA
Membership: $8 per annum.
Publication: Bi-monthly, *Paper Money* (for sale to non-members $10 per annum or $1.75 per copy).
Coverage: Principally North American paper money.

World Paper Currency Collector (formerly Maryland Foreign Paper Money Club)

Address:
Secretary, Alexander J. Sullivan
701 Hammonds Lane
Baltimore
Maryland 21225
USA
Membership: $5 per annum.
Publication: Quarterly, *The Currency Collector*.
Coverage: International paper money.

Currency Boards, Commissions

These issues generally have a common design for all countries indicated with an overprint to show the specific country in which they are to circulate.

Afrique Française Libre (French):
  Guiana
  Martinique
  New Caledonia
  New Hebrides
  Oceania (Second World War)

Allied Military Currency:
  For Italy (issued by
  Allied countries in
  Second World War)

Allierte Militärbehörde:
  For Austria and
  Germany (Issued by
  USA, USSR, France
  and Britain in Second
  World War)

Bahrain Currency Board

Banco Nacional Ultramarino (Portuguese):
  Angola
  Bolama
  Cape Verde
  Lourenço Marques
  Luanda
  Macão
  Mozambique
  Nova Goa
  Portuguese Guinea
  Portuguese India
  Portuguese Timor
  São Tome

Banque Centrale des Etats de L'Afrique Equitoriale (French):
  Gabon
  Cameroon
  Central African Republic
  Chad
  Congo

Banque de L'Afrique Occidentale (French):
  Dahomey
  Dakar
  Fezzan
  French Guinea
  French Sudan
  Ivory Coast
  Mali
  Mauritania
  Niger
  Senegal
  Togo
  Upper Volta

Banque de L'Indochine (French):
  French India
  French Indochina
  French Polynesia
  French Somaliland
  New Caledonia
  New Hebrides
  Siam
  Tahiti

Banque d'Emission de Rwanda et du Burundi (Belgian):
  (Active 1960–64).

Banque Impériale Ottomane (Turkish):
  Arabia
  Armenia
  Cyprus
  Iraq
  Palestine
  Syria
  Turkey
  Yugoslavia

Board of Commissioners of Currency Malaya (British) (Established 1939):
  Jahore
  Kedah

Kelanten
Malacca
Negri Sembilan
Pahang
Penang Island
Perak
Perlis
Selangor
Trengganu

Board of Commissioners of Currency Malaya and British Borneo (British) (Established 1952):
Brunei
Federation of Malaya
Sarawak
Singapore

Board of Foreign Missions of the Methodist Episcopal Church (American):
Liberia

British Armed Forces:
Austria
Germany
(for use in NAAFI canteens)

British Caribbean Currency Board (British) (Established 1959):
– See British Caribbean Territories

British Caribbean Territories, Eastern Group (British) (Active until 1965) – See also East Caribbean Currency Authority:
Barbados
British Guiana
British Leeward Islands
British Virgin Islands
Dominica
Grenada
Saint Kitts
Saint Lucia
Saint Vincent
Trinidad and Tobago

British East African Currency Board – See East African Currency Board

British Military Authority:
Greece
North Africa (Tripolitania)

Burma Currency Board (British): (Second World War Period only)

Caisse Centrale de la France d'Outre-Mer (Established 1944):
Cameroon
French Equatorial Africa
French Polynesia

Guadeloupe
Guyana
Martinique
New Caledonia
New Hebrides
Réunion
St Pierre et Miquelon

Caisse Centrale de la France Libre – See Afrique Française Libre

Cassa per la Circulazione Monetaria de la Somalia (Italian) (Established 1951):
British and Italian Somaliland

Central African Currency Board (British) (Established 1953–56):
Nyasaland
Rhodesia

Conseil Monétaire de la République de Congo (Belgian):
(Active since 1963)

Currency Commission Irish Free State (Irish) (Established 1927. Note issues until 1943)

East African Currency Board (British):
Aden
Kenya
Somaliland
Tanganyika
Uganda
Zanzibar

East Caribbean Currency Authority (British) (Active since 1965):
Antigua
Barbados
Dominica
Montserrat
St Christopher–Nevis–Anguilla
St Lucia
St Vincent

Gambia Currency Board

Institut d'Emission de l'Afrique Equatoriale Française
Cameroon
Central African Republic
Chad
Congo
Gabon

Institut d'Emission de l'Afrique Occidentale – See Banque de l'Afrique Occidentale

Institut d'Emission Malgasche (French)
(Established 1961):
  Comoro Islands
  Malagasy

**Institut d'Emission des Etats du
Cambodge, du Laos et du Vietnam** (French)
(Active 1952–53):
  Cambodia
  Laos
  Vietnam

Institut d'Emission d'Outre-Mer
  French Polynesia
  New Caledonia
  New Hebrides

Institut d'Emission de Syrie (French)
(Established 1956)

Kuwait Currency Board

Palestine Currency Board (British) (Active
1927–48)

Pan-Malayan Currency Commission – See
Board of Commissioners of Currency Malaya

Qatar and Dubai Currency Board

Saudi Arabian Monetary Agency (Saudi
Arabian) (Established 1952)

South Arabian Currency Authority

Southern Rhodesian Currency Board
(British) (Established 1930–53)

Sudan Currency Board (Sudanese)
(Established 1956)

Union Monétaire Ouest Africaine – See
West African Monetary Union

West African Currency Board (British):
  British Cameroons
  Gambia
  Gold Coast
  Nigeria
  Sierra Leone
  Togo

West African Monetary Union (British):
  Dahomey
  Ivory Coast
  Mali
  Mauritania
  Niger
  Senegal
  Togo
  Upper Volta

Military Issues of the Second World War

Issues for use by armed forces

To American forces:
  Military Payment Certificates; other issues for use in Denmark, Greenland, and Norway.

To British forces:
  British Armed Forces – Special Vouchers.

To German forces:
  Auxiliary Payment Certificates; Reckoning Notes.

To Japanese forces:
  Bird and Dragon notes.

Issues for use in occupied territories

American occupation notes for:

Austria,	Japan,
France,	Korea,
Germany,	North Africa,
Hawaii,	Sicily.
Italy,	

British occupation notes for:

Burma,	Italy (Southern),
Greece,	North Africa,
Faroe Islands,	Tripolitania.
Hong Kong,	

French occupation notes for:

Austria,	Germany,
German Austria,	Saar.

Russian occupation notes for:

Austria,	Manchuria,
China,	Poland,
Czechoslovakia,	Rumania,
Korea,	Hungary.

Italian occupation notes for:

Albania,	Italian East Africa,
Egypt,	Montenegro,
Greece,	Rhode Island,
Ionian Islands,	Sudan.

Japanese occupation notes for:

Burma,	Malaysia,
China,	Oceania,
Dutch Indies,	Philippines.
French Indo-China,	

German occupation notes for:

Austria (limited),	Jersey,
Belgium (limited),	Lithuania,
Croatia,	Poland,
Bohemia and Moravia,	Serbia,
Estonia,	Slovakia,
Greece,	Slovenia,
Guernsey,	Ukraine,
Hungary,	Yugoslavia.
Italy (limited),	

The following additional countries issued occupation notes:
  Denmark: used in German Territory;
  Holland: for Dutch troops stationed in Germany;
  Hungary: in Yugoslavia;
  Yugoslavia: in occupied Italian territories.

Liberation notes:

Belgium,	Morocco,
Corsica,	Philippines,
Dutch Indies,	Tunisia,
Holland,	Yugoslavia.
Luxembourg,	

Glossary

ACCOUNT, MONEY OF	Currency used for accounting purposes and existing in name only, e.g. the British guinea (21 shillings).
ADOPENGO STAMPS	Tax stamps, used as currency during the Hungarian inflation of 1945–46 (1 adopengo = 2 trillion pengo).
ALLIED MILITARY CURRENCY	See Military Currency.
ARMY BILLS	British note issues in Canada between 1812 and 1815.
ASSIGNATS	French Revolution issues backed by land expropriated from the Church. A system originally devised by John Law.
AU	Abbreviation for 'About Uncirculated' in grading. Highly criticised for its vagueness (see Grading, Appendix I)
AUXILIARY PAYMENT CERTIFICATES	Internal currency issued to German troops for use in military canteens during the Second World War.
AXIS MILITARY CURRENCY	See Military Currency.
BANK NOTE	A paper money issue emitted by an authorised banking institution under strict control of the government, which includes limitations on quantity printed and legal backing secured against future redemption of the issue.
BANKERS NOTES	Paper money issues of private individuals, private banks or unincorporated banking houses.
BANK TOKENS	USA merchants' and local governments' fractional paper money issues in 1812. These were also known as 'tickets'.
BILLS OF CREDIT	The American term applied to colonial and continental paper money issues. This is the derivation of the word 'bill' which is used in the US and Canada for 'bank note'.
BILLETS DE CONFIANCE	French fractional and low denomination emergency issues which were made during the French Revolution to alleviate a critical shortage of coinage due to hoarding.
BLUEBACKS	A nickname for paper money issues of the Confederate States of America which had a blue reverse; those of the Union were known as 'greenbacks'.
BONS	'Good Fors' (q.v.), particularly private issues in Canada in the early 1800s.
BRADBURYS	A term used to describe the first two series of the Treasury notes in England issued at the outbreak of the First World War. The term derives its name from the notes which were signed by the Treasurer, Sir John Bradbury.
BRANCH BANK NOTES	In the US, issues from different branches of the same bank; each branch had its own design. In the UK this applied to the

branch issues of the Bank of England. All the notes were identical except that the city of the branch was indicated.

BROKEN BANK NOTES The paper money issued by USA banks which became insolvent or defunct during the free banking period of the 1830s.

CANTEEN MONEY Notes, vouchers and tokens issued to British servicemen during the Second World War for use in military canteens only. Some, for use in vessels, were overprinted with the words 'HM Ships Afloat'.

CARD MONEY Emergency issues of the late eighteenth and early nineteenth centuries in Canada and France. Notes were made from playing cards due to a shortage of paper and the durability of card.

CERTIFICATES US paper money circulated in the form of a receipt for silver or gold coin. The US silver certificates are the best known. The redemption privilege was revoked by Congress on 14 June 1968.

CHARTERED BANKS US and Canadian term referring to banks which operated under a government and state charter as opposed to private banks.

COIN NOTES US Treasury notes of 1890 which were redeemable by law in specie and for which the Treasury put aside special reserves of silver dollars to cover redemption.

COLONIAL CURRENCY Paper money issues of the American states when they were still British colonies.

CONTINENTAL CURRENCY Paper money (by authority of the Continental Congress) which circulated in North America between 1775 and 1779. The Continental notes appear under three headings: 'United Colonies' from February 1777, 'the United States' from May 1777 and 'the United States of America' from January 1779.

CUT NOTES Paper money issues which have been officially bisected or quartered and each portion given its own value, normally indicated by overstamping. This was usually an emergency measure due to a shortage of coinage. See also Halved Notes.

DAMAGE NUMBERS Figures used in calculating the condition of a paper note, based on the Bramwell system. See Grading, Appendix I.

DEMAND NOTES The first series of the US currency which was put into circulation by authorisation of the Act of 1861, for denominations of 5, 10 and 20 dollars and so called because of the statement 'The U.S. promises to pay the bearer . . . on demand'. These were the original 'greenbacks' (q.v.).

DEMONETISED NOTES Issues that have been officially withdrawn from circulation, having no further redemption value. They must not be confused with notes that are no longer legal tender but still redeemable at the Central Bank or Treasury in question.

DEVIL'S HEAD NOTES Issues of 1954 made by the Bank of Canada which show a devil's head in the hair of Queen Elizabeth's portrait. The notes were withdrawn after some years and the plates reengraved, because the public objected to their use.

DOUBLE NOTES	Where the two sides of an issue are intentionally and completely distinct from each other. A practice in the USA during the Civil War, when new paper money was printed on the reverse of the earlier 'broken bank notes' (q.v.) no longer in use.
ENCASED POSTAGE STAMPS	Stamps encased in round metal frames and intended for use as small change. The idea was devised by John Gault in Boston in 1862 and they were issued as a result of the coin shortage during the American Civil War. The reverse of the cases was used for advertising. Issued also by some European countries.
ESSAY-PROOFS	The design of a paper money issue or part of such a design offered to the note-issuing authority by the printing company for approval. See also Proofs.
EXONUMIA	A numismatic term applicable to all non-governmental monetary issues.
FEDERAL RESERVE BANK NOTES	Paper money issues of individual banks in the Federal Reserve system of the US since 1915 for denominations of 1 to 1,000 dollars.
FEDERAL RESERVE NOTES	US paper money issues of the Federal Reserve system proper, since 1941, bearing the heading 'Federal Reserve Note'.
FIDUCIARY ISSUES	Bank notes placed into circulation without the backing of gold or any other security.
FLYING MONEY	Chinese money of the ninth century (*fei-ch'ien*) evidencing deposits in the Government Treasury. The term came from the easy transportability of these 'receipts' which often changed hands many miles away from the Treasury.
FORCED ISSUES	Notes placed into circulation under emergency circumstances, frequently by military authorities, as obligatory currency.
FOREIGN TRADE PAYMENT CERTIFICATES	Paper issues introduced by the US to Japan after their surrender in the Second World War, due to the absence of any currency other than military. The currency units used were the US dollar and the pound sterling and this made it possible for foreigners to trade in Japan.
FRACTIONAL CURRENCY	A paper money issue with a denomination which is part of the standard unit. In the US the reference is to paper money issues from 3 to 50 cents between 1868 and 1876.
GOLD BANK NOTES	The National Gold Bank notes of California. California and Boston banks were authorised to issue notes redeemable in gold during the Californian Gold Rush in 1848 in order to facilitate transactions which were in any case taking place almost entirely in gold.
GOLD CERTIFICATES	A series of US paper currencies issued between 1863 and 1933 (nine in all) for denominations from 20 to 10,000 dollars and based on deposits of gold coin in the Treasury of the US. The gold was payable to the bearer for the value stated on the certificate.
GOLD NOTES	Gold certificates (q.v.) or other paper money redeemable in gold.

GOOD FORS	A term usually applied to emergency low denomination notes, which state on the face that they are 'good for' a specified amount.
GREENBACKS	Commonly used term for the large-sized 1861 US issues and today's Federal Reserve Notes. The term derives from the green colour on the reverse, and originated with the 'Demand Notes'.
HALVED NOTES	Paper money issues cut into two equal portions as a precaution against loss during transportation from branches to the main offices of the bank. These are to be differentiated from 'cut notes' (q.v.).
HANSATSU	Chinese feudal paper money issued on cardboard by War Lords which first made its appearance in 1661 and continued until 1871; their long thin shape has led to the nickname 'book marks'.
HELL MONEY	Imitation paper money issued by the Chinese to be burnt with a corpse, as the Chinese believed the spirit would benefit from the currency in the next world. They are not considered proper notaphilic items.
HOUR NOTES	See Labour Notes.
HUNGARIAN FUND NOTES	The Lajos Kossuth notes issued in the 1850s to raise funds for the establishment of an independent government for Hungary.
INDENTED NOTES	Paper money which has an irregular or patterned cut along one side, introduced to prevent counterfeiting. By matching the note being redeemed with the stub which was retained, the issuing bank could keep a check on their issues.
INFLATION NOTES	Very high denomination notes issued by countries which suffered from hyper-inflation normally caused by war: Germany, Hungary and Greece are examples from this century. The issues were demonetised at the end of the inflationary period and the currency system completely reformed.
INTEREST-BEARING NOTES	Notes indicating an interest rate. In Colombia some such notes (Banco Lopez) had dates marked along the sides perforated after payment of the annual interest. In the USA, Treasury notes of various denominations carried interests of periods up to two years. Some Confederate issues bore an interest rate of 2 cents per day.
INVASION MONEY	Official notes of the Allied Military Forces issued to troops for use in the occupation of Germany, Italy, France and Japan. The term is also used to describe other military currencies of the Second World War.
JIM	Abbreviation for 'Japanese Invasion Money', issued by the Japanese forces in the Second World War during their occupation of the Philippines, Malaya, Dutch East Indies, Burma and Oceania. Strictly speaking these were occupation (q.v.) and not invasion (q.v.) notes.
LABOUR NOTES	Also known as 'Labour Exchange' notes, these were issued in

England in the 1830s by Robert Owen. The notes had a currency unit expressed in hours, and products valued according to the amount of labour which went into their production could be exchanged in a specially set up Bazaar.

LEATHER MONEY   A forerunner of paper money, some of which is on record as dating back to pre-Christian periods. See Skin Money.

LEGEND   The text on a bank note.

LIBERATION NOTES   See Military Currency.

LOW NUMBERED NOTES   Serial numbers on a note indicate the quantity issued and so those with low serial numbers were usually much more clearly printed. They can form the basis for a thematic collection.

MERCHANT NOTES   See Private Notes.

MILITARY CURRENCY   Paper money issued by official military authorities for use by troops. Military currency includes invasion, occupation (qq.v.) liberation and partisan issues.

MILITARY PAYMENT CERTI-FICATES   US Military currency solely for use in US military establishments.

MING NOTES   Chinese large-sized notes of the Ming Dynasty (1368–1644) printed on paper made from mulberry bark. These are the earliest specimens of paper money still obtainable.

MULTILINGUAL NOTES   Paper money issued with a legend in two or more languages, as in India, Belgium or Cyprus.

NATIONAL BANK NOTES   Notes issued in the US between 1863 and 1929, and normally secured by government collateral. Also known as National Currency.

NECESSITY MONEY   See Notgeld.

NEGOTIABLE NOTES   Redeemable notes.

NOTGELD   A German word meaning 'money of necessity', normally associated with the emergency issues made in German cities during the hyper-inflation of 1919–23.

OBVERSE   The front or face of a note, as opposed to the reverse or back (q.v.). It usually bears the value of the note, its date and the principal vignette.

OCCUPATION CURRENCY   Military currency (q.v.) introduced to a country by occupying forces.

OPERATION BERNHARD   The code name given to a German plan to forge the issues of the Bank of England. These forgeries were introduced to Britain in limited quantities during the Second World War.

OVERPRINTS   Official marks on a current paper money issue for the reason indicated: for purposes of revaluation, cancellation etc.

PAPER MONEY   The general term used for all forms of money inscribed on paper as a substitute for metallic coin. A bank note (q.v.) is only one type of paper money.

PARTISAN ISSUES   Paper money for limited circulation issued by partisans fighting the forces occupying their country.

PHANTOM NOTES   Paper money issued by unscrupulous Canadian enterprises who declared the notes redeemable upon presentation in certain offices in the USA. As soon as large quantities of them were presented, the offices were closed down and the 'bankers' disappeared. Similar to the American 'Wildcat Banks' (q.v.).

PILGRIMS' RECEIPTS
Currency on which an overprint indicated that the notes were for use by pilgrims in another country. Pakistan, for example, overprinted some notes 'for use in Saudi Arabia by pilgrims only'.

PLATE NUMBERS
Figures in extremely small print on the side of some notes, to identify the printing plate from which they were made.

POSTAGE CURRENCY
US fractional currency issued in 1862 and 1863, deriving its name from the facsimile of a postage stamp on the reverse. Not to be confused with 'stamp money'.

POW MONEY
Paper money issued during the war, valid for circulation in specified prisoner-of-war camps.

PRIVATE NOTES
Paper money issued by merchants, shopkeepers or anyone other than the legally authorised government agency.

PROMISSORY NOTES
Negotiable documents promising to pay on demand, or after a fixed period, a specific sum of money to the person named or the bearer. All bank notes are promissory notes, in that they 'promise to pay the bearer on demand . . .'.

PROOFS
The final design of an issue prepared by the printing company as a sample for the issuing authority, but not of course intended for circulation. Proofs are always printed on a different type of paper from that used for the final issue, whereas specimen notes are on the same paper. See also Essay-Proofs.

PROVINCIAL NOTES
British banks mostly outside London normally issued notes for circulation and redemption on presentation. Many of the provincial banks used Bank of England notes as backing when they were declared legal tender in 1833. The last provincial bank closed its doors in 1923. (The term is also used in Canada for note issues by a provincial or colonial government.)

PSYWAR NOTES
Propaganda notes without specific monetary value, issued in time of war by one side, and intended to demoralise the enemy.

RAG MONEY
American term for paper money.

RAG PICKERS
American term for paper money collectors.

RAISED NOTES
Those on which there is an overprint (q.v.) indicating that the denomination has been increased as a result of government revalidation.

RECKONING NOTES
Paper money issued to German troops about to depart to occupied territory. The notes were exchangeable for the local currency on arrival and were not intended for circulation.

REFUNDING CERTIFICATES
US ten dollar certificates of deposit, which had an interest of four per cent for an indefinite period of time. They were issued in 1879 and intended to induce people not to redeem them but keep them in circulation.

REISSUES
Issues replaced into circulation after a specific lapse of time and usually with an overprint.

REMAINDER NOTES
Unissued and unsigned notes of a bank which either closed

down or changed its design, leaving a supply of unwanted notes. With few exceptions these should be in absolutely perfect condition.

REPLACEMENT NOTES — Those which do not run in serial sequence and which have been issued to replace a damaged or misprinted note. They usually bear an indicative symbol of some sort, such as the asterisk used in the USA as part of the serial number.

REVALIDATED NOTES — Paper money bearing an official overprint (q.v.), stamp or other mark to indicate its renewed status as legal tender, despite the invalidity of the original note.

REVERSE — The back of a note. See Obverse.

SADDLE BLANKETS — Nickname for the large size US paper money before 1929.

SCRIPS — An American term relating to paper certificates entitling the bearer to redeem them for money, merchandise, services etc. as indicated. The scrip is not a currency note itself but is sometimes collected as auxiliary material.

SERIAL LETTERS & NUMBERS — The system of numbering notes assists the checking of quantities put into circulation as well as guarding against counterfeits. Misprints whereby different serial numbers appear on the same note are often collected as curiosity items, while others collect particular combinations of serial numbers. See Low Numbered Notes.

SHINPLASTERS — A name for the fractional 25 cent notes of Canada, due to their small size. Also applicable, in the derogatory sense, to the Continental notes of the USA, their values being in doubt.

SILVER CERTIFICATES — USA paper money since 1873 issued as receipts for the stated amount of silver in the US Treasury. The redemption privilege was revoked by Congress in 1968.

SKIN MONEY — Russian currency used in Alaska and originally printed on animal skins. See also Leather Money.

SMALL SIZE NOTES — American paper money since 1928, characterised by new designs and the smaller size.

STAMPED NOTES — Paper money issued with revenue stamps on it to add to the value.

STAR NOTES — US replacement notes. Also the name given to some interest-bearing notes issued in 1837 by the Treasury of Texas.

STATE BANK NOTES — An American term referring to issues by banks which are subject to the laws in the State, as opposed to Federal laws.

STATE NOTES — Issued by an American state, now forbidden.

SURCHARGED NOTES — See Overprints.

TICKETS — See Bank Tokens.

TOKENS — Unofficially issued for monetary purposes and usually made from cardboard or plastic.

UNCUT SHEETS — Sets of paper notes in sheet form prior to being cut for circulation. They are collectable in this form.

VICTORY NOTES — US paper money issued for the Philippines, overprinted with the word 'Victory' after the Allied forces liberated the islands from the Japanese in 1944.

WAR LORD NOTES   Chinese issues made by War Lords (Tuchuns) without any backing. They are usually very colourful and continued to be issued up to the beginning of this century when the Communists took power.

WILDCAT BANKS   A derogatory term applied to the US banks organised in the early 1800s by as few as five men. Many were founded on an extremely unsound basis and took advantage of the general public.

Bibliography

General
Angel, N. *The Story of Money*, London, 1930
Avebury, Lord *A Short History of Coins and Currency*, London, 1902
Beresiner, Y. and Narbeth, C. C. *The Story of Paper Money*, London, 1973
Griffiths, W. H. *The Story of the American Bank Note Company*, New York, 1958
International Criminal Police Organisation *Counterfeits and Forgeries Part II*, Amsterdam, 1971
International Secretariat of the League of Nations *Currencies after the War*, London, 1920
Jones, K. P. *The Money Story*, London 1972
Josset, C. R. *Money in Britain*, London, 1971
Keller, A. *Paper Money of the 20th Century*, IBNS, Texas, 1973
Lagerqvist, L. O. and Nathorst-Böös, E. *Sedlar*, (in Swedish), Stockholm, 1971
Lake, K. R. *Investing in Paper Money*, London, 1972
Narbeth, C. C. *Coins and Currency*, London, 1966
   Collecting Paper Money – A Beginner's Guide, London, 1968
   How to Collect Paper Money, London, 1971
Pick, A. *Catalogue of European Paper Money since 1900*, New York 1971
   Catalogue of Paper Money of the Americas, New York, 1971
   Papiergeld, (in German), Braunschweig, 1967
Reinfeld, F. *The Story of Paper Money*, New York, 1957
Sten, G. K. *Bank Notes of the World*, Vols I and II, California, 1967
   Encyclopaedia of World Paper Money, New York, 1965
Willis, P. H. and Beckhard, B. H. *Foreign Banking Systems*, London, 1929

Siege Notes
Bigsby, V. L. *French Assignats, Part III Monnoye de Siege*, WNJ, July 1966
Helfer, A. *The Mafeking Siege Note*, CC, Winter 1964
Howlett, P. *Siege and Blockade Notes*, C and M, February 1973
Ineson, J. P. *Mafeking Siege Notes*, IBNS, Christmas 1967
Parr, M. W. *History of the Gordon Notes*, IBNS, Part I, September 1972, Part II, December 1972
Philipson, F. *The Mafeking Saga*, IBNS, Summer 1963
Riach, R. D. *Coburg's Obsidional Currency*, C and M, February 1973
Sollner, G. *Italian Notes from 1848*, IBNS, March, 1969

Inflation Currency
Banyai, R. A. *The Legal and Military Aspects of German Money, Banking and Finance 1939–1948*, Arizona, 1971
   The Legal and Monetary Aspects of the Hungarian Hyper-Inflation 1945–1946, Arizona, 1971
   Note Issuing Bank Combats Inflation, NSB, Summer 1966

Kupa, M. *Paper Currencies of Hungary 1945–1946*, IBNS, September 1971
   The Second Hungarian Inflation, CC, Autumn 1963
   The Greatest Inflation in the World, 1945–1946, CC, August 1972
   World War II Hyper-Inflation, BNR, April 1973
Pick, A. *Interim and Inflation Notes*, CC, Spring/Summer 1964
Sollner, G. *19th Century Inflation*, BNR, April 1973
Talisman, M. R. *Paper Money Propaganda, The Ultimate Inflation*, WC, October 1971

Emergency Notes
Dickerson, R. E. *Notgeld and Politics in Arnstadt 1921*, CC, February 1972
Keller, A. *About German City Notes*, IBNS, Christmas 1961
Mason, W. L. *Fritz Reuter and Numismatics*, WNJ, May 1966
Meldrum, P. and Winkler, E. F. *Introducing Notmünzen*, C and M, December 1971
Milich, A. *Notgeld Outbursts*, WC, June 1970
Musser, D. L. *Notgeld Comes in all Shapes*, WC, August 1972
   Collectable Notgeld Need Not be Money, WC, December 1972
Philipson, F. *German Inflation Period City Notes*, IBNS, Christmas 1961
   Every Note a Story, IBNS, September 1969
   City Note Honours Hero of German East Africa, IBNS, August 1969
   Schleswig-Holstein Plebiscite, IBNS, June 1970
   German World War I City Notes, IBNS, December 1972
Rauche, B. *Notes of the Franco-Prussian War*, IBNS, March 1970
Rulau, R. *Allenstein-Marienwereder Note Catalogue*, WC, November 1971
   Upper Silesia Plebiscite Note Catalogue, Parts I, II and III, WC, April 1972, October 1972
   and January 1973
Siemsen, C. *German Emergency Currency of 1945*, IBNS, March 1971
Simes, C. J. *Collecting German Notgeld*, IBNS, Summer 1969
Talisman, M. R. *German Notgeld Errors*, CC, March 1973
Watling, L. F. *More in Notgeld than Meets the Eye*, C and M, November 1973
White, B. *The Currency of the Great War*, London 1922

POW Camp Money
Atsmony, D. *POW Camp No. 84*, CC, Summer 1965
Beresiner, Y. *A Monetary System that Never Was*, CM, August 1971
Donn, A. I. *World War II US POW Scrip*, WNJ, November 1968
Fisher, J. *Theresienstadt Notes*, IBNS, Spring 1969
Narbeth, C. C. *Banknotes of Concentration Camps*, C and M, March 1973
Pick, A. *POW Camp Certifications in the USA*, CC, Autumn 1963
Rauch, B. *Dunera Changes Course for Australia*, WC, December 1971
Sasburg, C. P. *World War II Concentration Camp Paper Money in Netherlands*, IBNS,
   Christmas 1968
Seibert, V. C. *Siberian POW Camp Notes*, IBNS, December 1969
Siemsen, C. *Certain Falsified POW Camp Issues*, CC, Summer 1961
Slabaugh, A. R. *POW Monies and Medals*, Illinois, 1966

Military Currency
Kann, E. *Notes on the Japanese Military Yen*, IBNS, Christmas 1965
Keller, A. *The Paper Currency of the First World War*, Berlin, 1957
Litwin, J. *World War II German Occupation Issues in Poland*, CC, Summer 1965
Narbeth, C. C. *When Banknotes Follow the Flag*, CCW, February 1972
Pick, A. *Allied Military Currency in Germany*, CC, Summer 1963

Shafer, N. *Military Replacement Notes*, WNJ, January 1968
   Subject: World War II, WNJ, August 1968
   Allied Military Currency – A Reappraisal, WNJ, November 1968
Slabaugh, A. R. *Japanese Invasion Money*, Illinois, 1965
Swails, A. J. *Booklet Chits of Army, Navy, etc. Units*, IBNS, Summer 1968
Toy, R. S. *Announcement on the Military Payment Certificate*, CC, Summer 1968
   The US Military Payment Certificates' Story, IBNS, Summer 1968
   World War II Currency, IBNS, Spring 1968
   World War II Axis Military Currency, Arizona, 1967
   World War II Allied Military Currency, California, 1969
Warren, C. F. *World War II Military Notes of New Hebrides*, IBNS, December 1971
White, B. *The Currency of the Great War*, London, 1921

Forgeries and Oddities
Allen, H. D. *Canada's Counterfeit Originals*, IBNS, Autumn 1963
Auckland, R. C. *Air-Dropped Propaganda Currency*, Hertfordshire, England, 1972
Beresiner, Y. *The Pound in your Pocket Could be Worth £75*, CM, November 1974
   Force of Circumstance, C and M, December 1974
   Bank of England Printers, WC, January 1974
Cuba, J. *Origin of Latin American Monetary Units*, LANSA, June 1973
Friedman, H. A. *Propaganda Currency of the Far East*, WNJ, April 1968
Graf, U. *Competition for the Creation of New Swiss Notes*, IBNS, Summer 1971
Hauck, R. *Devaluation Spurs Hong Kong Propaganda Notes*, WC, March 1968
Landeress, M. M. *I Made it Myself*, New York, 1973
Lindgrew, T. *Precautions Against Counterfeit of 18th Century Paper Money*, IBNS, March 1967
Mao, K. O. *Bandit Stunts on Propaganda Money*, WC, January 1971
Matz, A. C. *Overprints on South American Notes*, CC, August 1972
Milich, A. *Piazza Garibaldi Top Swap Shop*, WC, May 1971
Narbeth, C. C. *The Biggest Note Forgery Ever Undertaken*, IBNS, June 1965
   Nazi-Forged Bank of England Notes, IBNS, June 1961
   The History of Anti-Forgery Notes, IBNS, March 1971
   Bradbury, Wilkinson and Co., IBNS, Spring 1963
Rao, R. *Hyderabad Error Note*, WC, November 1966
Rauch, B. *Some Types of Forgeries*, IBNS, June 1972
Serxner, S. J. *A Serbian Overprint*, CC, Summer 1968
Walker, P. *Forgery of Banknotes*, CM, April 1971.

Index